"Entrancing and beautifully writt[...] the writer laureate of astronomy, wi[...] of stargazers to come. By the book's [...] to arrive quickly, so you can go outsi[...] nd behold these glittering wonders wit[...]

—*The Washington Post Book World*

"A love affair of the grandest proportions."

—*The Cleveland Plain Dealer*

"An outstanding new work from one of the finest science writers in the world, this is a real page-turner."

—*Toronto Star*

"Charming and lyrical . . . Ferris's bestselling books . . . set the standard for popular books about astronomy."

—*The Baltimore Sun*

"Discover my beloved with me, Ferris seems to say, in prose that at times has an almost erotic power. . . . Whether sitting in the chilly darkness of a backyard observatory or in the warmth of a well-lit room, with this book and its author as guide, readers will experience the passionate quest to discover one person's time and place in the universe."

—*Milwaukee Journal Sentinel*

"Ferris has an enviable talent for explaining the magnificence of the night sky, and you feel drawn to go outdoors, look up, and wonder."

—*New Scientist*

"If you think that experts have taken over the world, author Timothy Ferris has some surprises for you."

—*Minneapolis–St. Paul Star Tribune*

"Ferris's gift is making the complex understandable—the *Washington Post* dubbed him 'the best science writer of his generation'—and his pelucidation of the magic of the night sky will make you want to go out into your backyard and stare for a long, long time."

—*Austin American-Statesman*

"A first-rate science writer delves into his lifetime passion for stargazing, and the result is essential reading for kindred spirits and all would-be astronomers. . . . Mysteries, menaces, and thrills for the skyward eye."

—*Kirkus Reviews* (starred)

"With a glossary of terms and a guide for examining the sky, this book should turn many novices on to astronomy and captivate those already fascinated by the heavens."

—*Publishers Weekly*

"Lyrical and engrossing . . . highly recommended."

—*Library Journal*

"Has the power to convert a casual browser into an active observer."

—*Booklist*

SEEING IN

How Amateur Astronomers
Are Discovering the
Wonders of the Universe

THE DARK

TIMOTHY FERRIS

SIMON & SCHUSTER

New York London Toronto Sydney

SIMON & SCHUSTER
Rockefeller Center
1230 Avenue of the Americas
New York, NY 10020

First Simon & Schuster trade paperback edition 2003

SIMON & SCHUSTER and colophon are registered trademarks
of Simon & Schuster, Inc.

For information about special discounts for bulk purchases,
please contact Simon & Schuster Special Sales:
1-800-456-6798 or business@simonandschuster.com.

Book design by Ellen R. Sasahara

Manufactured in the United States of America

5   7   9   10   8   6

The Library of Congress has cataloged the hardcover edition as follows:
Ferris, Timothy.
Seeing in the dark : how backyard stargazers are probing deep space and guarding earth
from interplanetary peril / Timothy Ferris.
p.   cm.
Includes bibliographical references and index.
1. Astronomers. 2. Astronomy—Amateurs' manuals. I. Title.
QB35 .F49 2002
520—dc21   2002020693

ISBN-13: 978-0-684-86579-9
ISBN-10:   0-684-86579-3
ISBN-13: 978-0-684-86580-5 (Pbk)
ISBN-10:   0-684-86580-7 (Pbk)

To stargazers everywhere

Rapport of Sun, Moon, Earth, and all the constellations,
What are the messages by you from distant stars to us?

—Walt Whitman

Anywhere is the center of the world.

—Black Elk

CONTENTS

II:
BLUE WATER

PREFACE

We grow accustomed to the Dark —
When Light is put away —
As when the Neighbor holds the Lamp
To witness her Goodbye —
. . .

Either the Darkness alters —
Or something in the sight
Adjusts itself to Midnight —
And Life steps almost straight.

    —Emily Dickinson

To gaze is to think.

    —Salvador Dalí

T HIS BOOK IS ABOUT STARGAZING, which is at once one of the oldest and most ennobling, and one of the newest and most challenging, of human activities. It weaves together—or entangles, insofar as my efforts have been less than felicitous—three strands.

The first is an account of my own experiences as a lifelong stargazer, principally the experience of what happens in the moment when ancient starlight strikes the eye and incites the mind. These encounters, involving as they do the inconceivably remote and the deeply intimate, have meant so much to me that it seemed inadequate to describe them solely through the eyes of others.

The second is a report on the revolution now sweeping through amateur astronomy, where depths of the cosmos previously accessible only to professionals or to nobody at all have been brought within the reach of observers motivated simply by their own curiosity. Many of these amateurs are content to enjoy the aesthetics of the cosmic spectacle. A few labor at what amount to unpaid careers as research scientists, a situation that raises the question of just what constitutes an amateur. Perhaps the best definition was offered by the amateur turned professional astronomer George Ellery Hale, who defined an amateur as "one who works because he cannot help it."[1]

The third strand has to do with what's out there—what Saturn, the Ring nebula, the Silver Coin galaxy, and the CorBor cluster really are, as best we humans can tell at this early stage in our cosmic investigations. Much remains to be learned, but in studying the stars—as in attending a concert, or watching a baseball game, or talking with an old friend—it helps to know a few things about the objects of your attention and affection.

Although this book is not intended to be a how-to manual of amateur astronomy, I hope that it will encourage its readers to make the glories of the night sky a part of their lives. The universe is accessible to all, and can inform one's existence with a sense of beauty, reason, and awe as enriching as anything to be found in music, art, or poetry. No great familiarity with astronomy is required to read these pages, and readers encountering an unfamiliar term are encouraged to make use of the glossary. For those who wish to try their hand at stargazing, Appendix A, Observing Techniques, offers suggestions on getting started.

Casual observing of the stars and planets requires as little effort as casual bird watching, but serious amateur astronomers devote considerable time and effort to their avocation—not to mention lost sleep—and it seems reasonable to ask why they do it. I have asked many of them this question, only to find that most, like myself, simply got sensitized to stargazing at some point and stuck with it, for reasons they find as difficult to explain as why they married their spouse rather than someone else. Some mention the beauty of the planets, stars, nebulae, and galaxies. Others invoke the grandeur of the cosmos and their sense of belonging to it. A few mention that contemplation of the stars draws people closer to one another, by awakening us to our common status as fellow travelers on one small planet. As a Chinese amateur astronomer, Xie Renjiang, of Dalian, wrote me recently, "Astronomy is the most significant [way to] unify us. Although we have different skin colors and live in different countries, we should all be the family on this planet. No other cause is so noble in my eyes."[2]

As this is a narrative work rather than an academic review, many estimable astronomers and telescope makers go unmentioned here—not through any deficiency of theirs but because they didn't happen to fit into the story as I have told it. I can only apologize for these omissions with the plea that storytelling, like the Sun in the sky, obscures as much as it reveals.

In addition to the many stargazers whose cooperation was essential to my research and whose kindness and hospitality has been a boon, I am grateful to my wife and family, and to William Alexander, Andrew Fraknoi, Edwin C. Krupp, Owen Laster, Sara Lippincott, Alice Mayhew, Leif Robinson, Donna April Chua Sy, and Terra Weikel.

Seeing in the Dark was written in San Francisco, California; in Florence and Castiglione della Pescaia, Italy; and at Rocky Hill Observatory on Sonoma Mountain, California, from 1991 to 2001. Portions of it originally appeared, in a substantially different form, in *The New Yorker*.

— T.F.

I
THE SHORE

FROM THE OBSERVATORY LOG:
A PRIMATE AT DUSK

A T SUNSET ON A LATE AUTUMN DAY the skies have cleared, so I hike the two hundred paces from the house up the hill to the observatory. A two-day storm has scrubbed the sky clean of haze and left the fruit trees in the orchard bare, their dead leaves spread like vendors' wares in soggy puddles of russet and yellow at their feet. The vineyards flanking the path are soaked, their leaves hammered into sheets of glistening gold. Lined up toward the top of the dirt path stand three farm structures, each sided in railroad-red clapboard and roofed with corrugated steel. The first is a barn, the second a tractor shed, and the third, pitched out over the hillside, is the observatory. Its roof rolls off.

Inside, I step around the cylindrical concrete pier that supports the telescope one story above. Two feet thick, rooted in bedrock far below, the pier stands at the heart of the observatory but touches it nowhere, an isolation imposed to prevent its picking up vibrations. Upstairs, I am gratified to find that the telescope, hunkered down on its concrete lily pad and safe beneath the low roof, has weathered the storm bone dry. I release a big red safety latch and heave my weight against the nearest aluminum stud. The roof emits a groan and starts rolling on its twelve steel wheels, until it has cleared away and suddenly I am outdoors again, under a promising sky of unblemished and darkening blue.

I tilt the telescope skyward and reach in through the struts of its skeleton tube to uncover the concave primary mirror, which returns a fleeting, funhouse image of my face—big, goofy jowls and diminished brow—as if to underscore the preposterousness of my primate pretensions, an ape out to learn about the cosmos. The mirror, wide as a serving platter and thick as a phone book, is figured in a parabola to an accuracy of better than one-eighth of the wavelength of sodium light, but until it reaches the same temperature as the surrounding air it will be somewhat warped. While waiting for the mirror to cool down, I sit at the desk, switch on a small red night-vision light (red light, long in wavelength and relatively low in energy, minimizes dazzle to the dark-adapted eye), and make an entry on the ruled page of the observatory logbook: *Sky cloudless. Light southwest wind. Humidity sixty-seven percent and falling.*

I consult a favorite old star chart. Its vanilla pages, crossed by thin grid lines on which the stars and nebulae hang like grapes on a trellis, are dotted with inked-in notes on observations made over the years: the trails of comets past; a penciled dot locating a quasar, its light a few billion years old; and the numbered distances of stars and nebulae, part of a lifelong effort to comprehend our immediate neighborhood, out to just a couple of thousand light-years or so, in three dimensions. But most of the inscriptions refer to galaxies, each designated by an unfilled oval on the star chart, each home to a hundred billion stars. The darkness is deepening, and their hour is at hand.

1.

Beginnings

Poor boy—leaves
moon-viewing
for rice-grinding.
—Basho

When you see a drop of water,
you see the nature of all the waters of the universe.
—Huang-Po

A T DAWN ON A DESERTED FLORIDA BEACH in 1954, the first rays
of the Sun sent my father's long shadow and my shorter one rippling
like kite tails across the rumpled sands. We were out early to see
what had washed up during the night. In the past we'd found a gleaming conch
shell that whispered surf sounds like betrayed secrets; a dark, ancient wine
bottle, stout and heavy as a stonecutter's mallet; and a bottle with a note in it
from an English schoolgirl who tossed it from the taffrail of a cruise ship in the
Bahamas, a hundred miles away. The previous winter a freighter burned and
sank in the Gulf Stream, and for weeks afterward cases of its cargo washed up,
providing us with a set of new white wooden lawn furniture.

The rising Sun painted the seashore gold, lighting up the pear-amber flota-
tion balls clustered in tangles of seaweed, the indigo sails of beached Portuguese men-of-war, and the miles of unbroken Australian pines that
stretched south along the beach. My father in his faded swim trunks was
golden, too. He'd been a boxer and a tennis pro, had grown fat and famous in
the café society of prewar San Juan and Miami Beach, then gone broke and
gone to work driving a truck. Now he looked like an athlete again, tanned and
muscled, given to drawing lines on the beach so we could practice the standing
broad jump and the hundred-yard dash. Weekdays he loaded forty-pound

cement bags onto a flatbed truck and drove them to construction sites. On weekends he delivered wooden cases of soft drinks to gas stations and bait shops in the Everglades. Nights and early mornings, he sat at a typewriter propped on the rattan table in our tiny living room, writing short stories that he sold to magazines to bring in "extra" money. He'd stopped drinking when we fled the city, had escaped the boredom that stalked him when his name was in the gossip columns, and was full of wit and·wonder.

"Look," he said, touching my left arm lightly to bring me to a halt as we stepped gingerly around a delta fan of violet men-of-war tentacles. Up ahead, high on the beach, something strange was happening—a slow flurry in the sand, rhythmic and methodical. We advanced slowly, peering intently, trying to make sense of it. Scoops of sand flopped up and over, casting long shadows.

"It's a sea turtle," he whispered. "A loggerhead, I think. She's been laying her eggs."

Now I saw her—the enormous shell, so dusted with sand that it had been nearly invisible, and the big, powerful flippers, throwing sand into the hole beneath. My father explained that she had dug the hole five or six feet deep, laid something like a hundred eggs in it, and now was covering it up to keep them safe from predators. We backed away and watched as the turtle finished her labors. Then she heaved her gigantic form down to the shoreline and sank beneath the waves, her fins leaving strange, deep tracks on the beach.

My father found two fallen palm branches, handed one to me, and we swept the sand to erase the tracks, taking care not to step on the nest and pack it down, as this might hinder the little turtles when they hatched and clawed their way to the top. "It's against the law to dig up turtle eggs, but people do it anyway," he said, as we brushed. "If the nest is left alone, the baby turtles will dig their way out in a month or two and head straight back to sea. I don't know how they find the water, or get along on their own out there, but at least some of them must manage it or there wouldn't be any more sea turtles. Was it a full Moon last night?"

"I'm not sure."

"May have been. It's June, and they say loggerheads like to lay their eggs at the first full Moon in June. The tides are higher then, so the sea covers up more of the mother's tracks. But you'd think she'd prefer the dark of the Moon. They're usually done laying by dawn. This one was a bit late."

"How do they know when it's a full Moon—and the right full Moon?"

"I don't know. A female loggerhead can wander from here to the Azores and find her way back to the same beach where she was hatched when it's time to lay her eggs. Possibly they navigate by sensing the Earth's magnetic field lines."

OUR LIFE ON THIS all-but-deserted coast was a species of economic exile, punctuated by the petty embarrassments that afflict the poor. Each morning I siphoned gasoline from the gas tank of our used car to prime the carburetor so it would start, then with the taste of petrol in my mouth took a small yellow bus to an abject school where I was regarded as well off because I wore shoes and a shirt. (To this day I burn with humiliation at the memory of having been thoughtless enough to ask a classmate, "Gabe, how come you don't wear shoes to school?" to which he replied, in a wire-taut Appalachian drawl, "May-be if ah *hay-ed* some shoes, ah would *way-air they*-em.") At the market my mother took groceries back from the checkout counter when we hadn't the money to pay for them all, and through my flimsy bedroom door I could hear the strain in her voice as she contended with the landlord about when we would pay the rent.

But we lived in a beautiful part of this world. Our little house looked out across a windswept field of sea grape to the blue-green sea. At night the stars stood out so vividly that they seemed to crackle, and we would watch, transfixed, as the Moon rose the color of a blood orange, then changed into costumes of ermine and silver. My brother, Bruce, and I fell asleep each night to the subdued hiss and thunder of the waves—all alike, yet no two identical—and awoke to the same ceaselessly inventive sounds each morning, and did not understand that we were poor, and imagined that we were blessed.

Throughout those years, though we could afford no recreation beyond going to the drive-in movies two Friday nights a month, my parents always kept Bruce and me supplied with books. We got library cards and were thrilled when the librarian told us we could take home whole armloads of books at a time. For my ninth birthday I was given a chunky book with a green cover titled *A Child's History of the World*. The author, a New England schoolmaster named V. M. Hillyer, stated on its first page that his purpose was "to give the child some idea of what has gone on in the world before he arrived" and "to take him out of his little self-centered, shut-in life, which looms so large because it is so close to his eyes; to extend his horizon, broaden his view, and open up the vista down the ages past."[1]

It worked. Hillyer began by describing the formation of the Sun and its planets, at a time "long, long, long" ago, "when there was NO WORLD AT ALL!"[2] I found this astonishing, then, and have ever since. It meant that the world we lived in was not *the* world but *a* world, a planet, and that all of it—the rolling waves, the seagulls, the mud that oozed up between my toes when

Bruce and I stalked land crabs along the banks of the Intercoastal Waterway—was made of stuff that had not always been here but had once been *out there*, in the cosmos. The earth and mud had got here, Hillyer informed me, through astronomical processes. Those processes were in action in the tides and the phases of the Moon that the sea turtles seemed to know so well. If I wanted to understand this world, I'd need to broaden my view of space and time, as Hillyer implored. I'd have to learn astronomy.

Fortunately, astronomy turned out to be a wonderful subject. Soon I'd read all the astronomy books in the local library, along with dozens of science fiction novels that filled my head with notions of colonies on Mars and freighters plying routes to Ganymede and Titan.

My mother, who hadn't had a new dress in two years, decreed that once a month we would make the long drive to the nearest bookstore, where Bruce and I could each purchase one new book of our own preference. Soon my single bookshelf held cherished copies of astronomy popularizations by Patrick Moore, Dinsmore Alter, and Bertrand Peek. My father, who had long been interested in land crabs—when we were younger he invented a genie named Sam the Crab Man, who caused ice cream treats to materialize in the freezer on paydays—sold to *Blue Book Magazine* a novella-length story, "The Fifth Assault," in which giant crabs, mutated by radioactive fallout, attack the lone inhabitant of a desert island. (He was disappointed that the assailants in the pioneering monster film *Them* were not giant crabs but giant ants.) Our fortunes began to improve a bit. Dad got a white-collar job and we moved into a small house on Key Biscayne, an island off Miami that had been a coconut plantation, and bought a better car, and then a color TV.

The night skies over Key Biscayne in those days were inky dark and rock steady. The stars seemed close at hand, like the spangles inside a princely Bedouin tent. I learned the constellations from a book called *The Stars: A New Way to See Them*, written by H. A. Rey, coauthor and illustrator of the "Curious George" books. I would take a dining-room chair out to the front lawn, illuminate Rey's book with a little flashlight whose lens I'd painted with my mother's red nail polish, and trace the outlines of mighty Orion, Cygnus the swan flying south in the Milky Way, and the almost frightening Scorpius, swollen to gigantic size in the briny air along the southern horizon, its starry stinger lurking above the palm fronds.

Mars loomed in the east like a garnet out of Araby, getting brighter every night. I'd read that it was approaching opposition, the point when Earth lay in a direct line from Mars to the Sun so that the planets were close together, and that the opposition this year, 1956, was to be an especially favorable one. Mars would draw within 36 million miles of Earth, affording exceptional views of its polar caps and continent-like markings and prompting fresh debate over the reality of the famous canals, which the astronomer Percival Lowell believed to have been built by an ancient and parched civilization to ferry water to their cities from the poles. But to see these wonders required a telescope. I found a suitably cheap one in a tiny advertisement in the back of *Popular Mechanics*, and my parents gave it to me as an early Christmas present that fall.

Stargazers, like musicians, typically learn on inferior instruments, and my first telescope was suitably wretched. It consisted of a skinny tube made of Bakelite—a brittle and literally tacky substance that, like yogurt, is easier to recognize than to describe—mounted irresolutely atop a spindly tripod fashioned from wood so green that its legs bowed inward under their own meager weight. Into one end of the tube was glued a war-surplus objective lens with a diameter of 1.6 inches, giving it less light-gathering power than an ordinary pair of reading glasses. The other end held a cardboard eyepiece; you changed the magnifying power by taking it apart and reassembling its yellowing lenses into various bewildering combinations.

Nobody else could see much of anything through this telescope, nor did I have a great deal of initial success, lacking experience as I did—and having been a bit unnerved when, on one of my first attempts to use the thing, I looked through its four-power finder scope and was confronted by the grotesquely magnified image of a flying cockroach who had just landed on the tube and was scurrying my way. But I could see Mars—its polar caps, at least, and a few of the most distinct surface markings, especially the dark dagger shape of the northern hemisphere feature Syrtis Major—and the effect was transforming. Mars was, after all, a world, and even more mysterious then than it is today. Staying up late on cold, clear nights, out in the front yard watching Mars, I began to learn how to observe a planet. I came to realize that the air is rather like the lens of the eye, a curved membrane thinnest at its center—the zenith—and thicker toward the sides. That's why the sky on a sunny day looks deep blue overhead and pallid near the horizon, and it means that planets are seen most clearly when highest in the sky.[3] I learned that the highest powers of magnification do not necessarily produce the best results: Instead, for any given telescope, trained on a given object at a given time and place, there is an ideal power, a sweet spot. Once you've found it, the trick is to keep watching, waiting for moments when turbulence settles out of the air and the eye is treated to a gratifying and tantalizing instant of clarity—an instant as fleeting, yet as potentially significant, as the flash of insight that brings an original idea.

From the new TV inside streamed an advertising jingle: "Dream *caaar*, fifty-seven, *Murr*-cury," for an automobile that bore the name of a planet and looked like a spaceship. The future seemed pregnant with the promise of discovery. What lay before me was nothing less than the whole universe. But to see more of it, I'd need a better telescope.

I got a job, along with two of my best friends, sweeping up the sidewalks and parking places in front of stores in the local shopping center on Sundays. It was hard, sweaty work, but it paid well enough that I soon had the down payment on a superior telescope. It had a solid mount, an enameled white tube, and a 2.4-inch objective lens whose components were mounted in an aluminum cell with a thin layer of air between them, instead of being glued together as had been the case with the 1.6-inch. (The glue in the old lens had already begun to separate and turn cloudy.) I still remember the thrill of opening the package when, after an eternity of waiting, the new telescope was delivered to our house—the sharp, spar-varnish smell of its wooden cases, the gleaming chrome and black enamel of its eyepieces, the weighty, oily worm gears of its spring-loaded slow-motion controls. I came to cherish it as an instrument of deliverance, the keys to a kingdom vast, ancient, and spectacular. With it, I saw the sand-colored rings of Saturn, the blue-white stars of Orion, the golden glow of the Omega Centauri star cluster, and a thousand other things so big, so venerable, so hot or cold as to balloon one's sense of the plausible.

My father, meanwhile, became alarmed. Why was I sweeping sidewalks? I worked hard enough at school: *That* was my job. One had a lifetime in which to work; childhood was a time to dream. Late one hot Sunday morning I was wiping sweat from my eyes when he drove up in a borrowed convertible, the top down and the back-seat stuffed with brand-new beach toys—a football, a couple of beach balls, two inner tubes, and a rubber ball to bowl at stakes embedded in the sand, in a game he'd invented years ago. (My father could make a game out of anything.) He offered to pay me and my friends the rest of our month's earnings if we'd stop laboring away and instead devote our Sundays to having fun. He'd also help me make the payments on the telescope. We kids looked at one another, turned in our brooms, jumped in the big open car, and went to the beach.

Some of my friends got telescopes for themselves. The ablest observer among them was Charles Ray Goodwin III, a boy of sufficient seriousness of mind that he was teaching himself Russian so that he could read Solzhenitsyn in the original. Chuck and I learned from books how to make drawings of the Moon and the planets with charcoal and colored pencils. Later we got hold of a couple of old cameras and took time-exposure photographs that recorded the lurid orange hue of the eclipsed Moon and the tangle of gas clouds that engulf the constellation Orion. A few of us formed a club—the Key Biscayne Astronomical Association, or KBAA—with Chuck as president, and started keeping observing logs, portentously filling them with sketches and data like those we'd seen in the books of the august, full-grown members of the British Astronomical Association.

From the KBAA logbook:
June 14, 1958. Observed from 21 hour to 23 hr, made no drawings. Seeing excellent. All observations were of deep space objects, in Cygnus, Scorpius, Ursa Major, and Lyra. A very good night.
July 6, 1958. Seeing fair. Three drawings of Jupiter made, one in color.
July 11, 1958. Good seeing. Observed deep space objects in Scorpius, also Delta Cygnus, Omega Centauri. We had as our guest John Marshall, from Evanston, Illinois, who was made a member and president of the Illinois branch of the KBAA.
Aug. 1, 1958. Ferris made his standard two drawings of Jupiter, and observed a few common deep space objects. The rising of the full moon interrupted observing.
August 24, 1958, predawn. Goodwin and Ferris observed M34, the double cluster in Perseus, Haydes in Taurus, and Mars from 3h 30m to 5h 30m.

As happens if you spend a lot of time outdoors at night, we encountered some unexpected spectacles. One night we saw a mighty fireball—a meteor, a chunk of rock probably no bigger than a golf ball but spectacular when seen hurtling into Earth's atmosphere and burning up from friction with the air. I was bending to fetch a star chart from the lawn when suddenly the colors of the chart leaped into view—the blue of the Milky Way and the red oval galaxies on the white page, laid against an abruptly vivid carpet of bright green grass. I looked up and saw the whole neighborhood bathed in something approaching sunlight, with green coconut palms waving against a blue sky. Everything cast two shadows, one black and one red, and the shadows were shifting, clocking rapidly from north to south. In the sky I saw the fireball itself, silver and yellow with a red halo and brighter than the Moon, racing northwest and leaving behind a fading white trail flecked with gold.

Watching it fade I recalled a day, years earlier, when my mother had gone into a little rural grocery and I'd wandered over to the railroad crossing. All was quiet. The twilight sky was lavender and dark enough that Venus was out, hung above a freshly minted sickle Moon. Then the crossing-gate alarm gong started ringing, the big red lamps flashed, and the black-and-white-striped gates went down, blocking the dirt road. There was no train in sight, but the rails began to hum. I fished a penny from my pocket, laid it on the track, and dashed back a safe distance, having been warned that if you stood too close to a speeding train it could suck you in under the wheels. A yellow headlamp appeared in the distance and closed with incredible speed. The train flashed past—an express!—and shot through in a blast of noise, going so fast that my eye could capture only a few snapshots in the blur. There were tan cars with a blood-red stripe that rose from the diesel engine's bullet snout and extended down the cars under their windows, in the warm yellow light of which I thought I caught a glimpse of dining tables covered in white linen. Then the train was gone, in a waft of vacuum that sent sheets of newsprint spiraling in the warm, moist air.

I stood there frozen, my jaw agape, staring after it. I'd been wearing a black cardboard cowboy hat and found that I'd taken it off and was holding it over my heart. Years later I heard an old recording by the Mississippi blues singer Bukka White that captured the sensation:

> Got that fast special streamline,
> Leaving out of Memphis, Tennessee,
> Going into New Orleans.
> Be runnin' so fast the hoboes don't fool with that train,
> They just stand by the track
> With their hat in their hands . . .
> Play it lonesome, now, 'cause I'm a hobo myself sometimes.[4]

Visions like this one produced a sensation that I did not know how to express until, years later, I read what Einstein had to say about the lesson he'd learned from his first encounter with geometry, which, he recalled, provided a way "to free myself from the chains of the 'merely-personal,' from an existence which is dominated by wishes, hopes, and primitive feelings. Out yonder there was this huge world, which exists independently of us human beings and which stands before us like a great, eternal riddle, at least partially accessible to our inspection and thinking. The contemplation of this world beckoned like a liberation."[5]

Decades later, back in Key Biscayne to give a talk, I found that its once-dark skies had been turned to fish-gray by urban lights. Efforts were under way to reduce the light pollution, by limiting the size of advertising signs and encouraging the use of hooded lamps that illuminate the ground without wasting energy on the sky. The stargazers working on the light-pollution issue had found allies among marine biologists concerned with nesting sea turtles. The turtles, it seems, prefer their beaches dark at night.

2.

Spaceflight

Given only the ships and right sails for the heavenly space, there will also be men unafraid of the terrible distances.
— Kepler, to Galileo

We have nothing to hide: We have nothing, and we must hide it.
— Nikita Khrushchev, to his son,
on the Soviet space program, 1961

ON SATURDAY MORNING, October 5, 1957, in Key Biscayne, my father awakened me and thrust the front page of the *Miami Herald* up to my blinking gaze. It said the Russians had orbited *Sputnik*, the first artificial Earth satellite. We were astonished. Everyone had assumed that Americans would be first into space. Now, instead, the space race was on in earnest, and we were playing catch-up. At school Monday morning the principal addressed us on the PA system and said we were all going to have to work harder at science and math if we were going to beat the Russians.

The shock of *Sputnik* was good news to me and my friends in the astronomy club. We'd been in love with rockets and space for years, but nobody else much cared. Now, for a little while, pretty girls asked us questions about satellites and stared into our eyes while we explained how rocket engines work. We struggled to fashion stoic, jet-jockey replies to the cherished, breathless question, "Would *you* go into space?"

FLORIDA IN THE FIFTIES was a particularly futuristic part of a generally forward-looking society. Exotic military aircraft designated X for experimental, flying out of Homestead and Cape Canaveral, etched long, high contrails across the Kodachrome-blue skies, and we kids traded airplane cards bearing their pictures. The newspapers were full of UFO stories. Movies about alien invasions and men

15

flying to the Moon were playing at the drive-in. The Sunday morning TV lineup featured a smoothly produced propaganda series, *The Big Picture—Our Army in Action*, narrated in stirring tones by Master Sergeant Stuart A. Queen. It opened with a disquieting shot—intended, no doubt, to be reassuring—of a howitzer firing a nuclear shell that, after a long pause, detonated in the distance, producing a signature mushroom cloud. In evocatively illustrated books like *Across the Space Frontier*, the rocket scientists Wernher von Braun and Willy Ley promoted plans to build a big wheel-shaped space station in Earth orbit, establish a lunar base, then send astronauts to colonize Mars. I glued together a plastic model of the gigantic rocket that would get them there, and launched it, with elaborate countdowns, from a paving stone on our bright green lawn.

The space race heated up soon after we moved to Key Biscayne. Rockets launched from Cape Canaveral often passed almost directly overhead, skirting the coastline. (One night on a radio interview program an Air Force general was asked why rockets launched from the Cape didn't fly directly above Miami. "Well, it's always remotely possible that a missile launched on that trajectory might crash into downtown Miami, destroying buildings and setting massive fires," he replied, "and that would be bad for the morale of my men.") Rockets had been blowing up with dismaying regularity—notably Vanguards, the civilian missiles that were to have paved a peaceful way to space as part of the International Geophysical Year, but Jupiter, Thor, and Atlas missiles were immolating themselves, too—and when launches did go well they frequently happened hours or days late. The scheduled launch times were generally not made public, lest such embarrassments become even more of a spectacle than they were already, but occasionally a launch would be covered on live TV, providing us stargazers with a good chance of seeing it in the sky.

From the *KBAA logbook:*
Friday night–Saturday morning, December 5–6, 1958. At 12:45 watched a moon shoot on NBC TV, then went up on the roof and watched the first stage burnout. No vapor trail. Mission was to fly a Pioneer III probe near the Moon, but news reports say firing error will bring rocket back tonight after reaching an altitude of about sixty thousand miles. At 9:45 A.M. took photos of Sun using solar filter and 25-mm eyepiece.

Other launches came as a surprise.

January 30–31, 1959. Chuck and I observed Mars, but made no drawings. While we were observing, I looked up to see a Thor missile arcing across the sky, vapor trail streaming behind, lit up by a 1-million-candlepower lamp mounted on the rocket. We then went up on the roof to watch for an Atlas missile due to fire at 11:00 P.M. But the Atlas was not fired, and we descended from the roof at midnight. Observed Orion nebula. The Moon rose and was photographed. Went to sleep at 5:00 A.M. local time. In all, a fairly successful night and morning.

Having seen a couple of rockets roar past, we were hooked. My father would call a journalist friend, Ben Funk, who covered Cape Canaveral for the Associated Press. Ben would tip us off to scheduled launch times; then Chuck and I would go up on the roof, set up a borrowed camera, and wait. Knowing that rockets were bathed in powerful searchlights when a launch was imminent, we scrutinized the northern horizon, convinced that we could tell that the launch was coming if the horizon grew "warm." As the launches were often long delayed or canceled, we spent many hours on the roof for every time we actually got to see a rocket, so it made sense to pass the time by stargazing. Tiled and pitched, it wasn't the most accommodating roof, but we managed to haul my telescope up there and to set up my mother's ironing board as a chart table. One night a police patrolman, spotting our silhouettes on the roof with our gear past midnight, rang the doorbell and awakened my mother.

"Ma'am," he said politely, "do you know that there are two men with what appear to be rifles standing on your roof?"

"Oh, yes, officer," she replied, yawning as she cinched up her bathrobe. "It's perfectly all right."

On August 12, 1960, a couple of weeks before my sixteenth birthday, NASA orbited a gleaming silver Mylar space balloon, a hundred feet in diameter, called *Echo 1*. A radio message bounced off it traveled from coast to coast. President Eisenhower said this demonstrated the feasibility of passive communications satellites, but what mattered to me was that *Echo* could be seen with the unaided eye, a bright beacon in space as readily visible as a tall ship on the horizon. The faded pages of my log book for August 14–17, 1960, show that I would watch it glide across the sky at night, then get up before dawn to see it again, observing it seven times in four days. I felt like an ancient lungfish contemplating the land from the sea. We could get *up* there.

I got a driving permit the day I turned sixteen, borrowed my mother's car, and drove up to the Cape, where Ben Funk had reserved a motel room for me and my friends. Just down the hall from the AP bureau, it had a splendid balcony view of the launch pads. I was amazed, after years of reading descriptions of imaginary spaceports in the works of Robert Heinlein and other science fiction writers, to see what a real one looked like. The vastness of its intentions was reflected in its imposing scale—the gleaming rockets standing against primary-color blue skies and billowing white clouds just like those on the airplane cards, the blockhouses set back miles from the launch pads, and the security gates miles farther away still. What had been a dream was now reality, spread out across the Florida flatlands.

Having been advised by teachers and school counselors (accurately) that it would be difficult to earn a living as a writer, and (archaically) that I would have to learn Latin if I wanted to be a scientist, I instead studied English and communications, at Northwestern University, with the intention of becoming a lawyer. I'd been alienated from school since the middle of the fourth grade and my indifferent study habits persisted into college, where I seldom cracked a textbook, preferring to read poetry and philosophy at night and compose songs on a battered steel-string guitar. The only science course I took was taught by a bearded astronomer whom we freshmen regarded as an old man. When he said in class, "It is possible that one of us here in this room will one day fly in space," a skeptical murmur rose in the hall; most of us, I think, assumed that we'd been born too soon to entertain any such hopes. As it turned out, the one person in that room who assuredly did go into space was our venerable professor. His name was Karl Henize, and the year after I graduated he left teaching to join the astronaut corps, flying on the space shuttle *Challenger* in 1985. A few years later I ran into him at a banquet and took him for a late-night ride in a Porsche. When we hit high speed Karl laughed out loud, shouting, "This is the sort of car we astronauts are supposed to drive." Karl died in 1994, nearly seventy years of age, of hypoxia while climbing Mt. Everest. His body is buried there, at an altitude of 22,000 feet.

I'm ashamed to say that I studied little for Karl's class. Preoccupied with coeds and cars, I missed most of the labs and was on the verge of failing when, at the last lab meeting, held out under the stars on the Lake Michigan shore, I saw a satellite headed east.

"Excuse me," I said to the graduate student who was holding forth. "Sorry to interrupt, but I thought everyone might be interested to see that there's a satellite passing over. I'd have waited to mention it, but if I calculate correctly it's about to go into the Earth's shadow."

We all looked up and watched the satellite glide along in its perpetual Newtonian fall. Moments later, it vanished. The grad student grinned in the dark, and I passed the course.

3.

The Ozone

Sparrows in eaves
mice in ceiling—
celestial music.
—Basho

It is not always that there is a strong reason for a great event.
—Samuel Johnson

A T TWILIGHT ON KEY BISCAYNE in 1959, as the stars were coming out, I admired a brilliant crescent Venus through the telescope while Chuck tuned his handheld radio, searching through its shortwave bands and at the top of the AM commercial broadcast band and moving its rod antenna like a wand, seeking sounds from far away. The radio didn't look like much, but it could "bring in," as we said in those days, the voices of ham radio operators in what I imagined to be icebound cabins above the Arctic Circle and thatched huts in Mandalay, as well as the official "voices" of Moscow, China, and the BBC with the chimes of Big Ben sounding the hour. For timing events such as eclipses of the moons of Jupiter we tuned the radio to WWV, the Bureau of Standards atomic clock, and recorded our squeaky voices on an equally squeaky reel-to-reel tape deck over the throbbing beats that marked each second of the passing time that ushered the stars across the sky and ourselves toward the dimly anticipated pragmatisms of our approaching adulthood.

NIGHTS ARE PRIME TIME for radio listening, for it is then that the ionosphere firms up overhead, bouncing radio signals between continents. The radio pioneers used to call this reflective layer "the ozone," after the form of

19

oxygen (O_2 and O_3) that one smells after thunderstorms and around electric motors. (The ionosphere does contain some ozone, but is distinct from the "ozone layer," which lies less than a third as far up in the atmosphere, only twenty miles high.) At night, when the Sun stops stirring it up, the ionosphere congeals—in patches, like sargasso—and radios get legs. Signals from far away begin rising up out of the background noise, washing in and out with an alluring unpredictability.

Our favorite station during long nights of stargazing was WLAC, in Nashville, Tennessee. It was there that we first heard the blues.

Florida in the fifties was racially segregated by both law and custom, and Florida radio was segregated, too. All the big stations played white music only—a marketing situation common throughout the South, one that Elvis Presley's future manager, Colonel Tom Parker, had in mind when he predicted that he could make a fortune if he could find a white boy who sang like a black boy. The few stations that catered to Florida's black residents were mostly low-wattage operations, literally on the fringe out near the end of the dial, and they seldom played the blues. So we heard virtually no blues, and very little of the exciting new recordings through which black artists like Chuck Berry, Little Richard, and Fats Domino were transmuting the blues into rock and roll—until the night, listening to Chuck's radio out under the stars, that we discovered WLAC.[1]

WLAC, a "clear-channel," fifty-thousand-watt Nashville AM station, was owned by the Life & Casualty Insurance Company—a fact I heard repeated so often on the air that the phrase "life and casualty" became my youthful mantra, a summing up of the mysterious unpredictability of the human situation—and it treated black music as a happy fact of life, natural as moonlight and moonshine. The station's motto was "Health, Thrift, Entertainment, and Education," and its advertisements were almost as entertaining as its music: "Live baby chicks delivered right to your door! One hundred ten of the finest baby chicks you ever saw, for the low, low price of just $2.95!" "Royal Crown hair dressing and Royal Crown pomade, the light, bright, modern way to keep your hair in style." "Ernie's Record Mart! Man, they got 'em, records galore at that store!"

In this fashion, huddled around the radio in twos or threes, we white boys were initiated into the rootstock of black American music, a glimpse that was all the more enticing because it was so difficult to obtain: The sound swelled and faded with the vicissitudes of the "ozone," raising groans when it deprived us of a B.B. King guitar solo or a Sonny Boy Williamson verse. Stargazing all night with WLAC playing fitfully on Chuck's radio, I came to think of astronomy and music as woven together, though I did not yet know that this was already an old story back when Kepler wrote of the music of the spheres to Galileo, a musician's son.

I got a car—a raw, street-legal racer, bought in steaming rain off the track at Palm Beach Raceway—and became one of the millions for whom the cult of the highway is married to that of the radio. Late at night on long, high-speed drives on two-lane blacktops through the South and across the Mississippi into the West, in the stark cockpit with its mingled smell of leather, oil, and scorched primer paint, over the rolling thunder of the car's dual exhausts came WLAC, WWL in New Orleans, and cherished scraps of the black disk jockey Early Wright's blues broadcasts out of Clarksdale, Mississippi. ("That was a beautiful record I dropped on you for your listening plea-sure," Early Wright would say, often without otherwise bothering to identify it, on the assumption that his listeners knew their Delta blues.) Then, heading west, came the big, clear-channel R&B stations out of Chicago, which across the great reaches toward California faded to static, a whispered dialogue of Earth and Sun murmured along the magnetic field lines that arced to the pole, interrupted by pirate stations across the Mexican border carrying the bleak pleas of revival-tent evangelists: "Ladies, if you're pregnant, be sure to send for our 'Expectant Mothers Prayer Booklet.' It only costs a nickel. One woman didn't, and *her* child was born with one big eye in the center of its forehead!"

One star-filled night west of Abilene, a few minutes after passing a long freight train that ran parallel to the road, billowing smoke from its twin steam locomotives, I heard from somewhere off the ozone an old blues recording so powerful that it brought tears to my eyes, obliging me to pull over and stop:

> Let the Midnight Special
> Shine its light on me.
> Let the Midnight Special
> Shine its ever-loving light on me.

The Midnight Special was the train that an escaped convict might hope would carry him to freedom. The singer was Huddie Ledbetter, better known as Leadbelly, a convicted murderer whose songs appealing for clemency had twice helped him win release from prison. Hearing songs like "The Midnight Special" and Blind Willie Johnson's peerless "Dark Was the Night" got me tangled in the blues, and I became one among many young white guys who learned to pick out on the guitar tunes written in circumstances so remote from our own that they might as well have happened on another planet. I came to think of the blues singers of the American South as akin to orphaned stars, points of light in the wider darkness.

If you're young and don't know where you're going, the highway is an excellent place to be. The police officers I met along the way would sometimes ask, "What's the hurry?" but there was no hurry. A dedicated high-speed driver isn't anxious to get somewhere; he's already *there*, where he wants to be—at speed, with the car seemingly shrunk to the size of a motorcycle, or the motorcycle to something not much larger than his hands and wrists, screaming down a road that also has shrunk, to the thickness of one pounding vein, in which somehow there is always just enough room to get by, with nothing in his ears and mind but the scream of the engine and the sound of good strong music, bounced off the "ozone." Endlessly flying up the road—and for some reason, late at night, two-lane blacktops always seemed to be leading upward—peering into the headlights' yellow ellipse, I was as alone as some future astronaut adrift in the hard vacuum past Titan. But I never felt lonely. I was in just the right place.

The highways back then were dark, and the stars kept you company. They'd already been banished from downtown New York, Chicago, and even St. Louis, but out on the open road they were almost always with you, in pinpoint sprays like bouquets of spring flowers, decorating the periphery of your vision. If you pulled over to stretch your legs or take a sip from a pint of bourbon, the stars would fill the windshield as soon as you killed the headlights. Leaning back into the worn leather seat and gazing at them through the tempered glass, listening to cosmic rock or blues crackling out of the speaker in the center of the dash while the cooling engine headers and exhaust pipes crackled back in reply, you felt at home. You knew what the elderly Chuang Tzu meant when, asked by his disciples what sort of funeral arrangements he preferred, he said, "My coffin will be Heaven and Earth; for the funeral ornaments of jade, there are the Sun and Moon; for my pearls and jewels, I shall have the stars and constellations; all things will be my mourners. Is not everything ready for my burial?"[2]

Because the ionosphere is full of holes, not all the music broadcast by the big radio stations bounces back to Earth. Some of it goes right through, escapes into space, and speeds on to the stars. Bits and pieces of the blues tunes we heard in the fifties are still out there, moving through space at the velocity of light. Fats Domino's "Blueberry Hill" and Sonny Boy Williamson's "Nine Below Zero" have reached hundreds of stars by now, some of which have planets. They could in theory have reached an extraterrestrial audience, if there's anyone there glued to a radio set sensitive enough to pick them up.

In principle we could hear their broadcasts, too, as well as stronger signals intentionally dispatched, which with existing radio telescopes could be detected from star systems thousands of light-years away.

This is not a new realization—Marconi and Tesla, pioneers in radio engineering, knew that radio signals could cross space, and had listened for signals from Mars—but technology had by now advanced to the point where such dreams could be turned into real possibilities. In the fifties, astronomers got hold of microwave receivers developed for radar installations during World War II and hooked them up to dish antennas to form radio telescopes. These "radio astronomers" studied the natural radio energy emitted by galaxies and nebulae, and bounced radar signals off Venus and the Moon. The journals that we amateur astronomers pored over began to publish not just photographs of galaxies but radio maps of them too, in which the dark, unseen outer reaches of these gigantic cities of stars were shown to extend far beyond their visible parts, like melting butter around an egg in a frying pan. In 1960, a young astronomer named Frank Drake aimed a radio telescope at two nearby stars that resemble the Sun and listened for artificial signals, inaugurating what came to be called SETI—the Search for Extraterrestrial Intelligence.

By the end of the twentieth century, a half-dozen privately funded SETI projects were under way, and more than a million people in a couple of hundred nations were donating idle time on their home computers to comb through SETI data, looking for signals. The data, collected mostly by the giant radio telescope at Arecibo, Puerto Rico, were dispatched via email to personal computers in Germany, Sweden, the Netherlands, Mongolia, and the Congo, and the processed results were automatically sent back the same way. High-tech corporations formed teams, each hoping to detect the first signal, that competed with one another over who could analyze the most SETI data the quickest, on work stations left humming through the night. In a matter of months after its inception, this novel experiment in amateur volunteerism, called "SETI@home," had built itself into the world's largest supercomputer, carrying out the largest computation project ever achieved on planet Earth.[3]

Seth Shostak of the SETI Institute was observing at Arecibo one night in 1999 when I called to ask how it was going. He said that SETI@home, by scrutinizing data that had been collected weeks before, nicely complemented the coarser, real-time analysis his team does at the telescope. His explanation of the rationale behind the project sounded a lot like what Chuck and I used to do with that little radio under the stars:

"We process our data more or less in real time," Shostak told me. "Right now we're monitoring twenty million channels. We pick up signals that look like ET all the time—several every minute—and when we get a good candidate, which happens about once a minute, we send it to a second telescope and check it right away. This is important, because any signal coming to us from a kiloparsec [3,260 light-years] or more away is subject to interstellar scintillation—changes in the signal strength—caused by the hot gas clouds between the stars. It's in many ways analogous to what you hear on your AM radio. In California, you might pick up WLS in Chicago for a minute or two, and then a minute later not get it, because of changes in the ionosphere. Something similar can happen over long distances in the galaxy. You might hear something and then go back to it and find that it's no longer there, because you went back when it was down in amplitude, owing to interstellar scintillation."[4]

Detection of an alien radio signal presumably would constitute the greatest discovery in the whole history of science, but nobody really knows what it would be good for. Despite a certain amount of loose talk about advanced aliens teaching us to cure cancer or end war—which reflects, I suspect, a rather parochial view of what they'd have on their minds—we don't know whether an alien message could be decoded, or, if so, whether the worldview it evinced would overlap with our own enough to permit comprehension, much less communication. My own expectation is that the message, although not terribly difficult to translate, will be so long and complex that to understand and absorb it would involve establishing entire institutions, in something like the way that European universities were founded to interpret the works of Aristotle.

But that's because SETI is genuine exploration. Columbus didn't know what he was going to find, either. It's said that Michael Faraday, experimenting with electrical generators in the 1830s, was asked by the British chancellor of the exchequer, William Gladstone, what use such a thing might be. "I know not," he replied, "but I wager that one day your government will tax it."[5] If we can predict the outcome of an enterprise, it's not exploration.

Which bears on the question of why an advanced civilization on a planet orbiting a distant star, presumably possessed of vast learning and a technology that beggars that of an infant civilization like ours (which has been on the air for merely a century), should bother sending signals that we can detect, or listening for our old radio broadcasts. If we envision alien civilizations as monoliths, there would seem to be little practical value in such an endeavor, except perhaps to enrich their studies of various biological and sociological systems—peering at us coldly, as through a dissecting microscope. But as our own planet demonstrates, intelligent societies need not be monolithic. They may instead thrive on the diversity that permits individuals to do things for no better reason than that they feel like it. In that sense, the first message we receive may have been transmitted, not by the Grand Pooh-Bah in charge of a galactic federation, but by the alien equivalent of a high-school ham club. Nor do the first aliens to pick up *The Jack Benny Show* and *I Love Lucy* need to be scientists, their tentacles manipulating dials of a big radio telescope on Upsilon Andromedae 3. They could be just kids like Chuck and me, searching the "ozone" for something new.

Much has been made of how we ought to feel chagrined by the possibility that the first impression humans make on aliens will be based on *Amos 'n Andy* or Bill "Hoss" Allen's nightly show on WLAC. But there's something to be said for humor and music, as opposed to the earnest intonations of official proclamations issued by politicians, and in a sense the old radio shows are more honest for being incidental, and lie closer to the wild spirit of the real explorers.

In the 1970s I produced a phonograph record that was launched aboard the twin Voyager interstellar space probes, a sampling of Earth's cultures for the benefit of anyone who might encounter the spacecraft in the course of their billion-year wanderings through the stars. Among the twenty-seven pieces of music on the *Voyager* record—which range from Bach and Beethoven to Javanese gamelan music, an ancient Chinese ch'in piece, and Blind Willie Johnson's "Dark Was the Night"—we included one rock and roll song, Chuck Berry's "Johnny B. Goode." It depicts a rural youth's hope that his playing can win him fame in the big city: .

> He used to carry his guitar in a gunny sack
> And sit beneath the trees by the railroad track.
> Oh, the engineers would see him sitting in the shade,
> Strumming to the rhythm that the drivers made.
> People passing by would stop and say
> Oh my that little country boy could play.[6]

The record occasioned a skit on the television comedy show *Saturday Night Live* in which earthly scientists receive a radio signal from an alien civilization that has intercepted the Voyager spacecraft and played the record. Their message: "Send more Chuck Berry!"

In restless youth I crossed the country a half-dozen times, driving by night, tuned to the ozone, but eventually things changed. The highway back then was a symbol of freedom, and one still heard talk of "the song of the open road." Now there are few genuinely open roads. The two-lane blacktops are pinched off by detours, and traffic can clog even remote stretches of interstate in the depths of night. Most of the time you drive as if in one car of an endless train, with taillights in the windshield and headlights in the rearview mirror. Nowadays the only truly open roads go straight up—through the ionosphere and the ozone, and on to the planets and the stars.

Standing Watch:
A Visit with Mr. White Man

AT CHABOT OBSERVATORY in Oakland, California, on a Saturday evening in the early 1990s, having just given a lecture to benefit this public science center, I stopped in at the dome that housed its 20-inch John Brashear refractor— built in 1914 and still a fine instrument for studying the planets. A docent walked me over. Inside the big, cream-colored dome a hundred schoolchildren were waiting in line to climb a set of steps and have a look through the eyepiece. An elderly astronomer stood by, helping each child adjust the focus and whispering to each of them about Saturn, its size and distance and composition, and the fact that the dot of light to one side was not a star but its satellite Titan.

"Who is that?" I asked.

"Mr. Wightman," the docent answered. He pronounced it "White Man," and, indeed, Kingsley Wightman was one of the whitest men I'd ever seen. His hair was white, his skin was nearly as white as good bond paper, and he was dressed all in white. He seemed like an apparition, born with the telescope (actually he was thirty-three years younger than that), a living link with the astronomy of the prior century. The docent informed me that due to Chabot's funding problems—the observatory was at that time operated by the cash-strapped Oakland public-school system—Mr. Wightman had not been paid for his work in years. Yet he continued to show up for public sessions, insistent on showing children the planets and stars. I waited for a break, then introduced myself.

"How may I be of help to you?" Mr. Wightman asked, with a slight bow.

"I know Chabot has been having financial troubles," I said. "How are things going?"

He laughed. "That, sir, is the very history of Chabot Observatory. Every semester, the question confronting us is, 'How can we possibly keep Chabot open for just one more semester? If we can keep it open for just one more semester, we ought to be able to solve this budget crisis.' Every year I go and speak, here and there, about the value of Chabot, and why we should keep it open. They say, 'But, Kingsley, we don't have enough money to teach reading, writing, and arithmetic: Why should we be teaching astronomy?' But then, they get one look through the telescope . . ." He shook his head, and his voice trailed off.

Thereafter I often thought about Mr. White Man, and made a point of stopping in to see him when I was in the neighborhood on public nights. I learned that the observatory had been built at the instigation of the Oakland school superintendent, James C. Gilson, and was named for Anthony Chabot, a hydraulic engineer and philanthropist who had laid many of the city's first water pipes. Wightman told me that he had graduated from Berkeley with a degree in education and had then gone back and studied astronomy in order to get a job teaching public school in Oakland. "I was thrilled to teach at Chabot Observatory," he recalled. "I got hooked the first time I looked through the telescope, and evidently it had the same effect on other people."

Saturn literally moved him to tears. "Saturn, to my mind, is just a beautiful object," he said. "The proportions, the symmetry, the colors. Imagine my enthusiasm upon first seeing that planet! And it's just as great now."

Over the years, while Wightman and his colleagues struggled to keep Chabot open, civic leaders formed a foundation to raise money for a Chabot Science Center that would house the telescopes and provide the community with a new planetarium and teaching facilities. One day the word came through that they had succeeded in obtaining a $17-million grant, toward what would eventually become a $74.5-million facility that opened in the year 2000.

On a Saturday night after the grant was announced, I stopped in at the old observatory and was gratified to see Mr. Wightman there, under the vanilla dome, showing Jupiter to a line of schoolchildren. When he had finished I congratulated him on the new grant.

"We did it!" he said, laughing gleefully. "For years, it was, 'How can we possibly keep Chabot open for just one more semester?' Now we have money pouring out of our ears! It's, 'Do you need a little more money, Kingsley? Would a million do it? Would you like two million?'

"I'll tell you one thing," he confided, leaning close and lowering his voice. "They don't talk about closing you down anymore, once you've got seventeen big ones!" A new group of schoolchildren entered the dome, and Mr. White Man bade me a courtly goodbye. I watched as he made his way to the wheeled steps, his left arm trembling slowly with palsy. He took the first student by the hand and led her up to the telescope.

4.

Amateurs

Amateur, from the French, *amateur*, Latin, *amator* . . . to love.
—*Oxford English Dictionary*

To know something is not as good as loving it; to love something is not as good as rejoicing in it.
—Confucius

A T SUNDOWN AT A STAR PARTY on the high Texas plains near Fort Davis, west of the Pecos, the parched landscape was crowded with telescopes. Reared against the darkening skies to the west rose a set of rolling foothills known jocularly as the Texas Alps. To the east of us lay dinosaur country, with its wealth of oil.

The stars came out with imposing clarity—Orion fleeing toward the western horizon, pursued by the dog star, brilliant white Sirius, the square of Corvus the crow to the southeast, the scythe of Leo the lion near the zenith. The planet Jupiter stood almost at the zenith; scores of telescopes were pointed toward it, like heliotropes following the Sun. As the gathering darkness swallowed up the valley, the sight of the observers was replaced by land-bound constellations of ruby LED indicators on the telescopes' electronics, the play of red flashlights, and voices—groans, labored breathing, muttered curses, and sporadic cries of delight when a bright meteor streaked across the sky. Soon it was dark enough to see the zodiacal light—sunlight reflected off interplanetary dust grains ranging out past the asteroid belt—stabbing the western sky like a distant searchlight and looking like what Omar Khayyam and his translator Edward FitzGerald referred to as "Dawn's Left Hand." When the Milky Way rose over the hills to the east, it was so bright that I at first mistook it for a bank of clouds. Under skies this transparent, the Earth becomes a perch, a platform from which to view the rest of the universe, rather like the foothold one achieves at the top of a rickety stepladder in order to peer into the eyepiece of one of these giant Newtonian telescopes.

I HAD COME HERE to observe with Barbara Wilson, legendary for her sharp-eyed pursuit of things dark and distant. I found her atop a small ladder, peering through her 20-inch Newtonian—an instrument tweaked and collimated to within an inch of its life, with eyepieces that she scrubs with Q-tips before each observing session, using a mixture of Ivory soap, isopropyl alcohol, and distilled water. On an observing table, Barbara had set up *The Hubble Atlas of Galaxies*, the *Uranometria 2000* star atlas, a night-vision star chart illuminated from behind by a red-bulb light box, a laptop computer pressed into service as yet another star atlas, and a list of things she hoped to see. I'd never heard of most of the items on her list, much less seen them. They included Kowal's Object (which, Barbara informed me, is a dwarf galaxy in Sagittarius), the galaxy Molonglo-3, the light from which set out when the universe was half its present age, and obscure nebulae with names like Minkowski's Footprint, the Red Rectangle, and Gomez's Hamburger.

"I'm looking for the jet in M87," Barbara called down to me from the ladder. M87 is a galaxy located near the center of the Virgo cluster, sixty million light-years from Earth. A white jet protrudes from its nucleus. It is composed of plasma—free atomic nuclei and electrons, the survivors of events sufficiently powerful to have torn atoms apart—spat out at nearly the velocity of light from near the poles of a massive black hole at the center of this giant elliptical galaxy. (Nothing can escape from inside a black hole, but its gravitational field can slingshot matter away at high speeds.) Astronomers study the structure of the jet to map dark clouds in M87—clouds whose locations and densities can be inferred from the way the jet stacks up where it has collided with them—and to reconstruct, from knots and clumps along the line, the differing amounts of stuff ejected from the vicinity of the black hole in its recent history. To do this they use the most powerful instruments available to them, including the Hubble Space Telescope, the twin ten-meter reflectors at Keck Observatory in Hawaii, and the Very Large Array, a Y-shaped deployment of twenty-seven radio dish antennas splayed across the desert in New Mexico. I'd never heard of an amateur's having seen it.

There was a long pause. Then Barbara exclaimed, "It's there! I mean, it's *so* there!"

She climbed down the ladder, her smile bobbing in the dark. "I saw it once before, from Columbus," she said, "but I couldn't get anybody to confirm it for me—couldn't find anyone who had the patience that it takes to *see* this thing. But it's *so* obvious once you see it that you just go, 'Wow!' Are you ready to try?"

I climbed the ladder, focused the eyepiece, and examined the softly glowing ball of M87, inflated like a blowfish at a magnification of 770x. No jet yet, so I went into standard dim-viewing practice. Relax, as in any sport. Breathe fairly deeply, to make sure the brain gets plenty of oxygen. Keep both eyes open, so as not to strain the muscles in the one you're using. Cover your left eye with your palm or just blank it out mentally—which is easier to do than it sounds—and concentrate on what you're seeing through the telescope. Check the chart to determine just where the object is in the field of view, then look a bit *away* from that point: The eye is more sensitive to dim light just off-center than straight ahead. And, as Barbara says, be patient. Once, in India, I peered through a spotting telescope at a patch of deep grass for more than a minute before realizing that I was seeing the enormous orange-and-black head of a sleeping Bengal tiger. Stargazing is like that. You can't hurry it.

Then, suddenly, there it was—a thin, crooked, bone-white finger, colder and starker in color than the pewter starlight of the galaxy itself, against which it now stood out. How wonderful to see something so grand, after years of admiring its photographs. I came down the ladder with a big smile of my own. Barbara called a coffee break and her colleagues departed for the ranch house cafeteria, but she remained by the telescope, in case anyone else came along who might want to see the jet in M87.

Amateur astronomy had gone through a revolution since I started stargazing in the fifties. Back then, most amateurs used reedy telescopes like my 2.4-inch refractor. A 12-inch reflector was considered a behemoth, something you told stories about should you be lucky enough to get a look through one. Limited by the light-gathering power of their instruments, amateurs mostly observed bright objects, like the craters of the Moon, the satellites of Jupiter, the rings of Saturn, along with a smattering of prominent nebulae and star clusters. If they probed beyond the Milky Way to try their hand at a few nearby galaxies, they saw little more than dim gray smudges.

Professional astronomers, meanwhile, had access to big West Coast telescopes like the legendary 200-inch at Palomar Mountain in Southern California.[1] Armed with the most advanced technology of the day and their own rigorous training, the professionals got results. At Mt. Wilson Observatory near Pasadena, the astronomer Harlow Shapley in 1918–19 established that the Sun is located toward one edge of our galaxy, and Edwin Hubble in 1929 determined that the galaxies are being carried apart from one another with the expansion of cosmic space. At Palomar, Allan Sandage age-dated the stars, Halton Arp probed the structures of "peculiar" galaxies, and in 1964 Maarten Schmidt and Jesse Greenstein discovered that quasars are remote—the fiery nuclei of young galaxies whose light has taken billions of years to reach us. Professionals like these became celebrities, lionized in the press as hawk-eyed lookouts probing the mysteries of deep space.

Which, pretty much, they were: Theirs was a golden age, when our long-slumbering species first opened its eyes to the universe beyond its home galaxy. But observing the professional way wasn't usually a lot of fun. To be up there in the cold and the dark, riding in the observer's cage and carefully guiding a long exposure on a big glass photographic plate, with icy stars shining through the dome slit above and starlight puddling below in a mirror the size of a trout pond, was indubitably romantic but also a bit nerve-racking. Big-telescope observing was like making love to a glamorous movie star: You were alert to the honor of the thing, but aware that plenty of suitors were eager to take over should your performance falter. Nor did academic territoriality, jealous referees, and the constant competition for telescope time make professional astronomy a day at the beach. As a brilliant young cosmologist once told me, "A career in astronomy is a great way to screw up a lovely hobby."

So it went, for decades. Professionals observed big things far away, and published in the prestigious *Astrophysical Journal*—which, as if to rub it in, ranked papers by the distances of their subjects, with galaxies at the front of each issue, stars in the middle, and planets, on the rare occasion that they appeared in the *Journal* at all, relegated to the rear. Amateurs showed school-children the rings of Saturn at 76 power through a tripod-mounted spyglass at the State Fair, and sent snapshots of themselves (smiling, standing next to a homemade reflector in the yard) to *Sky & Telescope*—a fine publication, but not the eminent *ApJ*. Inevitably, a few professionals disdained the amateurs. When Clyde Tombaugh discovered Pluto, the astronomer Joel Stebbins, usually a more charitable man, dismissed him as "a sub-amateur assistant," adding that since the existence of Pluto "was predicted by the amateur [Percival] Lowell" and was confirmed with a telescope "made largely by another amateur, the Reverend Joel H. Metcalf," it followed that if Pluto was a planet, its discovery amounted to "the greatest joke that ever happened to the astronomical profession."[2] There were of course professionals who kept up good relationships with amateurs, and amateurs who did solid work without fretting over their status. But generally speaking, the amateurs lived in the valley of the shadow of the mountaintops.

Which was odd, in a way, because for most of its long history, astronomy has been primarily an amateur pursuit.

The foundations of modern astronomy were laid largely by amateurs. Nicolaus Copernicus, who in 1543 moved the Earth from the center of the universe and put the Sun there instead (thus replacing a dead-end mistake with an open-ended mistake, one that encouraged the raising of new questions), was a Renaissance man, adept at many things, but only a sometime astronomer. Johannes Kepler, who discovered that planets orbit in ellipses rather than circles, made a living mainly by casting horoscopes, teaching grade school, and scrounging royal commissions to support the publication of his books. Edmond Halley, after whom the comet is named, was an amateur whose accomplishments—among them a year spent observing from St. Helena, a South Atlantic island so remote that Napoleon Bonaparte was sent there, forty years later, to serve out his second and terminal exile—eventually got him named Astronomer Royal. The amateur astronomer John Bevis is still the only person to have observed an occultation of Mercury by Venus. (This rare event, a passage of Venus in front of Mercury, took place on May 28, 1737; the next one will occur on December 3, 2133.) James Ferguson, born in 1710, learned the stars as an illiterate shepherd boy in Scotland before going on to become a popular author, a fellow of the Royal Society, and the beneficiary of a royal pension from King George III—himself an amateur astronomer, who built an observatory to watch the June 3, 1769, transit of Venus across the Sun.[3] Ferguson's immensely popular 1756 book *Astronomy Explained Upon Sir Isaac Newton's Principles* captured the attention of the composer and organist William Herschel, who made his own telescopes and wielded them with sufficient skill to become one of the most acute observers of all time—but who didn't get paid for his research until he discovered the planet Uranus, in 1781. (George III rewarded him, too, with a royal stipend; Herschel used the money to concoct an enormous telescope, but it never worked very well, and he reverted to relying on the smaller ones he'd fashioned in his amateur days.) The first to "recover" Halley's comet—that is, to find it in the sky as it made its way inward from beyond the orbit of Uranus, brightening all the while—was Johann Georg Palitzsch, in 1758. For this feat, which ably demonstrated the validity of Newton's laws of motion, Palitzsch was lampooned by the Royal Academy of Science in Paris as "a simple farmer who did not realize the importance of his discovery."[4] Actually he was a sophisticated amateur scientist, who regularly corresponded with Herschel and correctly theorized that Algol is an eclipsing binary star—a pair of stars that periodically pass in front of each other.

Nineteenth-century astronomy was dominated by amateurs, most of them either wealthy gentlemen with private observatories or ordinary citizens whose skills sometimes won them royal or mercantile patronage. Friedrich Wilhelm Bessel, a high-school dropout working as a merchant's clerk in Bremen, did useful research on the orbits of comets and was made director of Prussia's new Königsberg Observatory in 1813. The first to have recognized his abilities was Heinrich Olbers, a physician who was also the leading comet observer of the day. Wilhelm Tempel, who discovered eight comets and the reflection nebula that entangles the Pleiades, was a lithographer with little formal education who ultimately garnered professional appointments in Marseilles, Milan, and Florence. In Dessau, Germany, the pharmacist Heinrich Schwabe studied sunspots on virtually every clear day for seventeen years, and discovered that the number of sunspots waxes and wanes in eleven-year cycles. The English amateur Richard Carrington established that sunspots are found increasingly close to the solar equator as each sunspot cycle evolves, creating a "butterfly pattern" when the latitudes of sunspots are plotted against time. Observing the Sun from his estate at Redhill, Surrey, Carrington was also the first to see a solar flare. Among those inspired by these solar discoveries was John Robertson, a porter who called out station names on the Caledonian Railway. Robertson scrutinized astronomy books in the local workingmen's reading room, spent his savings on a telescope, and eventually was publishing accounts of his observations of comets, meteors, and the relationship between sunspots, auroras, and magnetic compass-needle deviations. He was offered observatory posts but always declined, explaining that "the [railway] company are very kind to me, and I hope to serve them faithfully."[5]

In Ireland in the mid-1800s, Lord Rosse built a 72-inch reflector, with which he observed the spiral patterns in galaxies. Known as "the Leviathan of Parsonstown," it reigned as the largest telescope in the world until 1917, when the 100-inch reflector at Mt. Wilson was completed at the instigation of an amateur turned professional astronomer, George Ellery Hale. Henry Draper, who had completed his medical studies and was touring England while waiting to begin practicing medicine (he was still a minor), saw Rosse's telescope and became entranced by astronomy. Back home in New York, Draper built a series of telescopes and combined a spectroscope—which breaks down light into its component spectral lines, revealing which atoms are involved in making the light—with a camera to create the first reliable spectrograph, with which he helped in the creation of modern astrophysics and the classification of stars. Joseph Norman Lockyer, a civil servant in the War Office and self-taught in astronomy, discovered the element helium in the solar spectrum. (As it was as yet unknown on Earth, Lockyer named it after the Greek term for "of the Sun.") In India another amateur, Pierre-Jules-César Janssen, managed to observe a solar prominence—a streamer of glowing plasma arising from the Sun's surface—that had been spotted the previous day, August 18, 1868, during a total solar eclipse. This was the first prominence to be observed without the benefit of an eclipse, during which the Moon blocks most of the Sun's light. (Lockyer observed the same prominence a few days later; both he and Janssen went on to become professional astronomers.) Lockyer later established the British science journal *Nature*, in 1869, and determined that Stonehenge and the great pyramids of Egypt are oriented to the Sun and stars, a finding that inaugurated the science of archaeoastronomy. Meanwhile Warren De la Rue, a printer, had proved, using photographs taken of an 1860 solar eclipse from stations 400 kilometers apart, that prominences do indeed belong to the Sun and are not emitted by the Moon, as some had thought.

Harvard College Observatory was founded by an amateur astronomer, William Cranch Bond, who with his son George Phillips Bond discovered Hyperion, the eighth moon of Saturn. William Rutter Dawes, known as "Eagle Eye Dawes" for his acute observations of binary stars, was a physician and Nonconformist clergyman; "Dawes's limit" is still employed today to estimate the theoretical resolving power of telescopes. The silk merchant William Huggins made spectroscopic observations of planetary nebulae—shells of gas cast off by unstable stars—and measured the velocities of stars. The Liverpool brewing magnate William Lassell, who met his wife Maria King at a star party held by fellow amateur astronomers, built a 24-inch reflecting telescope with which he discovered Neptune's satellite Triton and Uranus's satellites Ariel and Umbriel.

Even in the twentieth century, while they were being eclipsed by the burgeoning professional class, amateurs continued to make valuable contributions to astronomical research. As the British author and amateur astronomer Patrick Moore recalled, prior to World War II, "if you wanted to know the longitude of Jupiter's Red Spot, the state of ashen light on Venus, or whether dust storms were developing on Mars, you went to one of the main amateur associations."[6] Arthur Stanley Williams, a lawyer, charted the differential rotation of Jupiter's clouds and created the system of Jovian nomenclature used in Jupiter studies ever since. Milton Humason, a former watermelon farmer who worked as a muleteer at Mt. Wilson, teamed up with the astronomer Edwin Hubble to chart the size and expansion rate of the universe. Grote Reber built the first true radio telescope in the back yard of his home in Wheaton, Illinois, and used it to chart the sky; for a time he was the only radio astronomer on Earth. The solar research conducted by the industrial engineer Robert McMath, at an observatory he built in the rear garden of his home in Detroit, so impressed astronomers that he was named to the National Academy of Sciences, served as president of the American Astronomical Society, a professional organization, and helped plan Kitt Peak National Observatory in Arizona where the world's largest solar telescope was named in his honor.

The fitful growth of professional astronomy in the late nineteenth and early twentieth centuries produced mixed careers like that of Sherburne Wesley Burnham, who combined amateur and professional work like a man stepping indecisively from one boat to another in midstream. A self-taught court stenographer who observed and cataloged double stars, Burnham was persuaded to spend two months on Mt. Hamilton in Northern California, testing the future site of Lick Observatory with his personal telescope, a six-inch refractor. When the observatory opened, Burnham took a pay cut to stay on as senior staff astronomer. Witty, cosmopolitan, and well-connected, he was equally at ease conversing with a hermit named Bennett, who stopped by the observatory regularly to pick up his mail, and with the chief justice of the United States Supreme Court. Eventually he quit, returning to Chicago and signing on as clerk of the United States Circuit Court at double his astronomer's salary. But he kept observing, on weekends, when he would take the train up to Yerkes Observatory, in Williams Bay, Wisconsin, to study double stars.

Why were the amateurs, having played such important roles in astronomy, eventually overshadowed by the professionals? Because astronomy, like all the sciences, is young—less than four hundred years old, as a going concern—and somebody had to *get* it going. Its instigators could not very well hold degrees in fields that didn't yet exist. Instead, they had to be either professionals in some related field, such as mathematics, or amateurs doing astronomy for the love of it. What counted was competence, not credentials: If you're raising the roof beam of a log cabin in the wilderness, you don't demand that your neighbors present contractor's licenses when they show up to help. That comes later, when the frontier has been turned into a subdivision.

The novelty of distinguishing between amateurs and professional scientists is reflected in the history of the terms themselves. The word *amateur* didn't enter the English language until around 1784, and *scientist* wasn't coined until 1840, when the English philosopher and mathematician William Whewell mused, "We need very much a name to describe a cultivator of science in general. I should incline to call him a scientist."[7] But once the distinction arose, there emerged, as the Oxford historian Allan Chapman notes, an "unfortunate dichotomy in which institutionally-funded expertise was supposed to set the standard to which 'amateurs' merely aspired."[8]

One capable amateur who felt the ragged edge of this dichotomy was John Edward Mellish. While serving as an unpaid observer at Yerkes Observatory in the summer of 1915, Mellish spotted what he took to be a comet, just before dawn. A Harvard Observatory telegram—the standard means of broadcasting new astronomical findings—was promptly dispatched, announcing his discovery. This proved embarrassing to the Yerkes director, Edwin Brant Frost, when it was soon ascertained that Mellish's "comet" was actually the diffuse nebula NGC 2261. What Mellish actually should have gotten credit for was discovering a nebula that varies in brightness.[9] But Frost, unwilling to trust an amateur further, took Mellish off the project, assigning it instead to a young Yerkes staff astronomer named Edwin Hubble. Hubble's Variable Nebula, as it has been called ever since, became the subject of Hubble's first published paper and the springboard of his distinguished career.[10]

Despite such frustrations, amateurs were back on the playing field by about 1980. A century of professional research had by then greatly increased the range of observational astronomy, creating more places at the table than there were professionals to fill them. Meanwhile the ranks of amateur astronomy had grown, too, along with the ability of the best amateurs to take on professional projects and also to pursue innovative research. "There will always remain a division of labor between professionals and amateurs," wrote the historian of science John Lankford in 1988, but "it may be more difficult to tell the two groups apart in the future."[11] That same year, Leif J. Robinson, the editor of *Sky & Telescope* magazine, reported that "amateurs and professionals have begun talking to one another with enthusiasm not seen in decades. There's a real freshness in the air."[12]

The amateur astronomy revolution, like the one that raised up the professionals, was incited by three technological innovations—the Dobsonian telescope, CCD light-sensing devices, and the Internet.[13]

Dobsonians are reflecting telescopes constructed from cheap materials. They were invented by John Dobson, a populist proselytizer who championed the view that the worth of telescopes should be measured by the number of people who get to look through them. As Dobson enunciated his credo, to thunderous applause from a group of amateur telescope makers assembled in Stellafane (literally "shrine to the stars"), a hilltop observing site near Springfield, Vermont, on July 25, 1987, "To me it's not so much how big your telescope is, or how accurately your optics are figured, or how beautiful the pictures you can take with it; it's how many people in this vast world less

privileged than you have had a chance to see through your telescope and understand this universe. That is the one thing that drives me!"

Dobson was well known in San Francisco as a spare, ebullient figure who
would set up a battered telescope on the sidewalk, call out to passers-by to
"Come see Saturn!" or "Come see the Moon!" then whisper astronomical lore
in their ears while they peered into the eyepiece. To the casual beneficiaries of
his ministrations, he came off as an aging hippie with a ponytail, a ready spiel,
and a gaudily painted telescope so dinged-up that it looked as if it had been
dragged behind a truck. But astronomical sophisticates came to recognize his
telescopes as the carbines of a scientific revolution. Dobsonians employed the
same simple design that Isaac Newton dreamed up when he wanted to study
the great comet of 1680—a tube with a concave mirror at the bottom to
gather starlight, and a small, flat, secondary mirror near the top to bounce the
light out to an eyepiece on the side—but they were made from such inexpensive materials that you could build or buy a big Dobsonian for the cost of a
small traditional reflector. You couldn't buy a Dobsonian from John Dobson,
though; he refused to profit from his innovation. Poverty was his habit, and
the mother of his invention.

Born in China, where his maternal grandfather was a founder of Peking
University and his father a teacher of zoology, Dobson studied chemistry at
Berkeley and in 1944 became a monk in the Vedanta Society, part of the Rama
Krishna order, and lived in monasteries in San Francisco and Sacramento.

Fascinated by astronomy since boyhood, he started making telescopes, but
his vow of poverty meant that he had to invent ways of making them much
more cheaply than had been done before. One of the significant expenses confronted by amateur telescope makers was buying a "blank," a piece of fine
optical glass that was then ground to the proper figure. Dobson couldn't afford
proper blanks, so he salvaged discarded portholes and jug bottoms and ground
them instead. For the telescope tube, he scrounged up cardboard sonotubes,
used for pouring concrete at construction sites.

Good telescope mountings are expensive, so Dobson cut a box out of scrap
plywood, lined it with anything slippery—PVC piping, scrap Teflon, or discarded vinyl phonograph records—and stuck the tube into the box, creating a
telescope that could be pointed anywhere in the sky just by giving it a shove.

Thus equipped, Dobson started "going over the wall" at night, as the
monks say, to deploy his telescope and show people the stars. When he encountered a youngster who seemed particularly passionate about astronomy, Dobson would give the kid his telescope and build another one. Since too much
going over the wall was presumed to indicate nocturnal preoccupations inap-

propriate for monks, Dobson was eventually banished from monastery life. (His first telescope, which otherwise might have wound up in the Smithsonian Institution, is said to have been flung into San Francisco Bay at the behest of an official at the monastery.) Much of Dobson's time thereafter was spent touring the West in a series of motor homes and rusted vans crammed with telescopes that he set up on street corners and in national parks, attracting long lines of neophytes. When a park ranger objected to his activities, saying, "The sky is not a part of the park," Dobson replied, "No, but the park is a part of the sky."[14] It is likely that John Dobson personally introduced more people to stargazing than anyone ever had before.

Dobsonians might be optically rather crude—few could deliver the sharp images of Mars and Jupiter traditionally cherished by amateurs—but they made it possible for stargazers of average means to own large telescopes. Observers armed with big Dobsonians didn't have to content themselves with looking at planets and nearby nebulae: They could explore thousands of galaxies, invading deep-space precincts previously reserved for the professionals. Soon, the star parties where amateur astronomers congregate were dotted with Dobsonians that towered twenty feet and more into the darkness. The British amateur astronomer Patrick Moore, describing the "Leviathan of Parsonstown," wrote, "It has been said that anyone who set out to use the great reflector had to be not only an astronomer, but also an experienced mountaineer."[15] Now, thanks to Dobson, the era of Lord Rosse had returned, and the greatest physical risk to amateur observers became that of falling from a rickety ladder high in the dark while peering through a gigantic Dobsonian. I talked with one stargazer whose Dobsonian stood so tall that he had to use binoculars to see the display on his laptop computer from atop the fifteen-foot ladder required to reach the eyepiece, in order to tell where the telescope was pointing. He said he found it frightening to climb the ladder by day but forgot about the danger when observing by night. "About a third of the galaxies I see aren't cataloged yet," he mused.

Meanwhile the CCD had come along—the "charge-coupled device," a light-sensitive chip that can record faint starlight much faster than could the photographic emulsions that CCDs soon began replacing.[16] CCDs initially were expensive but their price fell steeply: Obsolete CCD chips that a few years earlier cost more than mink coats are today being used in observatories as coffee-mug coasters. Amateurs who attached CCDs to large Dobsonians found themselves in command of light-gathering capacities comparable to that of the 200-inch Hale telescope at Palomar in the pre-CCD era.

The sensitivity of CCDs did not in itself do much to close the gap separating amateur from professional astronomers—since the professionals had CCDs too—but the growing quantity of CCDs in amateur hands vastly increased the number of telescopes on Earth capable of probing deep space. It was as if the planet had suddenly grown thousands of new eyes, with which it became possible to monitor many more astronomical events than there were professionals enough to cover. And, because each light-sensitive dot (or "pixel") on a CCD chip reports its individual value to the computer that displays the image it has captured, the stargazer using it has a quantitative digital record that can be employed to do photometry, as in measuring the changing brightness of variable stars.

Which brings us to the Internet. It used to be that an amateur who discovered a comet or an erupting star would dispatch a telegram to the Harvard College Observatory, from which a professional, if the finding checked out, mailed postcards to paying subscribers at observatories around the world. The Internet opened up alternative routes. Now an amateur who made a discovery—or thought he did—could send CCD images of it to other observers, anywhere in the world, in minutes. Anyone who took an interest could take a look, whether he was a back-yard amateur or a professional in a mountaintop dome. Global research networks sprang up, linking amateur and professional observers with a common interest in flare stars, comets, or asteroids. Professionals sometimes learned of new developments in the sky more quickly from amateur news than if they had waited for word through official channels, and so were able to study them more promptly. If the growing number of telescopes out there gave the Earth new eyes, the Net fashioned for it a set of optic nerves, through which flowed (along with reams of financial data, gigabytes of gossip, and cornucopias of pornography) news and images of storms raging on Saturn and stars exploding in distant galaxies.

For decades, professional astronomers had increasingly relied on computerized databases that could, for instance, instantly generate a list of every face-on spiral galaxy within a given brightness range that would pass within ten degrees of the zenith on a given night at a given observatory. The requisite software, processing power, and disk-storage capacity were expensive to acquire, but once the professionals started putting them online, amateurs were free to use them, too. And, since most professional telescopes were by now computer-controlled, the Net facilitated programs under which amateurs and professionals alike could order up automated observing runs on telescopes that they need never visit.

Amateur superstars emerged, armed with the skills, tools, and dedication to do what the eminent observational cosmologist Allan Sandage called "absolutely serious astronomical work."[17] Some chronicled the weather on Jupiter and Mars, producing planetary images that rivaled those of the professionals in quality and surpassed them in documenting long-term planetary phenomena. Others monitored variable stars useful in determining the distances of star clusters and galaxies. Amateurs discovered comets and asteroids, contributing to the continuing effort to identify objects that may one day collide with the Earth and that, if they can be found early enough, might be deflected to prevent such a catastrophe. Amateur radio astronomers recorded the outcries of colliding galaxies, chronicled the ionized trails of meteors falling in daytime, and listened for signals from alien civilizations.

The amateur approach had its limitations. Amateurs insufficiently tutored in the scientific literature sometimes acquired accurate data but did not know how to make sense of it. Those who sought to overcome their lack of expertise by collaborating with professionals sometimes complained that they wound up doing most of the work while their more prestigious partners got most of the credit. Others burned out, becoming so immersed in their hobby that they ran low on time, money, or enthusiasm and called it quits.

But many amateurs enjoyed fruitful collaborations, and all were brought closer to the stars. If they sometimes went too far, they could console themselves with the reflection that love is conjugal with excess. It's fine to strive for moderation in all things, as Terence advised in the second century B.C., but as William Blake noted, "You never know what is enough unless you know what is more than enough."[18]

How Much Can You See?
A Visit with Stephen James O'Meara

I MET STEPHEN JAMES O'MEARA at the Winter Star Party, held annually alongside a sandy beach in West Summerland Key, Florida. Arriving after dark, I was greeted at the gate by Tippy D'Auria, the founder of the Winter Star Party. An unfettered man, Tippy dropped out of school and left home at age thirteen to work as a charterboat captain, running moonlight tarpon-fishing expeditions, served in the Navy on an attack submarine, and raced cars before settling down as an electronics engineer in Miami. He and his wife, Patricia, were introduced to stargazing, as so many are, by their first sight of Saturn's rings. They bought a small telescope that same night, then a larger one, and had been observing ever since. Tippy led me through thickets of telescopes reared against the stars—eccentric home-made instruments fashioned from cardboard, plywood, aluminum, steel, or lovingly varnished cherry; a pair of Newtonian reflectors mounted together so that their paired eyepieces formed an enormous pair of binoculars; sleek refractors with cameras and CCD cameras mounted on them—past Bedouin-style tented portable observatories looming against the white beach sand, to a circle of aluminum lawn chairs where Tippy and the other tribal elders were gathered. He assigned me a chair and for the rest of the night, no matter how long I had been away or how many people were standing around, that chair was always empty and waiting.

"Steve's up there, drawing Jupiter through my telescope," Tippy said, nodding toward the silhouette of a young man perched atop a stepladder at the eyepiece of a big Newtonian that was pointing into the southwest sky. Comfortable in my lawn chair, I listened to the elders talk—a mix of astronomical expertise and self-deprecatory wit, the antithesis of pomp—and watched O'Meara drawing. He would peer at length through the eyepiece, then glance down at his sketch pad and draw a line or two, then return to the eyepiece. It was the sort of work astronomers did generations ago, when observing could mean spending most of a night making one drawing of one planet.

O'Meara likes to describe himself as "a nineteenth-century observer in the twenty-first century," and in meeting him I hoped to better understand how someone who works the old-fashioned way, relying on his eye at the telescope rather than a camera or a CCD, had been able to pull off some of the most impressive observing feats of his time. While still a teenager, O'Meara saw and mapped radial "spokes" on Saturn's rings that professional astronomers dismissed as illusory—until *Voyager* reached Saturn and confirmed that the spokes were real. He determined the rotation rate of the planet Uranus, obtaining a value wildly at variance with those produced by professionals with larger telescopes and sophisticated detectors, and proved to be right about that, too. He was the first human to see Halley's comet on its 1985 return, a feat he accomplished using a 24-inch telescope at an altitude of 14,000 feet while breathing bottled oxygen.

Time passed. O'Meara kept drawing. Someone called out that the Hubble Space Telescope could be seen gliding overhead in orbit, and hundreds of shadowy figures turned to follow the path of a dot of light that crossed the Milky Way and disappeared into the Earth's shadow—a big, distant telescope sailing above the smaller ones close at hand. From out of the darkness I could hear "Crazy Bob" Summerfield shouting encouragement to a dozen or so people lined up to take a look at the Orion nebula through his gargantuan 36-inch Dobsonian. "This is the world's largest portable observatory dedicated to public education," he exclaimed. "We've put in over one hundred thousand miles on this scope, and close to a quarter of a million people have looked through it. Once you get on the ladder, it's your telescope. Take as long to look as you like. The people waiting in line and bitching at you are just jealous. When it's their turn they'll take too long, too." He sounded like a carnival barker, although looking through telescopes at star parties is free of charge.

After nearly an hour, O'Meara came down the ladder and made a gift of his drawing to Tippy, who introduced us. Clear-eyed, fit, and handsome, with black hair, a neatly trimmed beard, and a wide smile, O'Meara was dressed in a billowing white shirt and black peg pants, and would not have looked out of place on the deck of a British man-of-war. We repaired to the red-lit canteen for a cup of coffee and a talk.

Steve told me that he'd grown up in Cambridge, Massachusetts, the son of a lobster fisherman, and that his first childhood memory was of sitting in his mother's lap and watching the ruddy lunar eclipse of 1960. "From the very beginning I had an affinity with the sky," he said. "I just loved starlight." When he was about six years old he cut out a planisphere—a flat oval sky map—from the back of a box of cornflakes, and with it learned the constellations. "Even the tough kids in the neighborhood would ask me questions about the sky," he recalled. "I'd see them on the street corner, sneaking a cigarette, and they'd call me over and instead of beating me up would ask, 'Hey, what's that star?' The sky produced a wonderment in them. I believe that if inner-city kids had the opportunity to see the real night sky, they could believe in something greater than themselves—something that they can't touch, control, or destroy."

When O'Meara was about fourteen years old he was taken to a public night at Harvard College Observatory, where he waited in line for a look through its venerable Clark nine-inch refractor. "Nothing happened for a long time," he recalled. "The line was stuck, and eventually people started wandering off, discouraged. The next thing I knew I was inside the dome. I could hear a whirring sound and see the telescope pointing up at the stars, and a poor guy down there at the eyepiece—searching, searching—and he was sweating. I realized that he was trying to find the Andromeda galaxy. I asked him, 'What are you looking for?'

"'A galaxy far away.'

"I waited a few minutes, then asked, 'Is it Andromeda?' There was a silence, and finally he said, 'Yeah, but it's difficult to get, very complicated.'

"'Can I try?'

"'Oh, no, it's a very sophisticated instrument.'

"I said, 'You know, nobody's behind me. I can get it for you in two seconds.' I got it in the field of view, and he said, 'OK, go bring the other people back here, but don't leave.' Everyone got to see the Andromeda galaxy through the telescope, and after they left he said, 'Show me what you know.' He was just a graduate student, and like a lot of astronomy graduate students he didn't really know the sky. I showed him around, acquainted him with Messier galaxies and all sorts of things. We stayed up till dawn. The next morning he took me to the business office and they gave me a key, saying that if I helped them out with open houses, in return I could use the scope any time I wanted. So now I was a fourteen-year-old kid with a key to the Harvard College Observatory!"

For years thereafter the observatory was O'Meara's second home. After school he would work afternoons in a Cambridge pharmacy, then spend his nights at the telescope, patiently making drawings of comets and planets. "Why draw at the telescope? Because what you get on film and CCD does not capture the essence of what you see with the eye," he told me. "Everyone looks at the world in a different way, and I'm trying to capture what I see, and encourage others to look, to learn, to grow and understand, to build an affinity with the sky. Anyone who wants to be a truly great observer should start with the planets, because that's where you learn patience. It's amazing what you can learn to see, given enough time. That's the most important and critical factor in observing—time, time, time—though you never see it in an equation."

In the mid-1970s, O'Meara studied the rings of Saturn at the behest of Fred Franklin, a Harvard planetary scientist. He began seeing radial, spoke-like features on one of the rings. He included the spokes in the drawings that he would slip under Franklin's office door in the morning. Franklin referred O'Meara to Arthur Alexander's *The Planet Saturn*, in the observatory library. There O'Meara learned that the nineteenth-century observer Eugene Antoniadi had seen similar radial features in another ring. But the consensus among astronomers was that they must be an illusion, because the differential rotation rate of the rings—they consist of billions of particles of ice and stone, each a tiny satellite, and the inner ones orbit faster than the outer ones do—would smear out any such features. O'Meara submitted a paper on the spokes to the Association of Lunar and Planetary Observers (ALPO, an amateur organization), but they wouldn't publish it. Undaunted, O'Meara persisted in studying the spokes for another four years, determining that they rotated with a period of ten hours—which is the rotation period of the planet, but not of the rings. ALPO wouldn't publish those results, either. "I did not find one person, honestly, who ever supported me in this venture," O'Meara recalled.

Then, in 1979, the Voyager 1 spacecraft, approaching Saturn, took images that showed the spokes. "It was an overpowering emotion, to have that vindication at last," O'Meara said. "I felt like William Herschel, when the reality of something he had been seeing and wondering about all his life was finally proved." The spokes are now thought to consist of dust particles electrostatically suspended in Saturn's magnetic field, which explains why they rotate in synch with the planet rather than its ring particles, as O'Meara observed but astronomers had dismissed as physically impossible.

I asked Steve about his determination of the rotation period of Uranus. This had long been unknown, since Uranus is remote—it never gets closer than 1.6 billion miles from Earth—and shrouded in almost featureless clouds. He told me that Brad Smith, the astronomer who headed the *Voyager* imaging team, "called me one day and said, 'OK, Mr. Visual Guy, *Voyager* is going to be at Uranus in a few years, and I'm trying to first obtain the rotation period for Uranus. Do you think you can do it visually?' I said, 'Well, I'll try.'" O'Meara first read up on the history of Uranus observations and then inspected the planet repeatedly, starting in June 1980. He saw nothing useful until one night in 1981, when "two fantastically bright clouds appeared. I followed them as they did a sort of dance over time, and from these observations, with some help, I determined where the pole was, modeled the planet, and got a rotation period for each cloud, averaging around 16.4 hours." This number was disturbingly discordant. Brad Smith, observing with a large telescope at Cerro Tololo Observatory in Chile, was getting a rotation period of 24 hours, and a group of professional astronomers at the University of Texas, using CCD imaging, were also getting 24 hours.

To test O'Meara's vision, Harvard astronomers mounted drawings on a building across campus and asked him to study them through the nine-inch telescope he had used as a teenager. Although others could see little, O'Meara accurately reproduced the drawings. Impressed, the astronomers vouched for his Uranus work, and his results were published by the International Astronomical Union, a professional group. When *Voyager* reached Uranus, it confirmed that the planet's rotation period, at the latitude of the clouds O'Meara had seen, was within one-tenth of an hour of his value.[1]

O'Meara's visual detection of Halley's comet on its most recent return began as an exercise in skepticism. One day at lunch in Cambridge, he was talking with two astronomers about how the faint, incoming comet—which had been recorded on a time-exposure CCD image taken at Palomar but was still far too dim to be seen with the eye, even through the world's largest telescopes—was certain to attract widespread attention among amateur observers. When *Halley* brightened to near the limits of visual observation, they wondered, how would they evaluate the reports of those who claimed to have been the first to see it? At the time, the faintest comets ever seen visually through a telescope were about 11th magnitude. (The dimmer an object is, the higher the number assigned to it on the astronomical magnitude scale. An 11th-magnitude object is a hundred times dimmer than the faintest star that can be seen with the unaided eye.) O'Meara knew from experience that with his hawk-eye vision, looking through a large telescope on an extremely clear, dark night, he could see stars as dim as 17th magnitude. *Halley*, warming as it grew closer to the Sun, was expected to get that bright by January. O'Meara reasoned that if he used the 24-inch planetary patrol telescope on Mauna Kea, near his home in Hawaii, he might be able to see the comet. If not, at least astronomers could confidently discount any claims by observers using lesser telescopes at lower altitudes that they had seen *Halley* that soon.

University of Hawaii astronomers initially rejected O'Meara's request on grounds that what he proposed to do was impossible. But they later relented, and O'Meara was given a crack at the telescope, at a time when three professionals—Dale Cruikshank, Jay Pasachoff, and Clark Chapman—would be taking electronic images of *Halley* with an 88-inch telescope on the same peak. That way, any claim of O'Meara's to have detected it visually could be checked against the comet's position as imaged by the CCD camera on the big scope.

Few astronomers, amateur or professional, have attempted to make serious visual observations at Mauna Kea's 14,000-foot altitude, where oxygen deprivation clouds the mind and reduces the eye's ability to perceive dim objects. At the telescope on that cold, clear January night, O'Meara breathed bottled oxygen before looking into the eyepiece, but when he set the mask aside he quickly grew befuddled, and he neglected to apply the proper correction for local time. "I converted Universal Time to Boston time instead of Hawaii—a six-hour error," he recalled. "In six hours a comet moves appreciably. It was a gorgeous night, and I star-hopped to the field readily enough, using the maps. But I then spent two hours straining to see the comet, with no luck. I kept putting Xs at the positions of stars that I could see that were not on the chart, and I would wait to see if any of them moved, but none did. Eventually I went over to the 88-inch telescope dome and showed Dale Cruikshank my map. He said, 'You're seeing to the limit of the Palomar telescope, but you're looking at the wrong place. We think you should go back and try again.'"

With his charts corrected to the proper time, O'Meara found *Halley*. "It wasn't even the faintest thing I saw," he recalled. "I could also discern a coma surrounding it, while the CCD images showed none. But when I told them, nobody believed me. It was rather devastating. I was mentally and physically beat up."

By then the mountaintop domes had been ordered closed owing to rising winds, but the astronomers gave O'Meara special dispensation to reopen the dome and keep observing. "He can go back out," one said. "This is history." O'Meara relocated the comet and tried to show it to Chapman and Pasachoff, but neither could see it. So to check his claim, they subjected O'Meara to blind field tests that night and the next, by having him draw stars that were then compared to Palomar images that he was not permitted to see. All these tests he passed, confirming that he had seen fainter objects than any visual observer in recorded history. "And by the way," O'Meara told me, "when those CCD images were cleaned up and processed, guess what? The comet *did* have a coma."

We finished our coffee and made ready to go back out into the darkness. "I've always been strictly a visual observer, researching the sky with an eye to finding something new there," Steve said. "It's no different than the way someone in the nineteenth century would, I assume, be looking through the telescope—with curiosity. If something seems interesting and needs watching, I just do it. When I observe, I'm constantly challenging tradition. Most of what was once believed to be the truth is not, and the whole point of science is to discern the truth.

"We're all star people, in the sense that we're all created from star stuff, so it's in our genes, so to speak, that we're curious about the stars. They represent an ultimate power, something we cannot physically grasp. When people ask, 'Why, God?' they don't look down at the ground. They look up at the sky."

5.

Professionals

In the beginner's mind there are many possibilities, but in the expert's there are few.
—Shunryu Suzuki

Our means of knowing are few, the number of bits of information is small, the universe is very strange.
—Astronomer Jesse Greenstein

D AWN WAS APPROACHING at Las Campanas Observatory in Chile, 1980. I'd been up all night in the control room of the 2.5-meter du Pont telescope with a wry, bearded astronomer whose comprehension of our environment extended for a hundred million light-years into space. He knew where we and our neighboring galaxies are located among scores of neighboring galaxy groups, clusters, and superclusters, and how it all moves as cosmic expansion unwinds space against the mutual pull of their trillions of stars, as intimately as you or I know how to find the bathroom light switch at night. Outside, the mountains—so metallic that a tossed rock rings out when it lands, and you hear earthquakes before you feel them—were bathed in gentle starlight, but we could see none of it. We'd remained in the warm, brightly lit control room, where our work consisted of punching buttons to keep guide stars centered on crosshairs on a computer monitor, making photographs of galaxies on big, square glass photographic plates attached to a telescope that we hadn't seen all night, either.

Now came the moment I'd been waiting for. We'd reached the onset of astronomical dawn—too late to start another time exposure, for fear the brightening sky would fog the plate before we'd finished, but still dark enough for visual observing. I took a weighty brass eyepiece out of a drawer—a dusty relic of an earlier epoch, big as a can of beer—and made my way into the inky blackness of the dome. There the massive telescope was silhouetted against clouds of stars visible through the dome slit. As I groped in the darkness it swung toward the Large Magellanic Cloud, a neighboring galaxy, and the dome rotated to follow its progress, the Milky Way crossing the slit like a river seen out the door of a banking helicopter. Then, with a descending bass note from the gears, all came to a halt. I inserted the eyepiece and prepared to do something almost unheard of in modern professional astronomy: I was going to actually look through a large telescope.

I focused on a rich field of stars and could see a waft of gray haze intruding into the left side of the field of view—the outskirts of the Tarantula nebula, a gigantic star-forming region in the Cloud. I pressed a button on the steel control box, and the telescope glided toward the center of the nebula. I gasped at the sight: Reefs of brick-red and pearl-gray gas clouds were parading by like drapery in a palace of dreams. The nebulosity became ever brighter until I arrived at the core, where sheets of gas entangled the stars of the cluster 30 Doradus. Their light had been traveling through intergalactic space for 180,000 years, dissipating all the while as it spread out, but it was still bright enough to make me squint. I recoiled, and found myself gazing at a stream of light that spilled out from the eyepiece like a flashlight beam. Looking up, I saw that it projected a fuzzy, circular image of the nebula on the inside of the dome.

The night assistant's voice crackled through the intercom. "Tim, you OK down there?" I tried to speak, but could find no words.

PROFESSIONAL ASTRONOMY has been propelled for more than a century now by the advent of ever larger telescopes, but big telescopes have their limitations. Typically they can image only small portions of the sky at a time: All other things equal, larger aperture means longer focal length, which means a more constrained field of view. They eat up time, too: Because they normally are used to go as "deep" as possible, taking long exposures to record the dimmest accessible objects and features, big telescopes often don't obtain more than a handful of images each night. And although this world is blessed with more large-aperture mountaintop telescopes than ever before, there are still relatively few of them compared to the thousands of smaller instruments in amateur hands. So the professionals, through no fault of their own, overlook a lot of what goes on in the sky. They cannot very well aim a five-meter telescope at the dark side of the Moon for a year, hoping to record meteoroid impacts, or keep it pointed at a few variable stars for months on end to record the "light curves" that show their changing brightness over time. Amateurs, however, are free to observe whatever they like. For many this may be nothing

SEEING IN THE DARK

more than gawking at the Jewel Box cluster or trying to take a pretty picture of the Trifid nebula. But for some it means making a real contribution to science, and the professionals take notice. The amateurs, in turn, are often grateful for the professionals' technical expertise, without which many would be hard-pressed to make useful observations in the first place, or at least to reduce their data competently and get them published. The result has been a flourishing of amateur-professional collaborations.

Attending a conference of the American Astronomical Society in Rochester, New York, I found the floor of the sprawling convention center dominated by rows of posters reporting on such work. Beside several of the posters stood the amateur astronomers themselves, on hand to answer questions and looking as pensive as wallflowers at a dance. (It's one thing to be a young postdoc with a poster, quite another to be a grown-up amateur who is painfully aware of the dangers of having one's ignorance exposed by a "real" scientist.) I talked with Doug West, an aerospace engineer whose poster concerned a project he'd been doing with a career astronomer, David Alexander, of Wichita State University, titled "Late-Type Star Research." ("Late" stars are stars cooler than the Sun.) Working with professionals and graduate students, West had been studying these stars in an effort to better understand their chemical composition and how their atmospheres change when they vary in brightness. He was using an off-the-shelf eight-inch telescope that he set up each night in his back yard.

"You'd think they had thousands of flux-calibrated spectra of late stars out there, but that's not the case," West said. (I nodded judiciously, rather than volunteer that I'd never given the matter any thought and have about as much ability to take a flux-calibrated spectrum—the chemical signature of a star, corrected for the absorbing effects of Earth's atmosphere—as to tune a grand piano.) "I've already calibrated spectra of several stars that aren't in the literature at all. I can't stay up all night—I have to go to work in the morning and function—so I've got it down to an art. It takes me just twenty minutes to haul my telescope out into the yard, align it to the pole, and start taking data.

"Astronomy is a black hole for time," he added. "You can pretty much dump your whole life into it. So why do it? I don't know. I was in the second grade when I told my dad I wanted to be an astronomer someday. It's just a nice thing to do. It keeps me out of the bars."

Another poster reported on discoveries made by the Center for Backyard Astrophysics, led by the Columbia University astronomer Joseph Patterson, who researches cataclysmic variable stars. These are, Patterson explained in an article for *Sky & Telescope:*

close binary systems in which a fairly normal, low-mass star gently pours a stream of gas toward a white dwarf [star]. The stream settles into orbit around the white dwarf and forms an accretion disk, releasing its gravitational energy as it gradually spirals its way down to the white dwarf's surface. Perhaps surprisingly, we don't usually see the white dwarf or its companion star directly. The system's light output is ordinarily dominated by the hot, brilliant accretion disk alone. . . . Small changes in the gas transfer rate, disk structure, or accretion pattern will manifest themselves as slight, rapid fluctuations in the system's brightness.[1]

The most dramatic fluctuations occur when gas dumped onto the surface of the dwarf star gets dense enough to ignite in a thermonuclear detonation that causes the system to suddenly flare up. Professionals lack the telescope time to wait for these unpredictable cataclysms, so Patterson organized amateurs to monitor them.

He found that recruiting amateurs to make repeated observations wasn't easy. "Only an unusual person gets excited about making five thousand sequential measurements of the same star," he admitted. "It takes an enduring love of stargazing and of science."[2] But he eventually managed to put together a global network of some three dozen amateur observatories, in locations ranging from Moscow and Manhattan to Adelaide, Australia; Bloemfontein, South Africa; Søndervej, Denmark; and Ceccano, Italy.

Such a network permits continuous monitoring of the night sky: To invert the old saying about the British Empire, the Sun never *rises* on a global observing team. In one important finding made through such monitoring, Patterson and his amateur collaborators identified the twenty-three-hour fundamental variability period of the cataclysmic variable star V803 Centauri. Astronomers had been studying V803 Centauri for decades but none had discovered its fundamental brightness cycle, simply because no single observer this side of the South Pole could monitor it for twenty-three straight hours before the star set or the Sun rose. The network, however, could keep a steady eye on V803 Centauri, and the cycle was found. The group also discovered the orbital period of what Patterson called "the strange but wonderful helium star AM Canum Venaticorum," a result that confirmed a 1993 theory concerning the architecture of that exotic binary system.

The conference was abuzz with news of other feats recently accomplished by amateurs. A group of middle-school and high-school students invited to a summer workshop in Tucson had used a telescope at Kitt Peak to identify seventy-three novae in the nearby galaxy M31. (Novae are stars that flare up suddenly, increasing several hundred to a million times in brightness.) Warren Offutt, an amateur in Cloudcroft, New Mexico, had made twenty-five hundred observations of a dwarf nova, V2051 Ophiuchi. Working from Offutt's data, Sonja Vrielmann, of the University of Cape Town, was able to determine the distance of V2051 Ophiuchi and the inclination angle and spin characteristics of its accretion disk, a swirling mass of surrounding material being drawn in toward the star.

In Buffalo, New York, four members of the Buffalo Astronomical Association had imaged the optical flash of a gamma-ray burster only thirty-four hours after the burst itself was detected by satellites. Gamma-ray bursters are mysterious objects that occur, entirely unpredictably, all over the sky, and quickly fade away. Optical images coinciding with their rapidly fading bursts had been obtained for fewer than one percent of them. The Buffalo amateurs used a 12-inch telescope to record the image of this one. The few professional astronomers who had managed to do the same relied on the Hubble Space Telescope and the giant Keck telescope in Hawaii. They determined that the object lay four-fifths of the way across the observable universe, making it the most distant burster yet imaged by amateurs.

Downstairs in a dim, windowless conference room, an amateur astronomer from Phoenix, Arizona, named Gene Hanson was reporting on his visual monitoring of the cataclysmic variable star U Geminorum. U Gem, as it is nicknamed, is an eclipsing contact-binary system, consisting of a red dwarf and a white dwarf orbiting close together around their common center of gravity along a plane that happens to intersect Earth. The white dwarf draws gas off the surface of the red star, forming a blue-white accretion disk that periodically eclipses, and is eclipsed by, the red star. The resulting light curve is a complex combination of the rhythmic eclipses and the unpredictable outbursts that occur when gas stolen by the white dwarf reaches critical mass and detonates. Hanson scrutinized more than a hundred such stars each clear night, relying on his experienced eye to spot when one was misbehaving.

One November night in 1997, Hanson saw that U Gem, which normally shines at magnitude 15 during eclipses and magnitude 14 the rest of the time, was suddenly getting much brighter. He sent an email to amateur observers around the world, warning them that U Gem was about to blow. He also contacted Janet Mattei, a professional astronomer who alerts NASA officials to suggest that they interrupt the scheduled observations made by costly satellites like EUVE (the Extreme Ultraviolet Explorer) and RXTE (the Rossi X-ray Timing Explorer) to turn their attention to a flaring contact binary.

Meanwhile the Sun rose in Phoenix, and hours went by without Hanson or Mattei hearing any word from observers on Earth's dark side. For Mattei, it was a time of rising tension. If she contacted NASA on the authority of just one amateur astronomer—who, since he works visually, had not even a CCD image to offer in support of his claim that "this star is going to flare"—she risked crying wolf. If she kept silent and Hanson proved to have been right, scientists would have squandered a chance to catch U Gem in the act. She decided to trust Hanson's vision. She contacted NASA and the costly satellites were diverted to look at U Gem.

Back in Phoenix, an anxious Hanson sweated out dusk, then aimed his telescope at U Gem through a gap in the trees. He was relieved to see that it had indeed flared, all the way up to magnitude 9.7. Thanks to his efforts, the NASA satellites had captured valuable data that would otherwise have been lost. Hanson later learned that the silence of his fellow terrestrial observers resulted from their all having been clouded out that night. "As it turned out, I was the only person with a clear opening in the entire world," he mused.

As director of the American Association of Variable Star Observers, or AAVSO, Janet Mattei annually compiled over three hundred thousand variable-star observations, most of them made by amateur astronomers, and had coordinated hundreds of observing programs involving collaborations between professionals and amateurs. These data provided the fodder for nearly a thousand research projects and had been employed to program twenty-three satellites. A notable example was the European Space Agency's *Hipparcos* satellite, which measured with great precision the position of stars at different times of the year and obtained their distances by use of the parallax method—that is, by noting how nearby stars shift against more distant ones, owing to the Earth's changing position in its orbit. By loading *Hipparcos* with data on thousands of variables monitored by amateur observers, scientists were able to let their computers distinguish between known variable stars and the new ones that the satellite's detectors were discovering.

Mattei told the conference that astronomers employing AAVSO records and the Hubble Space Telescope to study Mira, a pulsating red giant star orbited by a white dwarf, near the minimum of its 332-day brightness cycle, had discovered a tail-like structure extending from the red giant toward the white dwarf, possibly the first direct image ever made of material being exchanged in a contact-binary system. She added that amateur observations of a flare-up of the dwarf nova SS Cygni had alerted NASA to train two satellites on this odd star, with results that helped astronomers understand its complex structure. In gratitude, NASA assigned nearly three days of observing time on the EUVE satellite to the amateurs, to do with as they wished.

Which was an honor, of course, but amateurs granted observing time on a major telescope or satellite may find themselves in a sweaty-palms situation as challenging as that faced by a weekend auto racer who suddenly gets his hands on a Formula 1 car. William Alexander, a petroleum chemist and part-time astronomy teacher, was awarded time on the Hubble Space Telescope under a short-lived project intended for amateurs who came up with research projects of potential use to the professionals.[3] "I'm kind of a lazy amateur," Alexander said. "I don't like to drag my telescope out to the back yard too often. So when the HST amateur observer program was announced in 1993, I tried to think up something useful for it. I settled on studying the deuterium-hydrogen ratio in the interstellar medium. Using the Hubble telescope, I could work in wavelengths of ultraviolet light that are available only from space."

When Alexander's project was approved, he was invited to the Space Telescope Science Institute, in Baltimore, Maryland—which, he soon found, was geared to the demands of professional astronomers rather than the learning curves of amateurs. "I was trying to find the best way of doing these observations," Alexander recalled, but "although I was called the PI [the principal investigator, the head of a team of scientists] I didn't know much about it." He wound up taking spectra of two nearby stars, with results indicating that at least some deuterium is produced in them. If true, this would modify a standard cosmological assumption that almost all the deuterium in the universe was made in the big bang. Alexander eventually published the data—in the *Astrophysical Journal*, with two professionals, Jeff Linsky and Brian Wood of the University of Colorado, as coauthors—and was proud that the paper was subsequently cited at least a dozen times in the scientific literature. Nevertheless, his observations were incomplete by comparison to what a professional would have done—astrophysicists informed him, after the fact, that he should also have measured magnesium II lines in the stellar spectra he obtained—so the paper's impact was somewhat blunted.

Bill Aquino, an amateur who studies gamma-ray bursters, suggested at the conference that the *Hubble* amateur-observer project "fell short of its ultimate goals" and eventually was canceled because "amateurs and professionals are different. We amateurs have day jobs. We're willing to learn, but we need some help. Professionals have to be willing to work with and educate amateurs, and amateurs have to be able to rise to the challenges that the professionals set—to be willing to learn how to work at a professional level."

But that can require more time and effort than an amateur can muster.

Paul Boltwood, a Canadian computer systems analyst and amateur astronomer who frequently collaborated with professionals, had written that "I have less time to do research (I have to earn a living), yet I have to do everything involved from building and maintaining my observatory to sweeping the floor. I have no support staff and I have no time to 'keep up with the field,' nor do much of anything else other than make observations. Everyone should do what he or she is fitted to do—I produce good quality data. But I disappointed some professionals when I did not produce Ph.D.-level papers on astrophysics, or fully understand some of the papers that I coauthored." Boltwood complained of being "burned" more than once by professionals who, he said, used his data but failed to give him proper credit for it, either because they were out to hog the glory or, more often, because they feared that their peers would distrust the information, and the astronomer relying on it, if they knew it had come from "a seven-inch amateur telescope."[4] Still, such adversities did nothing to keep Boltwood away from the night sky. He did blazar photometry, measuring variations in the brightness of active galaxy nuclei (blazars are thought to be jets protruding from the vicinity of black holes); videotaped the occultation by Saturn's rings of the star 28 Sagittarii; and imaged galaxies down to 24th magnitude, winning a competition for the "deepest" CCD image ever made by an amateur. He accomplished this last feat by "stacking"—adding up—601 exposures, totaling more than twenty hours, made with a 16-inch telescope at his suburban Ottawa home. The image records starlight more than twice as old as the Earth.

When it comes to doing real science, what the amateurs have to offer includes their numbers—there are perhaps ten times as many experienced amateurs as professional astronomers—and their time.[5] As Aquino notes, "Amateurs are the best equipped to do long-term projects that take days, weeks, or decades." Discovering a nova, for example, takes an average of five hundred to six hundred hours observing. Hence, almost all of the brighter novae have been discovered by amateurs.

Amateurs equipped with a computer and modem can do proper research even if they lack access to a telescope, by studying the "virtual universe" of data generated by automated survey programs. One such project was described at the Rochester conference in an urbane talk given by Bohdan Paczynski of Princeton University. Paczynski reported on the All Sky Automated Survey (ASAS) directed by another astronomer, Grzegorz Pojmanski. ASAS had been running for two years, using nothing more than a 135-millimeter camera lens fitted with an off-the-shelf commercial CCD chip and mounted on a small drive in one corner of a runoff-roof shed at Las Campanas Observatory in Chile. "The ultimate goal of the project is to monitor the variability of all stars in the sky down to the magnitude limit we can afford," Paczynski said. The data were collected automatically, with no observer present, and recorded on tape. Every month or so a technician at Las Campanas would pull the data tape, mail it north, and insert a fresh tape in its place. Using this simple system, Pojmanski and his colleagues had, in just two years, identified nearly 3,900 variable stars, of which only 155 had been cataloged as variable in the standard literature and only 46 by *Hipparcos*, the mightiest variable-star detector among the space satellites.[6]

Paczynski estimated that there are "probably about one million variable stars waiting to be discovered" with modest equipment. "I don't think there is any natural limit to growth, even for a small instrument like this one," he said, noting that ASAS had to date covered only about one percent of the sky. "It's like buying a lottery ticket. You may discover a nova, discover an optical flash—who knows what. We just don't know what may be out there."

As nobody can accurately predict what might be found in the raw data obtained on thousands of variable stars by such surveys, Paczynski talked of future researchers probing the virtual universe via the Internet for years to come. "What I am proposing is a vacuum-cleaner approach," he said. "You get all of these data, put them in the database, then you look at them and you think. I used to be an amateur observer of variable stars myself, and I still have a warm spot in my heart for those things. When I retire I hope to be able to sit comfortably with a cup of coffee and watch variable stars on the Internet. Some may be reluctant to give up the romance of being cold in the dark. But I prefer to relax with a cup of coffee in a warm room and let the telescope sit in the cold and the dark."

Picturing the Universe:
A Visit with Jack Newton

WHEN JACK NEWTON WAS a boy, in Canada, he was fascinated by astronomy but discouraged about observing the stars for himself. "All the photographs I saw in the magazines and books were taken by the two-hundred-inch at Palomar," he recalls. "I was humbled. How could I expect to see anything with my little telescope, if all the pictures came from giant telescopes like Palomar?"

Then one night he stumbled on Saturn while sweeping the skies, and began to wonder what other astonishing sights might lie within his reach. His friends at school ridiculed his claims that he could see the craters on the Moon from his back yard, so to prove them wrong he used his paper-route money to buy a camera and attached it to his telescope. With that, he embarked on a life-long avocation in astrophotography, working with a combination of skill, perseverance, good equipment, and dark skies that set new standards in the field.

When Jack retired from his job working for a department store, he and his wife, Alice, built a mountaintop house with a dome housing a 25-inch telescope. There, pioneering new CCD techniques, Jack produced images of galaxies that rivaled those made by professionals using the world's largest telescopes. "I had a simple creed," he says today. "Go where nobody has gone before, and leave a path for others to follow."

To further spread the gospel, Jack and Alice eventually sold their dream house, donating the 25-inch telescope to the Lester B. Pearson College of the Pacific, and opened a pair of astronomy-oriented bed-and-breakfast inns—one in Osoyoos, British Columbia, and another in a dedicated dark-sky community that had been established by and for amateur astronomers seven miles south of the rural town of Chiefland, Florida. That community was founded by the amateur astrophotographer Billy Dodd, of St. Petersburg, who in 1985 drove thousands of miles in Florida, testing the night skies by pulling out lengths of unexposed film, holding them up to the stars, then rolling them back into the canister and shooting astronomical photographs on them. The skies outside Chiefland were so dark that they didn't fog the film, so Dodd parceled eighty acres of the land into an amateur astronomers' subdivision. Some built houses there. Others showed up in motor homes to observe for a few nights at a time, leaving money in a donation box to pay for their electricity and the use of the outdoor hot showers. Hundreds of visitors, part of a growing astrotourism movement, came to learn CCD imaging at the Jack Newton B&B Observatory.

I visited Jack and Alice in Chiefland on a crisp January afternoon when an enormous blue sky hung over the pine woods. Jack emerged from his house— a slim, balding man with bright, attentive eyes and a droll manner—showed me into the observatory, and pulled on a rope to roll back its roof. "I had to build it myself," he called over the rumbling steel wheels. "This is hurricane country, and I couldn't find a contractor who was willing to take it on. They said, 'We spend all our time trying to keep the roofs on buildings, not rolling them off.'"

With the roof out of the way, we had a clear view of Jack's array of telescopes, which included a seven-inch Maksutov, a sixteen-inch Dobsonian, and a sixteen-inch Schmidt-Cassegrain on a concrete pier. "It's amazing what amateurs can do," Newton said, as he switched on electrical power to the big Schmidt-Cass and the CCD cameras attached to both it and its five-inch guide scope. "I've taken images of the Hubble telescope deep field and found that in just a matter of minutes I can get about seventy-five percent of what *Hubble* can observe. I've imaged clusters of galaxies three billion light-years away, and got eighty percent of the galaxies that can be recorded by the big telescopes. I've seen galaxies that no human has ever seen before. I can observe six hundred stars in the daytime, including all the stars in the Big and Little Dipper. Double stars, like the double-double in Lyra, are more easily split in daylight, because there's less contrast and glare. You can simply see them sitting there, with the sky between them. It's a whole new ball game."

While the telescopes cooled down and the stars came out, we strolled down a dirt road past the houses that comprise Chiefland's dark-sky observing community. Their windows had blackout curtains and their porch lights were night-vision red. "They look like whorehouses, don't they," Jack said. "But the reason, of course, is that nearly everybody here has a telescope. The folks in this house here put Christmas lights out, but they were on a dimmer so they didn't bother anyone."

We stopped in at Tom Clark's house and toured the workshop where until recently he'd been making big Dobsonian telescopes that could be broken down and transported in the back of a station wagon. They sold briskly but Tom had recently retired and was winding down his business. "I've made a couple of hundred telescopes," he estimated, pulling off a tarpaulin to unveil one of his gargantuan 36-inch "Yard Scopes." Half the length of a European freight car, it consisted of two elegant hardwood boxes, one for the eyepiece and the other housing its primary mirror, connected by a set of black struts. Tom and Jack bantered about the best way to support primary mirrors—an endless bone of contention among telescope makers—with Tom adhering to the virtues of the conventional metal "floating" suspension system and Jack arguing that ordinary bubble wrap was cheaper and just as effective.

Although retired, Tom was still observing—he would roll the Yard Scope through a garage door in the workshop onto a concrete pad outside—and still serving as the editor of *Amateur Astronomy* magazine, a nuts-and-bolts period-ical unburdened by literary pretensions. In the current issue, he had responded to complaints about certain articles by noting, "We have 68 pages to fill each issue, and run just about every article received. Since *Amateur Astronomy* is written by its readers we can not dictate what they write."[1]

Back at his own observatory, Jack fired up the computers in the warm room and started taking images through the 16-inch Schmidt outside. He made a test exposure, loped out to the telescope to correct the focus—"I'll move it three- or four-thousandths of an inch"—and then took four one-minute shots of the spiral galaxy M77, each through a different color filter. His automated filter wheel was broken at the moment, so each change of filter required that he leave the warm room, fiddle with the telescope, and then return to the computer screen, a process that he accomplished at the patient, measured pace of a painter at work on a fresh canvas. As each image appeared, he examined it, processed it on the computer, stared at the results with bird-like concentration, cocking his head from side to side, and then processed it some more.

"Now I'll combine them," he said. He digitally overlaid the four shots and a spectacular color picture of the galaxy appeared. I knew the galaxy fairly well, but the image showed an outer halo unfamiliar to me. While Jack adjusted the colors, I looked up M77 in *The Carnegie Atlas of Galaxies* and found that its photo, shot on film with the 100-inch Hooker telescope on Mt. Wilson, showed no sign of the outer halo. The Carnegie photo was over a half-century old so the comparison wasn't entirely fair, but clearly Jack in a few minutes was outper-forming the published work of what was, back then, the largest telescope in the world.

After imaging a few more deep-space objects, Jack inserted a Barlow lens into the telescope—which extended its effective focal length, so that a larger image was projected onto the CCD chip—and moved on to Jupiter, taking four short exposures of the giant planet in each color filter and promptly discarding the ones that didn't measure up to his cocked-head scrutiny. "I'm trying for that moment of good seeing," he said. "There, that's better."

Once he had four clean images, he combined them to produce a sharp, colorful portrait of Jupiter as it had looked just minutes earlier. "That's good," he said happily. "In fact, that's beautiful! But it's not as much fun as galaxies, is it?"

Fog was creeping in from the Gulf of Mexico, less than ten miles to the west. Jack's antidewing system—wires that he rips from electrical heating pads, covers with foam insulation, and wraps around the objective lenses of his refractors and the corrector plates at the skyward end of his Schmidt telescopes—kept the optics clear, but by midnight the sky had turned milky white all the way to the zenith, so we powered down the telescopes and shut the roof.

I spent the night in one of the B&B's four guest bedrooms. The walls were adorned with large color prints of Jack's astrophotographs and the bedside table was stacked with copies of *Amateur Astronomy*. In the morning, we lingered over breakfast in the dining room while the climbing Sun burned off the fog and Alice extolled the virtues of amateur astronomy ("In how many areas of science can you still make an important discovery without a ton of funding?") while refilling my mug of coffee.

When the sky cleared, Jack returned to the observatory and imaged the Sun, using a hydrogen-alpha filter attached to the front of the five-inch refractor. Such filters block the entire solar spectrum except for one ruby-red line produced by hydrogen atoms. I had a look through the eyepiece and was rewarded with a spectacular view of gigantic solar prominences hovering in space above a blood-red solar disk etched with dark, swirling magnetic field lines of sunspots. Jack flipped a small mirror to direct light away from the eyepiece and into the CCD camera. Then, in the darkened control room, he set to work taking exposures, combing through them as he had the night before to select only those that, having been made at moments of the steadiest air, revealed the greatest detail. In minutes he had produced a striking composite portrayal of ruddy prominences hanging over the limb of a golden, tumultuous Sun. For comparison we consulted a solar photo made recently at Big Bear Solar Observatory in Southern California. Big Bear is a leading professional facility—a tall white dome set out in Big Bear Lake, whose waters serve to cool the environment and minimize atmospheric turbulence—and the astronomers there monitor the Sun through three research-grade telescopes, the largest of which is a 26-inch reflector. Jack's latest solar image wasn't quite as sharp as Big Bear's, but it came close.

A couple of months later, one of Jack's solar pictures was published as a full-page spread in *Newsweek*. "If you're lucky," Jack said, "if you get the focus right, and the seeing right, and the tilt of the solar filter right—you get *magic!*"

6.

Rocky Hill

Four and fifty years
I've hung the sky with stars.
Now I leap through—
What shattering!
  —**Dogen**

Who would believe that so small a space could contain all
the images of the universe?
  —**Leonardo da Vinci, on the eye**

WHILE THE SKY DARKENED at Rocky Hill Observatory I checked the western horizon for signs of fog. We're in the California wine country forty miles from the coast, but fog can cover that distance before you know it. It may arrive suddenly, in a gray wall that obliterates the stars as decisively as a closing eyelid, or insinuate itself gradually, in flat, bedsheet-white fingers that fan across the valley below before creeping up the hillsides. The zenith is the last to go, and since I tend to work near the zenith I sometimes don't realize what's happening until the view through the eyepiece fades to gray and I look up to find the star charts and desktops slick with water, the hydrometer reading ninety-nine percent, and the observatory virtually submerged. But there was no fog in evidence tonight, so I could simply enjoy the view—a small valley with a chuckling creek and a swaybacked vineyard in it, set against the purple north slope of Sonoma Mountain.

This is earthquake country—the San Andreas Fault lies just off the coast—and these hills are on the move, restless as cumulus clouds. They're basalts, mostly, vomited up by active volcanoes less than ten million years ago, but over by the coast lie slabs of radiolarian chert that started out on the floor of the central Pacific a hundred million years ago, and sandstones that once belonged to Central American beaches. Geological timescales, although infamously long by ordinary human standards, are astronomically unexceptional. Using nothing more than the eye and the 18-inch telescope at Rocky Hill, you can have spectacular views of galaxies whose light is ten million years old, and can see many at distances of over a hundred million light-years.

A few bright stars shone in the sky. To orient the telescope I aimed it at them, marveling as always at their colors—orange Aldebaran, yellow Capella, blue Vega. Once the sky was fully dark I had a look at the Triangulum galaxy, which at a distance of less than three million light-years from Earth is a local object by intergalactic standards. Its rangy spiral arms, tangled with glowing clouds of gas, spilled out beyond the field of view. As often happens, I was struck by the fact that all these things, unimaginably big or small or hot or cold as they may be, really are out there. Like giant squid or loaves of French bread—and unlike, say, postmodernism or public opinion polls—they confront us with the regality of the materially real.

I opened a loose-leaf binder full of galaxy photographs and checked the photo against what I was seeing. If a star has exploded there, it will look like a new (*nova*) star, one that is seen in the telescope but not on the photograph taken previously. There are two classes of such stars—novae, which flare up on occasion and may do so repeatedly, and supernovae, stars ending their careers in a fatal and titanic explosion. In a matter of days a supernova can swell in brightness until it outshines all of the hundred billion or so other stars in its galaxy combined; then, for months or years, it slowly fades. Tonight I was looking for supernovae. The trick is to catch one while it's still getting brighter, since this provides astronomers with the best opportunities to study it. The odds that a supernova will occur in a given galaxy on a given night are longer than ten thousand to one, so trying to discover a supernova is like fishing for trout in a stream where the average angler never once catches a trout. But even if you never find one, searching for supernovae provides a splendid excuse for gaping at galaxies.

Stately, self-possessed, a murk of mingled stars and gas clouds presenting itself to the eye in hues of silver to charcoal to India ink, a galaxy is so commodious as to contain, I should think, more stories than anyone, anywhere, shall ever come to know. Although there was no sign of a supernova in this particular galaxy on this night, I lingered a moment before moving on, just to look. I felt absurdly happy, like the early French balloonist who, once aloft, refused to come down.

THE WORD *TELESCOPE* COMES FROM THE GREEK for "far-seeing." Distant objects generally are dimmer than nearby ones, so seeing far requires gathering lots of light and bringing it to a focus. In refracting telescopes, of which spyglasses are an example, the light is gathered by a large "objective" lens. In reflecting telescopes a big "primary" mirror does the job. Big telescopes may be better than small ones, but they are also more expensive, harder to transport, and more vulnerable to turbulent air, called "bad seeing." The two most important quantities of any given telescope are its aperture (the size of the light-gathering lens or mirror) and its focal ratio, which is the focal length—the distance that the gathered light travels before coming to focus—divided by the aperture. Telescopes with large focal ratios generally produce the sharpest images, but also have relatively narrow fields of view: An F/12 refractor may be excellent for observing planets, but be less well suited to looking at extended objects like nebulae and nearby galaxies. Eyepieces magnify the image, to a degree measured by dividing the focal length of a given eyepiece into that of the telescope. Different eyepieces work best for different objects, and most observers have a collection of them, much as photographers use several lenses on one camera body. Low-power, wide-field eyepieces afford the best views of galaxies, nebulae, and other extended objects, while medium and high powers are preferred for planets and remote galaxies.

All telescopes introduce distortion—noise in the signal—starting with the accuracy to which the lens or mirror comports to the ideal (usually parabolic) curve required to bring a wide swath of light to a pinpoint focus. The tolerance for error is extremely small. The physicist and amateur astronomer Harold Richard Suiter notes that a premium-quality eight-inch telescope mirror, of the sort that many skilled amateurs have made by hand, if enlarged to a diameter of one mile, would be curved to an accuracy of better than a quarter of a millimeter, "a playing-card thickness error on a disk a mile across and three hundred yards high."[1] On the same scale, an extremely fine piece of metal machinery, such as a bearing used in a jet engine, would have undulations bigger than softballs. As Suiter puts it, a good telescope mirror "contains the most accurate macroscopic solid surfaces yet shaped by humans."[2]

Reflecting telescopes bounce the light right back the way it came, so there has to be some means of getting the light beam off to one side, lest the observer's head get in the way. (Some of the giant mountaintop reflectors, starting with the Hale telescope at Palomar, are big enough that the observer can actually ride in a "cage" at prime focus. Now that such telescopes are almost always used for electronic imaging, all that has to fit at prime focus is the camera and its cooling equipment.) In Newtonian telescopes, the most popular kind of reflector, this is accomplished by suspending a flat secondary mirror near the mouth of the tube. The suspending struts, usually made of metal and fashioned thin to minimize their interference with incoming starlight, make up what is called the spider. The secondary mirror bounces the light to the side of the tube, where the focused image hits a camera or is magnified by an eyepiece. To get to the eyepiece of a Newtonian telescope pointed near the zenith one climbs a ladder, as if on a stairway to heaven.

Intergalactic space is admirably clear, and ancient starlight with hundreds of millions of light-years on the odometer routinely arrives at the solar system in admirable condition, a bit ruffled by interstellar and intergalactic clouds but otherwise almost as good as new. What most distorts and degrades starlight is its passage, during the final one-ten-thousandth of a second of its journey, through Earth's atmosphere. Air turbulence makes stars twinkle, and telescopes magnify every twinkle. To minimize it, observatories normally are located at high altitudes, to get them above as much of the atmosphere as possible, and at sites where the air is stable.

Rocky Hill Observatory is on a hill, not a mountain, at an altitude of only four hundred feet above sea level, but it's still a dream come true. To find a good site for it I tested various locations with a portable telescope, arriving at a nice, level patch of ground with good local seeing, as it is called, resulting from a fairly smooth flow of clean Pacific air wafting in from the west. The only problem was that a pair of tall old oaks obscured most of the eastern sky. While I was pondering this problem, I got a call from an architect named John Miller, inviting me to address the American Institute of Architects. "We don't pay an honorarium," he said, "but we can offer you a nice case of wine."

"I don't need any wine," I replied, looking out over the vineyards.

But we got to talking about my hopes of building an observatory, and Miller offered to trade an architectural consultation for the talk. We met for lunch in Kenwood, then went out to the site. Miller quickly settled into a seemingly aimless, pointless working manner that I had fortunately—by virtue of having witnessed it in a few major artists, among them the physicist Richard Feynman and the painter Donald Kaufman—come to recognize as a badge of creativity. He ambled about, idly picking up a leaf here and a scrap of bark there, chewing on a twig. Then he announced that I would have to chop down the trees.

"Some consultation," I said. "Those trees are more than a century old. I don't want to cut them down."

Miller smiled, poked around some more, then picked up a weathered piece of cardboard, took out a pencil, and sketched on it. The thing to do, he said, is to put the observatory not on the flat ground but out over the adjacent hillside, like this.

The sketch alarmed me. What had been a simple one-story structure was now, at its westward end, a three-story building. But I got an orchard ladder, put it where the telescope pier would be, and climbed up to have a look. Everything fell into place. It was perfect.

"This is going to be more expensive than I'd planned," I grumbled, climbing off the ladder. Miller smiled again. "Of course," he said. "What do you think architects are for?"

With the site settled on, I started questioning veteran observatory builders about the specifics of design. Their advice was mostly ecological: If you've got a good site, build on it with as little impact as possible.

Use a roll-off roof, not a dome; domes are expensive, and they tend to trap warm air. Avoid tar paper roofing and exposed concrete floors, which soak up heat from the Sun during the day and release it slowly by night, scrambling the seeing. When you must use concrete, coat it; otherwise vapor from the lime in it may attack the glass in the telescope.

The most concerted advice I got came from Clyde Tombaugh, the discoverer of Pluto. I went to see him in Mesilla, New Mexico, in 1991. He took me out behind his house, hard by a trailer park, and showed me his two telescopes, both constructed by hand.

Neither had any enclosure at all; they had stood out there in the desert air, day and night, for years. No dome effects. "The polar axis on this one came off my dad's 1910 Buick," Tombaugh said, patting his nine-inch telescope. "I finished the mirror in the spring of 1928. The Smithsonian Institution wanted this telescope, but I told them, 'I can't really do that; I'm still using it.'"

He climbed up into the scaffolding of his 16-inch telescope, a steel-and-glass affair that loomed up into the dark like an inked equation. Tombaugh was in his mid-eighties, a bent old man, but he swarmed over the instrument's ladders, platforms, and steps with spiderlike agility, winding its gravity-driven clockwork drive and hoisting the tube into place while I struggled after him, furiously taking notes.

"You should be at least eight feet above the ground," he called to me. "Up here, your feet are sixteen feet above ground when you're observing the zenith. The best thing to have around a telescope is wood. Concrete is murder. A skeleton tube is best, to avoid air currents trapped inside. If you must have a closed tube, line it with cork. All this is hand-bolted, not a weld on it. Six tons of foundations, a ton of steel. I tap-fitted the bolts; drilled the holes a little too small for them, so they'd stay tight. The whole thing cost about five hundred dollars.

"It's a super telescope," he shouted. "A humdinger. With this scope, I've seen markings on Ganymede"—a satellite of Jupiter, about the size of Earth's Moon but two thousand times farther away. "To do that, you have to have some awful good seeing, and some awful good optics."

When he saw that I'd finally caught up with him, he lowered his voice. "Very few people," he said quietly, "have seen as much on the planets as I've seen with this telescope."

Taking Clyde's advice, I built my observatory mostly out of wood, with a minimum of concrete anywhere near the telescope, and it worked fine. Heavily insulated, it stayed cool inside on all but the hottest days—when an exhaust fan could be turned on to draw cold air up from the basement—and it rapidly equalized with the ambient temperature once the roof was opened. The observing deck, two stories high, was free from ground-effects air turbulence and its low walls afforded sweeping views like those enjoyed by ancient stargazers atop ziggurats.

One day, soon after it was finished, I was loading shelves on the floor below the telescope deck with some old astronomy books that my mother had sent me from the house on Key Biscayne. Among them was a faded blue loose-leaf binder containing observing notes and sketches of the Moon, Mars, and Jupiter I'd made as a boy. The cover was stained from the floodwaters of Hurricane Betsy, in 1965, but the contents were well preserved. I was paging through it nostalgically when a leaf slipped out from the back and fell to the floor. On it was a pencil drawing that I'd made at about the age of thirteen of an observatory floor plan. It envisioned a one-story structure, not canted out on a hillside as in John Miller's innovation, but otherwise it was almost identical to the building I was standing in. All dreams, as they say, begin in childhood.

Farsighted:
A Visit with Barbara Wilson

THE ALLIGATORS AT THE OBSERVATORY weren't a big problem for Barbara, but they did sometimes discourage visitors. Barbara would be waiting for a busload of schoolchildren to show up at the George Observatory—a public facility amid the swamplands of Brazos Bend State Park, an hour's drive south of Houston, where she is staff astronomer and assistant manager—and they'd be late. Considering that they might be trapped on the bridge by an alligator, she would get her rake and go down there, and sure enough an alligator would be lying across the bridge, sunning himself and refusing to move, with the kids stopped on the far side, their faces pressed to the windows of the yellow bus. Barbara would chase the alligator away so the kids could come and see the telescopes.

"I've learned to use a regular yard rake," she told me. "I scrape it in front of the alligators. They don't like that sound, so they'll move, usually. The kids love it—they think you're cool—but their parents and teachers aren't always so enthusiastic about it.

"The men are the worst. What is it about men and alligators? A six-foot alligator will show up in the parking lot and the men won't do a thing about it. Instead they'll call me on their cell phones and wait for me to come over with my rake. The other day there were five or six grown men just standing there, looking at the alligator and asking me to deal with it. So I chased him off into the woods. The bigger alligators aren't all that much of a problem, anyway, since they're more experienced and a little afraid of human beings. But the little ones will snap at you when you try to shoo them away, and they have a tendency to turn around and come right back. It's hard to herd them. Alligators are interesting creatures, though. They've been around for several hundred million years, and the mothers are very gentle with their young."

Modest and easygoing, a roly-poly figure in soft old clothes and athletic shoes, Barbara could be mistaken for just another Houston housewife—which is what she was, as well as a business executive and a real estate broker, before she went to work at the George Observatory, which she had helped establish.

During her initial tenure there, amateurs using the observatory's three telescopes discovered scores of asteroids and obtained a rare, optical image of the
light from a mysterious gamma-ray burster. Meanwhile Barbara continued
with her own astronomy projects, one of which involves observing all of the
twenty-five hundred deep-space objects that William Herschel ever cataloged
and spotting every one of the Milky Way's 147 known globular star clusters.
(The 147th, designated IC 1257, had long been identified as an open star cluster, until Barbara raised the possibility that it was a globular instead. The distinction is significant: Open clusters are much smaller, less dense, and younger
than globular clusters. She and three other amateur astronomers coauthored a
paper with several professionals on IC 1257.)

In the course of such endeavors, Barbara earned a reputation in amateur
astronomy circles as one of the world's most skilled observers—a voracious
explorer of deep space capable of seeing, and showing others how to see,
objects that had traditionally been assumed to lie beyond the reach of the
human eye. Even Stephen James O'Meara, her chief competitor for the title of
the most acute visual observer alive, described her accomplishments as "amazing." We sat at a picnic table at the Texas Star Party one hot, sunny afternoon
and talked, over a background of birdsong and the rustle of hundreds of
stargazers removing Mylar shrouds from their telescopes in preparation for a
night's observing.

Barbara told me that she was "an army brat, basically," born in Gorizia,
Italy, who grew up in Germany, Switzerland, and the United States, "wherever
Dad was stationed. At dusk one evening in 1956, in Green Bay, Wisconsin, we
were taking clothes off the line and the sky was turning dark blue, and I
remember seeing a brilliant red object in the east. My father said it was the
planet Mars. It wasn't until years later that I realized that this was the great
1956 opposition of Mars, when it came within thirty-six million miles of
Earth. Later, when we were living in Monterey, California, Dad bought me a
telescope. In the officers' quarters there was a water tower with a clear area. I
would take the telescope up there with some star charts and a little lantern,
and find bright stars and Saturn, things like that. Always alone. I didn't have
any contact with anybody who did amateur astronomy, but in the seventh
grade I read a wonderful little book about how Halley's comet would return in
the year 1986, and I remember thinking, 'Wow, that's a long time to wait!'
This was in the early 1960s.

"I wanted to become a professional astronomer, but once I got into high
school, I'd be the only girl in the math and science classes, and with all the
teasing and everything, I became self-conscious and discouraged. So I gravi-

tated away from science and math, although with great regret. I ended up moving here to Texas after high school. I married a fellow army brat, we had children, and I deferred going to college until after my children were in school. Then that marriage failed. It was doomed to fail: We got married way too young. My first husband was of the old school: He felt that women should stay home, and that they shouldn't do this and couldn't do that. My second husband is a completely different kind of man, very encouraging. One day he came home with a used telescope, a six-inch reflector, and I rediscovered astronomy. Before that, with the business of raising a family and surviving in the world—finding jobs and all—I had lost track of my love of science for a long, long time. I felt quite left behind, actually."

Inspired by the work of William Herschel, whose observing logs she read in the rare book room of the Rice University library, Barbara went after dim and distant objects that few if any observers had seen before, tracking dark rivers of dust in the Milky Way and laying eyes on dim galaxies, nebulae, and star clusters that previously had been detected only in long-exposure photographs and CCD images. She was often frustrated—"about half the time I fail"—but she succeeded often enough to become something of a legend. I asked her about some of the tales I'd heard. Was it true, for instance, that she had inhumanly good night vision, which she augmented by eating carrots and taking vitamin A?

"Hell, no," she laughed, lighting a cigarette. "My vision is, like, 20/15, so I *am* a bit farsighted, but I don't think I have particularly good vision. Patience and the desire to see are the most important things—that, and knowing exactly where to look. To realize that light that has traveled for eons across the universe is entering your eye—that you are *allowed* to see those kinds of things—is amazing to me."

When some of her amateur astronomer friends nicknamed her "The AIN'T NO Queen," the anagram standing for an imaginary "Association of Invisible Nebulae and Things Nobody Observes," Barbara responded by drawing up a wish list, "The AIN'T NO 100," of visual observations all but impossible to make. Its entries include footprints on the Moon, a transit of an asteroid across the Moon, an extrasolar planet, the gravitationally lensed arcs of light found in some clusters of galaxies ("I would give my arm to see one of those blue arcs"), the Black Widow pulsar, which is smaller than the city of Omaha, Nebraska, and lies thousands of light-years from Earth, and the "strange dog-leg jet in the galaxy NGC 1097."

"Have you observed any of these one hundred objects?" I asked her.

"No," she replied cheerfully. "There are a lot of things that I try for that I can't see. I've been kicked in the butt many times. The universe *will* kick you in the butt. The first time that you think you know something, the sky will knock you down to size. But it's all so beautiful. You look at the spiral structure of a hurricane, and a spiral of water swirling down the drain in a bathtub, and the spiral shapes of galaxies—all the continuity and repetition of pattern in nature—and it makes you feel good to be alive. Everything is so detailed, and the closer you look the more detailed it gets. It's a wonderful world—a wonderful universe—we live in. To me it's a visual world, and I just want to see."

II
BLUE WATER

7.

The Realm of the Sun

I feel the ocean and the forest—somehow
I feel the globe
    itself swift-swimming in space
    —Walt Whitman

The words of Mercury are harsh after the songs of Apollo.
    —Shakespeare

THE SUN IS A STAR, the only star whose phenomena can be studied in detail."[1] The words—almost a mantra, really—are those of George Ellery Hale, who fell in love with the Sun as a teenager. With his father's support, Hale built a solar observatory, equipped with a 12-inch refracting telescope, at their home in Hyde Park, Illinois, a Chicago suburb. He studied physics at MIT and in 1892 was appointed professor of astronomy at the University of Chicago, although he never got around to earning his doctorate. A spectacular career followed: Hale founded and directed Yerkes, Mt. Wilson, and Palomar observatories, recruiting astronomers who extended the dimensions of the known universe farther than had all their predecessors combined, establishing that the Sun is one among more than a hundred billion stars in one galaxy among billions of others, in an expanding universe billions of years old. But for all these soundings of deep space, Hale never lost his enthusiasm for the nearest star. When he was finally forced by his physicians to retire, after suffering a nervous breakdown under the combined workload of conducting his own research, administering the observatories, and raising funds for new ones, he amused himself by building a solar observatory behind his home in Pasadena, California. He equipped it with a spectroheliograph—a device he'd invented, through which solar prominences, previously visible only during a total solar eclipse, could be photographed every day—and began taking data. Soon he was working as hard as ever, a Sun-worshipper to the end.

Few stargazers share Hale's wholehearted dedication to the Sun, but even casual solar observing makes a lot of sense. It doubles your available observing time, and small telescopes can produce useful results. Before we go into that, though, it's time for the warning that astronomy writers are obliged to include when raising the prospect of observing the Sun: *Never look through an unshielded telescope pointed at the Sun! Serious eye damage, including blindness, can result!* Telescopes are designed to concentrate lots of light on a tiny focal point: To put your eye or any other part of your anatomy at the focal point of a telescope pointed at the Sun is like volunteering to be an ant fried under a magnifying glass. A telescope in sunlight should be handled with much the same care as a loaded pistol.

I was told by an astronomer—I'll call him Jack—that he once watched the robustly unpopular director of a major observatory being photographed for a picture-magazine spread. The director elected to pose in the observer's cage at the prime focus of the observatory's largest telescope, the throne room of celebrated astronomers like Hubble and Sandage, although he did mostly administrative work and seldom used the telescope at all. To avoid squandering valuable observing time, the photo session was done by day. As Jack looked on, the photographer had the dome slit opened and began shooting, ordering the telescope and dome moved to various orientations as he strove for the most dramatic lighting. The observatory director, eager to oblige, overrode a series of computer warnings and continued to slew the telescope and rotate the dome to comply with the photographer's wishes. Watching this scene unfold, it dawned on Jack that the photographer unwittingly was trying for a shot that could be obtained only by aiming the telescope directly at the Sun. Jack weighed the outcome if he remained silent. On one hand, sunlight concentrated by the giant mirror would quite possibly kill the despised director. On the other hand, they'd finally be rid of him. Ultimately Jack's humanitarian instincts prevailed, and he called out a warning before it was too late.

As a rule, no large telescope should ever be aimed at the Sun. The Hubble Space Telescope is not even permitted to observe the planet Mercury, which never strays more than 28 degrees from the Sun, for fear of its accidentally picking up direct sunlight. Smaller telescopes can, however, be used for solar observations, if proper precautions are taken. One method is to project the Sun's image on a screen. If you try the projection method, keep the finder scope capped, lest someone carelessly look through it, and be aware that there is a danger, if the telescope is more than a couple of inches in aperture, that solar heat may damage the eyepiece. The preferred method is to filter out most of the sunlight by putting a solar filter in front of the objective lens and other optics of the system.[2]

The first things one notices on the Sun are sunspots. Although they're actually quite hot and bright, they look dark because they are cooler than the surrounding solar surface. Each has a black central region, the umbra, surrounded by a gray penumbra. As Hale theorized, sunspots are created by swirling magnetic storms. They are usually found in pairs, one at each end of the magnetic loop that forms them. One spot in each pair is magnetically positive and the other negative, and their priority is mirrored by solar hemisphere: When the leading spot (in terms of the Sun's direction of rotation) in each Southern Hemisphere pair is positive, the leading spot in the Northern Hemisphere is always negative. The Sun rotates on its axis once every twenty-seven days, so sunspots can take almost two weeks to troop across the solar disk, during which time observers can chart their changing appearance. Sunspot counts, many of them done by amateurs, have confirmed that the number of spots waxes and wanes over an eleven-year cycle: At solar minimum there are sometimes no sunspots at all. At the end of each cycle the Sun's global magnetic polarity reverses, swapping magnetic north and south, and a new cycle begins. At the start of each new cycle, sunspots are found mostly at higher solar latitudes, and as the cycle goes on they come to prefer locations nearer the equator: This is the butterfly pattern first identified by the nineteenth-century English amateur Richard Carrington.

Solar flares and prominences boil up off the Sun's surface into space, launching sprays of high-speed particles. (Both types of eruptions are generated by the Sun's magnetic field, but flares are generally hotter and faster-moving than prominences.) When these particles reach Earth, they can create electromagnetic storms that disrupt radio communications and in some cases interfere with electronic and electrical gear. A gale-force solar storm that hit the Earth on March 13, 1989, sent electrons arcing across power lines, railroad tracks, and expanses of iron-bearing rock in eastern Canada, where alarmed engineers tried but failed to keep the Hydro-Quebec power grid operating. Power surges from the solar storm eventually overloaded the system at 2:44 A.M., putting out the lights of the city of Montreal and virtually all of Quebec. In space, meanwhile, friction induced by contact with the solar particles herded six thousand Earth-orbiting satellites into lower orbits. Ironically, the worst hit was the Solar Maximum satellite, designed to study the Sun during the peak of its eleven-year cycle. Crippled by the star it had been launched to observe, Solar Max was bludgeoned to low orbit and soon entered the atmosphere, crashing into the Indian Ocean southeast of Sri Lanka.

Solar storms incite auroras: Molecules of oxygen and nitrogen in the upper atmosphere are set aglow by the incoming charged particles in much the same way that the gas in a neon tube is lit up by an electrical current. When "space weather" stations post news of a major solar flare—particles from the Sun take three days to reach Earth, so there's plenty of warning—stargazers prepare to photograph and videotape the dancing auroral curtains of red, blue, and green. Concentrated toward the magnetic poles, auroras are seen at median latitudes only when Earth is struck by a particularly powerful storm. The 1989 storm that knocked out the Quebec power grid generated auroras that lit up night skies all the way from the poles to Bolivia and the Florida Keys.

The most spectacular views of the Sun are obtained through hydrogen-alpha filters like the one Jack Newton employed for his CCD solar images. The narrower the spectral line admitted by the filter, the better the view: The best filters have "subangstrom"-level resolution, meaning that they transmit less than one angstrom, or one ten-millionth of a millimeter, of solar light. An amateur at a star party once treated me to the view through a long-focus, gleaming silver refractor equipped with a 0.1-angstrom filter he'd made himself. The sight was riveting: Magnetically induced black swirls around sunspots, set against a mottled background of solar granulation consisting of small convection cells (diameters typically about five hundred miles), and arcing flares and prominences climbing off the solar limb. You could also see flares and prominences on the disk itself, like looking down from an airplane at the tops of tall trees. Flares can evolve in a matter of hours, but large prominences change more slowly, and so may seem frozen throughout the course of a day. In any event, the sheer dynamic energy of a star seen close-up was stunning to behold. Hydrogen alpha filters are expensive (you can pay for them either in money or, if you make one yourself, in time), but amateurs using them have seen and photographed sights as spectacular as any to be found in the solar system. When "Grandpa"—the largest eruptive prominence ever seen, spanning a quarter of the solar disk—rose into space in June 1946, only a handful of observers were equipped to see it, but its successor should be accessible to hundreds of amateurs.

Given the Sun's importance to humanity, as sustainer of life, generator of solar storms, and a ready laboratory for the study of astrophysics, it's not surprising that professional astronomers employ scores of space probes, satellites, and solar telescopes to study it. But it would be a mistake to presume that amateur observations of the Sun are therefore pointless. Observatories get clouded out, satellites falter or fail, and it is always possible that you will be the first—or only—person in the world to see a gigantic prominence appear or a new sunspot grow from what is called a "solar pore." "Never assume that you can miss a day of observing because 'others are probably observing the same thing anyway,'" advises the American amateur astronomer P. Clay Sherrod. "Always assume that your observations might be the only ones."[3] As the British amateur Gerald North notes, "There are significant gaps in the professionals' monitoring programs [and] it could just be that you might watch the unfolding of an eruption on the Sun's surface at a time when no professional telescope is trained on our daytime star."[4]

For all its tempestuousness, the Sun is quite even-tempered as stars go. It has been shining for nearly five billion years and is thought to have sufficient thermonuclear fuel in its core to maintain its equanimity for about another five billion, whereupon it will slowly swell to become a red giant star whose outer atmosphere may embrace the Earth. (The Sun has so much thermal momentum that when its central fires eventually do falter, several million years will pass before any change occurs at the solar surface.)[5]

Above the Sun's surface stands the solar atmosphere, composed of the chromosphere and the corona. The chromosphere ("color sphere") glows with a pink hue produced by excited hydrogen atoms. Beyond that extends the much larger, pearl-gray corona, its delicate tendrils tracing magnetic field lines far out into space. Both can be seen during total solar eclipses, when the Moon passes directly in front of the Sun, blocking out its light for a few minutes. Owing to a lucky coincidence, the Sun, which is four hundred times larger in diameter than the Moon, is also four hundred times farther away, so that the disks of both the Sun and the Moon have the same apparent size in the sky. Hence the Moon neatly blocks out the solar disk—except when it happens to be near the far point of the lunar orbit, creating an "annular" eclipse in which some direct sunlight spills around the edges. The Moon's shadow has two components—a dark, central umbra surrounded by a gray penumbra. Only those who view the eclipse from within the umbra see it as total, while those in the penumbra witness a partial eclipse, and those beyond the penumbra see nothing unusual at all. Eclipse enthusiasts travel thousands of miles to get themselves into the umbra's path. If skies are clear, they are rewarded by one of nature's most memorable spectacles.

Adept eclipse observers watch for several sorts of phenomena—if they watch at all: Some are so devoted to photography or to otherwise gathering data that they may neglect to look up from their equipment and take in the scene. Baily's Beads, bright globs of light created by solar rays shining through lunar valleys in the moments before totality, were first recorded by Edmond Halley in 1715 but are named in honor of Francis Baily, a businessman turned astronomer, who vividly described them in 1836. "A row of lucid points, like a string of bright beads . . . suddenly formed round that part of the circumference of the Moon that was about to enter on the Sun's disk," he wrote. "Its formation, indeed, was so rapid that it presented the appearance of having been caused by the ignition of a fine train of gunpowder."[6] The final rays, shining through a single lunar valley, can produce what is called the diamond ring effect. In the last moments before totality, if enough surrounding countryside is visible, you may see the umbral shadow of the Moon rushing toward you at a velocity of a thousand miles per hour. During totality, which typically lasts only about three minutes, the pink chromosphere, the red prominences protruding from the edge of the Sun, and the pearl-gray corona blaze forth, while planets and bright stars appear in the blackened sky.

I saw my first solar eclipse in North Carolina on March 2, 1970. I caught a ride down from New York with two friends, and when we reached the path of totality we found the farmlands dotted with glistening white telescope tubes and camera arrays, as if the path of the coming eclipse had been chalked in white across the land itself. We stopped at a small farm and introduced ourselves to the owners, a middle-aged couple whose forebears had farmed there for five generations. They kindly invited us to set up my eight-inch telescope but they refused to look through it, having been warned on the television newscasts that it was dangerous to stare at a solar eclipse. (The dangers are real enough, but can be overemphasized. While setting up to observe a solar eclipse on February 26, 1979, alongside a Montana highway, I encountered a school bus full of local children who lamented that although their school lay directly in the path of totality they would have to view the eclipse indoors, on TV, rather than going outside to see it for themselves.)

During the early stages of the North Carolina eclipse, as we watched the Moon carve the Sun down to a crescent in the eyepiece of the solar-filtered telescope, the dimming of sunlight was not particularly evident. The sky remained a brilliant blue, and the vivid greens and browns of the surrounding farmland looked much the same as when we'd arrived. Then, in the final minutes, the world began to get chilly and dark at an alarmingly accelerating rate. Cows lowed uneasily in the gloom, chickens flocked home to roost, and we began to feel apprehensive. We found ourselves saying things like, "Uh, you know, you don't really appreciate the Sun until it goes away."

Suddenly the sky collapsed into darkness and a dozen bright stars appeared. In their midst hung an awful, black ball, rimmed in ruby red and surrounded by the doomsday glow of the gray corona. No photograph can do justice to this appalling sight: The dynamic range from bright to dark is too great, and the colors are literally unearthly. (The ionized gas of the solar corona is hotter than anything gets on Earth except, momentarily, in the detonation of a hydrogen bomb, and is thinner than a laboratory vacuum.) I staggered back a few steps, like a drunken man—or like the Medes and Lydians, who stopped fighting and made peace when a solar eclipse interrupted their battle in 585 B.C. Observers more disciplined than myself have taken leave of their senses at just this moment. The astronomer Charles A. Young of Princeton University berated himself for falling into a trance during the 1869 solar eclipse in Iowa and failing to carry out his scientific tasks: "I cannot describe the sensation of surprise and mortification, of personal imbecility and wasted opportunity that overwhelmed me when the sunlight flashed out," he recalled.[7] Recovering, I snatched the solar filter off the telescope and had a look. A vivid red prominence was arcing up from behind the Moon, standing against the filmy silver backdrop of the Sun's outer atmosphere.

A minute or two later, as the Moon began to move off the Sun, a dot of direct sunlight, pure white as a welder's arc, poked through a valley between two mountains on the edge of the Moon. The "diamond ring" had returned. Soon it was joined by other spots to form Baily's Beads. My friends called out a warning and I knew I should pull away from the eyepiece, but the sight of sunrise on the Moon was so overwhelmingly beautiful that I couldn't stop looking until this peerless light had flooded my field of view. There was no pain—the retina has no nerves—but I suspected that I'd damaged my right eye.

Once sunlight returned to the North Carolina fields, our hosts asked us in for lunch. We said grace and ate grilled cheese sandwiches. Afterward, chatting at the table, I happened to remark that although sunlight was streaming in through the window as if everything was normal, the eclipse was actually still going on. When this met with skeptical responses, I demonstrated that it was so by making a pinhole with my crooked forefinger and projecting an image of the crescent Sun on the place mat in front of me. I instantly regretted having done so. The farmer and his wife turned pale and recoiled. They saw us to the door with decency and tact, but it was clear that they had been frightened by what they regarded as something akin to witchcraft.

Back in New York an ophthalmologist confirmed that, by continuing to look through the unfiltered telescope when the first rays of direct sunlight appeared between the mountains of the Moon, I'd burned a small hole in my retina and knocked loose a few "floaters," black bits of detached retina that I've been able to see ever since, moving like out-of-focus fish in the background of an undersea film. The retina as a whole remained intact, though, and no surgery was required. This came as a relief, especially since my eyes were none too good to start with (I'm nearsighted), and I'd previously been advised (wrongly, it turned out) that I had a fifty-fifty chance of suffering a retinal tear and going blind in one or both eyes. Perhaps the prospect of blindness was in the back of my mind when I persisted in peering through the telescope past the point of safety: If you think you're going to lose your sight anyway, you might as well see something memorable while you can. In any event I had few regrets. Love affairs can make you reckless and scar you for life, but what is life without love?

I watched the total solar eclipse of July 11, 1991, from a golf course on the island of Hawaii. Thousands had come to see it, and the hotel bars were serving up concoctions like the Solar Flare (rum, blackberry brandy, sweet-and-sour mix, schnapps, and grenadine) and the Total Eclipse (vodka and an Oreo cookie). Lead-gray clouds crowded the morning skies, and in our anxiety over the weather I reflected that waiting for totality is like waiting to be executed, except that all the emotions are reversed: You want it to hurry up and happen, and fear that it may not. Then, just in time, a hole opened in the clouds as the jet-black Moon imposed itself in front of the Sun. A great moan emerged from the crowds assembled on the fairway, and through my small portable telescope I could see a solar prominence perched on the rim of the black disk. A chilly wind blew down from the slopes of Mauna Kea, and off in the distance a sterling-silver glow arose from the wider, penumbral world. Then the sunlight returned, filling us with an atavistic sense of gratitude.

That night I set up the telescope by the seashore just after sunset to have a look at the white crescent of Venus, so bright it cast shadows, standing high in the western sky close to the bright star Regulus. Beneath it, near the western horizon, lay the planet Mercury. Like Venus and the Moon, Mercury displays phases. On this occasion it was gibbous—that is, a bit more than half-illuminated—but I could see nothing much on its surface. This is often the case: The *Mariner 10* space probe imaged half of Mercury in 1974, showing it plastered with lunar-style craters, but few such features can be seen from Earth, given Mercury's tiny apparent size and the fact that it never strays far from the Sun: To observe Mercury by night means looking through turbulent air close to the horizon in a dusk or dawn sky. Better views can sometimes be had by day. The Canadian amateur astronomer Terence Dickinson describes Mercury by day as having a creamy color similar to that of the daytime Moon, and presenting "an appearance of being vaguely textured, like fine sandpaper."[8] Two nineteenth-century observers—William Frederick Denning, an English amateur, and the great Italian astronomer Giovanni Virginio Schiaparelli, then director of Brera Observatory in Milan—made extensive daylight observations of Mercury but were unable to discern its rotation period (which turns out to be 58.6 days, two-thirds of its 88-day orbital period, a spin-orbit resonance caused by the tidal friction of the Sun). Intriguingly, the most prominent feature on Schiaparelli's Mercury maps—a large marking shaped like the numeral 5—corresponds to nothing in the *Mariner 10* photographs, and may lie on the side of the planet that has remained uninvestigated by spacecraft. "This is one of the increasingly rare cases where the amateur astronomer equipped with a modest telescope can still see to the very limits of our knowledge," advise William Sheehan and Thomas Dobbins of ALPO.[9] It was a treat to catch sight of this weird little world before it disappeared behind a cloud bank hovering over the ocean.

A few minutes later, I was looking at galaxies in Virgo when a friend turned up, apologized for being late, and asked if she could still see Mercury through the telescope. I explained that Mercury had disappeared behind the low-lying clouds, but I scanned the horizon anyway and found it peeking out from under a set of billowing thunderheads. She had a look, and when I returned to the eyepiece I saw something remarkable: So clear was the mid-Pacific air that as I watched, Mercury simply sank beneath the sea, bit by bit, like a miniature Moon. I would not have anticipated such a sight: The atmosphere at the horizon is usually much too hazy to afford a view of a tiny planet actually setting. (Legend has it that Copernicus never saw Mercury in his life, owing to mists rising from the rivers near his home.) But this memorable observation demonstrated anew that it never hurts to have a look. As the fishermen say, you can't catch a fish unless you get your line wet.

The most memorable episodes in Mercury's dance with the Sun come when it transits—passes in front of—the solar disk. I stayed home from school to observe the transit of November 7, 1960, from Key Biscayne. The morning skies were a deep, cloudless blue as I deployed my little telescope with its solar filter, a tape recorder, and a shortwave radio tuned to WWV, the Bureau of Standards station that beeps out the exact time. I'd read in *Sky & Telescope* that amateur astronomers could help the professionals measure the diameter of Mercury, which in those days was known to an accuracy of only about ten percent, by carefully timing the transit and sending in their data. The magazine planned to print the transit timings they received from their readers, and I was excited about my first chance to publish scientific data.

Tape recorder running, I gazed eagerly through the telescope and called out, "Mark! First contact!" when I saw the diminutive disk of Mercury appear, a small black bite eating its way into the edge of the yellow Sun. The next big moment was second contact, when the trailing edge of Mercury clears the edge of the Sun. Timing it was problematical, owing to the "black drop" effect: The disk of Mercury doesn't clear the Sun's limb cleanly but instead seems to drag a tail behind it. (Among the likely causes of this phenomenon is the interference of light waves coming from two slightly different directions. To see it, hold your thumb and forefinger up to the daytime sky, separated by a tiny space. A small dark zone seems to connect them even though they're not in contact.) The black drop showed up, all right, and when the first sliver of sunlight parted it, I called out "Mark! Second contact!"

For the next four hours Mercury glided across the Sun. At 3:09 P.M., when my friends were just getting out of school, it touched the other edge of the solar disk and I called out "Mark! Third contact!" followed by a final "Mark!" when Mercury disappeared. Then I mailed off my timings and awaited my entry into the annals of scientific research. As promised, the magazine published them, along with those of more than a hundred other observers—in Australia, New Zealand, South America, Europe, and across the United States—and the data did help refine the value of Mercury's diameter. But my heart sank when I compared my results with those of more skilled observers. Their timings fell within a few seconds of the consensus values, while my errors were in double digits. I wasn't the most bungling amateur of the bunch, but I'd done a thoroughly mediocre job. Better luck next time, I told myself, but the next Mercury transit was almost forty years away, on November 15, 1999. I resolved to see it, if I lived that long.

As it turned out, keeping my date with the Sun and Mercury was easy. When Monday, November 15, 1999, rolled around I was living in San Francisco, and all I had to do was set up a portable telescope on the roof of my house with a solar filter fitted safely over its objective lens. As I had back in 1960, I trained it on the expected point of Mercury's transgression onto the solar disk and waited. In the intervening decades, scientists had learned enough about the Sun and Mercury that there was no longer much to be gained by collecting amateur transit timings, but no matter. I was on a sentimental journey.

It was a spectacular, windy day. Flags swatted in the breeze, whitecaps dotted the green waters of the bay, and the Sun scudded in and out from behind great billowing clouds. A cloud obscured my view of first contact, but at 1:22 P.M. it moved away from the Sun, and I could see Mercury—a conspicuous dot, distinct as a BB pellet, its sharp edges confirming that the planet has little or no atmosphere—with a black drop connecting it to the edge of the solar disk. I felt oddly gratified that Mercury had kept its appointment, showing up on schedule after all these years. Life is uncertain, but the cosmic clockworks grind on.

Rock Music of the Spheres:
A Talk with Brian May

BRIAN MAY WAS ONE of the founders of the British rock band Queen. By his fiftieth birthday he'd written twenty-two hit songs; even a tune he did for a car commercial, called "Driven By You," wound up in the U.K. top ten. He wrote the Queen hit "We Will Rock You," which has become the anthem of sporting events throughout the English-speaking world, and played its famous opening bars—a sustained note that seems to gather itself in the sky like a tornado descending to Earth—on "The Red Special," a guitar that he and his father made by hand when he was a boy.

Given that Queen was viewed more as a flamboyant arena-rock band than an intellectual ensemble, I was surprised to learn that May had majored in astronomy and mathematics, and that other members of Queen had degrees in medicine, art, and physics. When I phoned May at his estate in the south of England, he told me—in unexpectedly soft, almost hesitant tones—that when he was about seven years old he "got passionate about astronomy and music, and they have never left me. My first telescope I made with my dad, at about the same time we made the guitar that I still use. It was just a little four-inch reflector on an altazimuth stand, but it's amazing what you can see with it. My dad could turn his hand to anything, and he was a genius in electronics, a fantastic craftsman. I wish he was still around.

"If it weren't for the tug on my heart that music has, I'd have been an astronomer for sure. I did postgraduate work at Imperial College in infrared astronomy—my field of research was interplanetary dust—and functioned as a professional astronomer during the years that I was doing my Ph.D. I built the first English hut in Tenerife, the basis of the observatory that's there now—well, I didn't build it single-handedly, but I organized the building of it."

May was disappointed by his fellow astronomers' lack of awe at the beauty of the night sky. "Once astronomers have set up their equipment, they don't bother to look up and go 'Oooo' anymore," he said. "But I feel about astronomy the way I feel about music, instinctive rather than analytical. First there's the 'Oooo!'—the pure emotional enjoyment factor, in music and astronomy, just allowing the beauty of things to wash over you. After that you can get analytical, but if you don't first allow yourself to be overwhelmed, I think you've missed the best part of it. Scientists tend to get wrapped up in what they think are explanations but actually are just relationships between facts. There's a lot to the world outside of that, having to do with the beauty of nature and our small place in the universe. I spend a lot of time just appreciating trees, flowers, and wonderfully clear, crisp night skies. I hope that doesn't sound pompous or critical.

"People sometimes ask if my interest in astronomy has made its way into my songs. I did write one song, called '39,' which is about a guy who travels in space at relativistic velocities. Owing to the time-dilation effect of special relativity, when he returns home it's a hundred years later, but he's still the same age. It's on a Queen record called *A Night at the Opera*.

"I have a small telescope and I stargaze—which is something that the professional astronomers don't do, generally; they don't look up much—and I have a hefty pair of binoculars that I use for comets. I get very excited if there's a comet. We've been robbed of the beauty of the night sky, you know? When you see those old pictures of the great comets over London, it's amazing what people on the street could behold, before there were so many lights. If there was a great comet now, I'm afraid that relatively few people would see it. Hope we get one, though! We're overdue, don't you think?"

8.

The Morning and Evening Star

Here come more stars to character the skies,
And they in the estimation of the wise
Are more divine than any bulb or arc,
Because their purpose is to flash and spark,
But not to take away the precious dark.
> —Robert Frost, "The Literate Farmer
> and the Planet Venus"

. . . I turned away to thee,
Proud Evening Star,
In thy glory afar,
And dearer thy beam shall be;
For joy to my heart
Is the proud part
Thou bearest in Heaven at night . . .
> —Edgar Allan Poe, "Evening Star"

THE PLANET VENUS is one of the finest sights in the sky for the unaided eye. It outshines everything but the Sun and the Moon, and does so with a pure white light that has long impressed people with its beauty. Small wonder that the Sumerians, the Pawnee, and the ancient Greeks and Romans all named Venus after feminine figures of allure.

Its uniquely changing appearance and its complex motions across the sky add intellectual appeal to Venus as well. Closer to the Sun than we are, Venus at sunrise and sunset never strays much more than halfway from the horizon to the zenith. Hence its nicknames: Venus is called the morning star when it rises before the Sun, and the evening star when it sets after the Sun. Its apparent path over the six months or so of each such apparition describes an elegant set of patterns, produced by the combination of its orbital motion and that of

Earth. Some of these patterns are as narrow as awls, others as beamy as sails and guitar picks. Some contain retrograde figure-eight loops in which Venus crosses back over its tracks.

The orbits of Earth and Venus obey an almost exact 5-to-8 resonance: Venus appears as the morning star every 584 days; multiplying 584 by 5 yields 2,920, which divided by 8 is 365, the number of days in one Earth year. The upshot is that once every 52 years these two cycles synchronize with each other, whereupon Venus appears in the same spot in the sky, on the same date, that it did 52 years earlier. This happens because Venus, orbiting the Sun every 224 days, takes a total of 584 days to catch up with Earth, but early observers discovered the relationship without knowing anything about orbits. The ancient Mayans—who built Caracol, in Chichen Itza, an observatory whose ruins bear a striking resemblance to a modern observatory dome, to study the motions of Venus—called the 52-year cycle the Calendar Round, and attached great significance to it. Quetzalcoatl, a light-skinned god identified with Venus in Mayan mythology, was said to have vanished into the East to found another society, predicting that he would one day return. His return was expected to coincide with the culmination of a Calendar Round cycle. Their devotion to this legend proved disastrous for the Mayans when, in 1519—which happened to be just such a year—the light-skinned but otherwise distinctly ungodlike conquistador Hernando Cortez landed in the New World. Montezuma, believing Cortez to be Quetzalcoatl, offered no resistance. In the wanton destruction that ensued, only a single book of Mayan astronomy survived.

TRACES OF VENUS WORSHIP HAVE ENDURED—the Mayan calendar is still in use in rural parts of Central America, and the Skidi Pawnee of Nebraska sacrificed a teenaged girl to Venus in the predawn hours of April 22, 1838—but increasingly the devotees of this brilliant planet are amateur astronomers rather than shamans. Visible to the naked eye in broad daylight during its periods of maximum brightness (if you know where to look) Venus is best observed telescopically by day: It can be studied while high in the sky, where there is less air turbulence, and the bright background makes for easier viewing than does the harsh contrast between a dark sky and the planet's brilliant disk. Eugene O'Connor of the Astronomical Society of New South Wales found that if he chose a cycle when Venus neither approached the Sun too closely nor passed directly behind it, he could view it through binoculars on every day of the year. (He was careful not to point his binoculars at the Sun.) "Anyone, with a little care, practice and moderate eyesight, can observe Venus for a full year—if you pick the year," he reported. "Allow eyes time to focus on infinity when viewing and expect to be distracted by Venus-like shapes—floating seed pods, not to mention the constant distraction of planes, migrating pelicans and—would you believe it—drifting cobwebs."[1]

The moonlike phases of Venus impressed Galileo, who noted that they supported the Copernican theory that the planets orbit the Sun. Observing Venus in 1610, Galileo conveyed the potentially heretical news to Johannes Kepler in the form of a Latin anagram: *Haec immatura, a me, iam frustra, leguntur—o.y.*—"These thing not ripe are read by me"—which, suitably rearranged, reads *Cynthiae figuras aemulatur Mater Amorum*, "The Mother of Love [Venus] imitates the phases of Cynthia [the Moon]." As Venus draws closer to Earth it swells in apparent diameter, growing from a nearly full, gibbous disk only 10 seconds of arc across to become, at inferior conjunction—the point at which it is closest to Earth—a slender crescent fully 48 arc seconds wide, even larger in our skies than mighty Jupiter.

Early telescopic observers saw evidence that Venus has a dense atmosphere. Sometimes the tips of the crescent extend into the dark side of the globe—forming "cusp extensions"—and various observers have reported that the dark side sometimes is illuminated with a faint "ashen glow," resembling earthshine. As Leonardo da Vinci recognized, earthshine on the Moon is caused by sunlight reflected off the Earth: When the Moon presents a crescent to us, the Earth is nearly full as seen from the Moon. But the ashen glow of Venus, if real, must be produced in some other fashion, and its cause remains in dispute. Some theorize that it is caused by "airglow," a release of energy stored when sunlight ionizes the upper layers of Venus's atmosphere. Others dismiss it as an optical illusion.

Twice in a long while Venus transits the Sun. The transits come in pairs, with eight years intervening between the members of a pair and more than 120 years separating pairs. A pair of transits occurred in the seventeenth, eighteenth, and nineteenth centuries; there was none in the twentieth, and there will be two in the twenty-first century—on June 8, 2004, and June 5–6, 2012. "When the last transit occurred the intellectual world was awakening from the slumber of ages," wrote the astronomer William Harkness of the U.S. Naval Observatory, in 1882, "and that wondrous scientific activity which has led to our present advanced knowledge was just beginning. What will be the state of science when the next transit season arrives God only knows."[2]

Mikhail Lomonosov, observing the transit of 1761 from the University Observatory in St. Petersburg, Russia, saw a pronounced "black drop" effect, which he attributed to Venus's having "an atmosphere equal to, if not greater

than, that which envelops our earthly sphere."[3] Observers subsequently
sought to discern patterns in the clouds, hoping to learn the planet's rotational period and perhaps to peer through a hole in the clouds and get a
glimpse of the surface, but most such projects proved ill-fated. All the published estimates of Venus's rotation period were wrong, and while the clouds
were easy to see—looking at Venus, one sees nothing *but* its cloud cover—the
putative cloud patterns were mostly illusory. Indeed, able observers were char-
acterized less by seeing patterns than by having the discipline to study Venus
at length without seeing things that weren't there. After the wealthy German
amateur astronomer Johann Schröter said he had watched sunlight strike the
summit of a mountain peak poking through the clouds of Venus, William Herschel had a long, cool-headed look and declared Venus to be featureless. The
planetary observer Bertrand Peek, alert to the dangers of telescopic illusions,
once complimented the young Alan Heath, a fellow member of the British
Astronomical Association, on his observations of Venus.

"But I hardly ever see anything," Heath objected.

"Precisely my point," said Peek.[4]

A notable exception was the French amateur astronomer Charles Boyer.
Trained as an attorney, Boyer held a government post from 1955 at Brazzaville
in the Congo. Finding the air there to be exceptionally steady, he constructed
a ten-inch reflector and wrote to an astronomer friend, Henri Camichel, at Pic
du Midi Observatory, asking him to propose worthwhile observing projects.
Camichel suggested that Boyer try photographing Venus in ultraviolet light.
Astronomers had found that the clouds of Venus, opaque at visual wavelengths, showed dark features in the ultraviolet, although the patterns were
too indistinct and confusing to yield a clear rotation period. Boyer's telescope
lacked the clock-driven equatorial mount that normally would be required to
take time-exposure astronomical photographs, so he fashioned an Erector-set
device that moved the camera across the focal plane instead. He used slow film
and a violet, rather than ultraviolet, filter, and his images of Venus were just
tiny dots on the emulsion. Others had trouble discerning anything at all on
them, but Boyer claimed to see evidence of a four-day rotation period.
Camichel did, too, and the professional and the amateur coauthored an article
on the four-day period, which appeared in a popular magazine, *L'Astronomie*, in
1960, plus two papers in technical journals.

Meanwhile another amateur, Rodger Gordon, was observing Venus visually through an ultraviolet filter. Although he knew nothing of the Boyer-
Camichel findings, Gordon, too, obtained a four-day period, and he published
a paper saying so in his astronomy club newsletter in November 1962. "I don't

have much confidence in my findings," he wrote, modestly. "It would not be surprising if I am considerably off the mark."[5]

As far as the rotation period of Venus is concerned, Boyer, Camichel, and Gordon were indeed far off the mark. Radar echoes bounced off Venus by American and Russian scientists in 1962 revealed that Venus spins on its axis once every 243 days—in retrograde, the opposite direction from most planets. (The reason for this strange situation is unclear; perhaps a large object hit Venus in the early days and reversed its spin.) "I felt like crawling into a hole," lamented Gordon. Boyer and Camichel, undaunted, submitted a paper on their four-day cloud-rotation value to *Icarus*, an American journal of planetary studies, but it was rejected after Carl Sagan, then a young astronomer at Harvard, was asked to referee the paper and scoffed that "the four-day rotation is theoretically impossible, and shows how foolish the work of the inexperienced amateur can be."[6] Boyer, Camichel, and another French astronomer, Bernard Guinot, continued observing Venus anyway, and kept coming up with the four-day period during which the same cloud patterns crossed the disk.

There matters rested until February 1974, when the Mariner 10 spacecraft flew past Venus, photographing it in the ultraviolet. Movies made by assembling *Mariner*'s still photos showed that the atmosphere of Venus, unlike the solid planet beneath, does in fact rotate once every four days. The problem was that the clouds' rotation does not match that of the planet beneath the clouds. The amateurs had been right after all.

Venus is Earth's twin in terms of gross anatomy. Its diameter is 95 percent, its mass 85 percent, and the gravitational force at its surface 90 percent that of Earth; the radius of its orbit is 70 percent of ours; and it has clouds that superficially resemble those of Earth, if only by virtue of their bleached-white color. Knowing this much (and not much else, owing to the clouds), some scientists and scholars presumed that the resemblances went further. This hyperbolic chain of reasoning was stretched to the breaking point by Camille Flammarion in 1884. Flammarion, a former theology student whose work mixed amateur and professional astronomy, had a romantic enthusiasm for the night sky that made his books extremely popular, but he sometimes got carried away. In *Les Terres du Ciel*, he reasoned that since Venus has "nearly the same dimensions, the same weight, the same density, [and] the same surface gravity" as Earth, it must also have "the same length of day and night, the same atmosphere, the same clouds, the same rains."[7] The first four similarities were factual. The fifth had some scientific basis—it came from the wildly inaccurate rotation estimates being made at the time—but Flammarion's conjectures about atmosphere, clouds, and rain were entirely unsupported and clangorously wrong.

So were the speculations of those who reasoned that a cloud-covered planet closer to the Sun must resemble the tropical rain forests of Earth. Jacques-Henri Bernardin de Saint-Pierre (1737–1814), a disciple of the philosopher Jean-Jacques Rousseau, published in one of his books a fanciful engraving showing the inhabitants of Venus as resembling "the happy islanders of Tahiti." The Nobel Laureate chemist Svante Arrhenius in 1918 described Venus as a "dripping wet" nursery of giant, short-lived plants. Garrett Serviss, in his popular 1888 book, *Astronomy with an Opera Glass*, suggested that the atmosphere of Venus might "furnish the breath of life to millions of intelligent creatures, and vibrate with the music of tongues speaking languages as expressive as those of the Earth."[8] The influential science fiction illustrator Frank R. Paul (1884–1963), in *Fantastic Adventures* magazine, called Venus "a planet of almost tropic richness . . . a young world, full of danger, and possible monster forms of life."[9] As late as 1960, the science journalist G. Edward Pendray, a founder of the American Rocket Society, was proposing that "Venus may turn out to be a wonderful place to live . . . like Florida all over."[10] The white clouds presented a blank slate on which the human imagination could project almost any fantasy.

Rude reality intruded once Russian and American space probes began taking a closer look, in a series of unmanned missions beginning in the 1960s. Coincidentally, the mission strategies of the two competing Cold War superpowers complemented each other splendidly, with the Russians dispatching a series of Venera landers while the Americans imaged the planet from space, using Mariner and Pioneer flyby probes and finally, in 1990, mapped Venus with the Magellan orbiter.

The Venera missions, though given short shrift in the American press, ranked among the most innovative in the history of robotic spaceflight. For nearly a quarter of a century, beginning in 1961, the Russians sent a new and improved *Venera* on its way at almost every launch window. (Such windows come at nineteen-month intervals; the craft is launched before inferior conjunction, when Venus is an evening star, and arrives after conjunction, reaching Venus the morning star.) The early ones failed utterly. *Venera 1* and *Venera 2* died en route. *Venera 3* hit Venus, but had already stopped transmitting. *Venera 4*, parachuting in, reported alarmingly high atmospheric pressures and temperatures before falling silent while still sixteen miles high. The next three missions failed for various reasons, but *Venera 8* managed to land safely, on July 22, 1972, and survived for nearly an hour. The photos and data it sent back, combined with studies by American flyby probes and subsequent Venera missions, made it clear that Venus is no Tahiti. Its surface is not wet, but dry—at least a hundred times dryer than Earth. Its temperature is a blistering 864° Fahrenheit—hotter than the surface of any other planet; hot enough to make rocks glow; "seven time hotter than it ought to be," to borrow a phrase from Louis Armstrong's "Shadrach." The atmosphere is 97 percent carbon dioxide, with too little oxygen to measure, and its pressure, 90 times that on Earth at sea level, approximates that experienced by terrestrial sea divers at a depth of 900 meters, enough to crush a deep-diving suit. The clouds are made not of water vapor but of battery acid. The ambient color is dark red, the prevailing odor that of sulfur. Even a fervent nonbeliever who found himself on Venus could be excused for thinking, during the few moments of life remaining to him, that he was in Hell.

Subsequent studies based on *Magellan*'s radar mapping of the surface centered on the unsettling question of how a planet outwardly so similar to Earth could be so alarmingly different. Astronomers theorize that the appalling heat and dryness of Venus resulted from a runaway greenhouse effect, in which the abundant carbon dioxide in the atmosphere trapped solar heat under the cloud cover. Venus and Earth are presumed to have begun their careers with about the same mix of chemical elements, since both condensed out of the same solar nebula at similar distances from the Sun, so why did they not evolve similarly? There are a number of theories, but the short answer is that nobody yet knows.

Adding to the piquancy of the question are two other considerations. The first has to do with the ratio of hydrogen to deuterium on Venus, which suggests that Venus once had oceans comparable in extent to those of Earth. Planetary scientists calculate that those oceans could have survived for almost half of the planet's 4.5-billion-year history, before something—the runaway greenhouse process, presumably—boiled them off, sending water vapor into the upper atmosphere where solar ultraviolet light bound the oxygen into carbon dioxide molecules and blew the hydrogen into space. The other is that the surface of Venus is geologically young, having formed only about 600 million years ago. This can be inferred by counting impact craters: Although the thick atmosphere of Venus protects the surface from being struck by meteoroids smaller than about a kilometer in diameter, big impactors should have gouged out many times more than the 963 craters found on the *Magellan* radar maps. Possibly Venus underwent a spasm of volcanic activity and "turned itself inside out," as the astronomer David Grinspoon puts it.

Venus remains a beautiful sight in the sky for stargazers, but its beauty is attended by a new version of the tinge of terror that prompted past human sacrifices to the morning and evening star. If Venus had oceans and a temperate climate for billions of years, there could have been life there, too, with ensuing volcanic upheavals wiping out virtually all traces of it. And if a runaway greenhouse effect was responsible for the climatic catastrophe, that fact should be of intense interest to humans, living as we do on Venus's twin and finding ourselves in an industrial era when the temperature of our planet is rising—due, it is thought, at least in part to growing levels of industrial carbon dioxide in the atmosphere. As Grinspoon and his colleagues write, "The conditions on Venus and its path of development serve as a warning to humanity. . . . Studying Venus, however alien it may seem, is essential to the quest for the general principles of climate variation—and thus to understanding the frailty or robustness of our home world."[11] The worrying question when looking at Venus from our polluted planet is whether it presents us with a vision of our potential future.

Founding Father:
A Visit with Patrick Moore

PATRICK MOORE, THE PATRIARCH of astronomy popularizers, wrote the books from which I first learned about observing the Moon and planets. I followed his friendly, crisply informative advice on using telescopes, emulated his drawing style—inky blacks for lunar craters, charcoal and pastels for the planets—and traced the squat template of the oblate disk of Jupiter from his book *The Amateur Astronomer* so many times that it threatened to fall out of the page, leaving a Jupiter-shaped hole behind. His urbane yet enthusiastic writing style helped me connect astronomy with the wider culture—specifically, with the British tradition of accomplished amateurism, a driving wheel of the scientific and technological revolutions that transformed the world. At age sixteen I took one of Moore's books with me on an ocean voyage with my parents to England, and was reading it on the day that the green coast near Southampton came into view from the deck. Forty years later, in the summer of the year 2000, I arrived in the village of Selsey along that same south coast to meet Moore at last.

"Ah, there you are!" he boomed, opening the door of the seventeenth-century house where he has lived since 1966. "Care for lunch?" He limped to my car, and we drove a few hundred meters to a neighborhood restaurant.

"It's a temporary limp, but it makes me quite grumpy, I'm afraid," he growled, as we got out of the car in front of the Seal Hotel. "Ordinarily I play cricket—I'm a spin bowler—but obviously these days I can't. I'm walking around like somebody aged seventy-seven. The fact that I *am* seventy-seven is neither here nor there. Normally I can keep up with any forty-year-old."

The sight of the monumental Moore lumbering down the sidewalk slowed village traffic almost to a halt. A big man whose monocle, shock of white hair, and rolling, Churchillian voice—a splendid broadcast voice, at once emotive and cracklingly precise—were national icons, Moore had long been one of the most famous men in Great Britain. He was soon to be knighted, and his BBC television show, *The Sky at Night*, which had aired every month for forty-three consecutive years, was listed in the *Guinness Book of World Records* as the world's longest-running TV program with a single host. We entered the hotel and sat at the bar. Moore ordered me a double vodka and himself a double gin, solicitously offering the bartender a drink as well. I inquired about his childhood, having read that he'd never really gone to school.

"I've always had a crocked heart," Moore said cheerfully. "Had to be kept home from school from age six to sixteen. Still, I had more education than most boys. I couldn't do much of anything else. Got started in astronomy by reading a book of my mother's—*Story of the Solar System*, by George F. Chambers, FRAS, published 1898, price sixpence. I bought a star map, got a pair of binoculars, and learned my way around the sky. Had to teach myself everything. Taught myself to type, on my grandfather's 1892 Remington. I still have it, but my main typewriter these days is a 1908 Woodstock. On it I can type accurately at ninety words per minute, whereas on the computer I can type only fifty or sixty words per minute.

"I joined the British Astronomical Association at age eleven, and published my first paper at age thirteen in the *BAA Journal*, about craterlets on the Moon. I did that work with a three-inch refractor. When the war came along I knew that if I went into the army or the navy I'd not last ten minutes—but I could fly, I expected, so that's what I did. I had to fiddle my age on the RAF application, and fiddle the physical examination, but they took me. Served as a navigator in bombing command, out of the north of England mostly, and I could also fly a bit when I had to. I once had to land a plane with five men aboard, after the pilot was hurt. The Germans were potting at us—I thought it most unfriendly—and the pilot was wounded. They found out the truth about me when I was eighteen. The officer who confronted me said they'd learned that I'd fiddled my age and the physical. 'But,' he said, 'you've been a commissioned officer in the RAF since age seventeen, so I suppose the only thing to be done for it is to ask you to come along to the mess and have a drink with me.'

"Top you off?" Moore gestured almost imperceptibly to the bartender, and another round of drinks appeared.

His studies interrupted by the war, Moore never attended university. Instead he was approached by the British associate of an American publisher about writing a popular astronomy book on his favorite subject, the Moon. *The Moon* captured a wide readership, launching his career as an author. He'd since written more than sixty books, by my count. (Moore, when I asked him, said, "I don't know—a substantial number, I suppose.") Rereading a few of them, I was impressed by his encyclopedic command of the scientific literature and his firm grounding in the art of coaxing useful, reliable observations out of smallish telescopes.

Lodging a monocle in his right eye, Moore consulted the menu. The monocle promptly fell out, and Moore replaced it unhurriedly, like a pipe smoker relighting his pipe. "I've always had a monocle," he said. "It's part of me. I started wearing it when I was sixteen. They wanted to give me a pair of glasses with one plain lens. I said, 'Why a plain lens?'"

We made our way to a table, the focus of a dozen sidelong gazes. A young man came up, thrust out a piece of paper, and said, "Excuse me, Mr. Moore, would you mind signing this for me? Thank you so much. Sorry for the interruption. You are a *star*, sir!" A bottle of chilled white wine appeared, closely followed by two plates of Dover sole.

We discussed Moore's lunar studies. His maps, although based largely on back-yard observations, were sufficiently reliable that the Russians used them to coordinate the first photographs of the far side of the Moon, taken by the automated *Lunik 3* probe, and American navigators consulted them when preparing for the manned *Apollo* landings. "I'm not a researcher," he mused. "I haven't the brain. I am a lunar cartographer—my main work was mapping the Moon—and my research was done years ago. But if I can give other people a hand, I will." Befuddled by his deference—and by the wine; Moore had produced another bottle—I asked a dumb question: "You're not particularly discovery-oriented, are you?"

"I have had one discovery," Moore said mildly. "Mare Orientale, on the Moon's eastern limb."

I slapped my forehead. Mare Orientale is the largest impact crater on the Moon, a baleful-looking bull's-eye larger than Nicaragua. Centered just off the Moon's visible side, it is usually assumed to have been discovered from space probe images of the lunar far side. But its outer edges impinge on territory that comes into view occasionally, owing to lunar librations: Because of our varying perspective on the Moon, and its intrinsic bobbing motions (it nods a bit each month, both north-south and east-west, like someone indicating "yes" and "no"), we on Earth can see nearly sixty percent of the lunar surface, if we look at all the right times. In the 1930s, Moore, mapping libration areas near the edge of the Moon with his 12.5-inch reflector in Sussex, charted the outer edges of Mare Orientale, recognized it as part of an undiscovered feature, and even proposed the name (it means "Eastern Sea") by which it is known today. I'd forgotten that.

"And the crater known as Einstein," Moore added. "I discovered it, too—right on the limit. Top you off?"

"Astronomy is one of the few sciences in which amateurs can be useful," Moore added. "The great advantage that amateurs bring to astronomy is in the continuity of their observations. If a dust storm happens on Mars, or a new white spot appears on Saturn, it will be an amateur who finds it. I found a white spot myself—a very small one—in 1961. I also saw the one Will Hay saw in 1933. Coffee?"

"Yes, thank you."

"Irish?"

I'd intended to pay the check but none appeared. "Quite impossible," Moore said amicably. "This is Selsey. You're on my turf." We emerged from the restaurant into brilliant sunshine and a bracing sea breeze.

"'Selsey' means 'Seal Island,'" Moore expounded. "I've never seen a seal off this coast, and it's not an island, but there you are. I'll never leave here. I have the house I want, in the place I want to be."

"Do you sail?"

"No. I live by the sea, but I don't really understand the sea. The *air* I understand. I live in the air, with my feet on the ground."

Back at Moore's house, we passed through the foyer, which was cluttered with small telescopes and four Russian lunar globes, toured rooms piled high with books, scientific journals, cameras, binoculars, and the memorabilia of Russian, European, and American space projects, and sat in the study, a place alive with the sounds of clocks ticking and chiming, where the few expanses of wood-paneled walls not covered with Moore's books were crowded with citations, plaques, and honorary degrees. Moore lit a pipe, offered me a cigar, and stacked before me a sampling of his early observing logs, bound ledgers neatly titled "Mars," "Jupiter," and "Venus." Their pages contained carefully drawn illustrations, in colored pencil on black backgrounds, accompanied by concise notes inscribed in black ink. They produced a curious sense of recurrence, since I'd taught myself to draw in a style very much derived from his—which, in turn, he had derived in part from his boyhood reading of George Chambers's *The Story of the Solar System*, a book notable for its lucid drawings. And here were Moore's drawings and commentaries on the Mars opposition of 1956, the event that had got me started as a telescopic observer. August 8, 1956: "The best view . . . I have ever had. Phase now 95%. The cap has a border . . ." October 18: "Probably the best view of the opposition."

Moore was making me a present of one of his CDs (he had composed two operas and a number of military marches) when two young men strolled in and were introduced by Moore as amateur astronomers with expertise in electronics. They were Chris Reid, a cheerful blond who smiled and said little, and

Tim Wright, a rail-thin lad dressed in jeans, a dark blue T-shirt, and enormous aviator glasses who talked in the forceful staccato of a hands-on achiever. They explained that they were there to test a dipole radio telescope, set up on Moore's property, part of which was soon to be transferred to the South Downs Planetarium, a public institution being built around a star projector that Moore had wrangled from the Armagh Planetarium in northern Ireland, which he'd established and directed years earlier.

"Where's my syringe?" boomed Wright. "I'm not a junkie, but I need to clean my syringe." He found it in a box of odds and ends, washed it at the sink, then filled it with ink that he injected into the printer cartridge of the radio telescope chart recorder in a corner of the study. He scrutinized the night's chart recording, a long sheet etched with lines in blue ink. "Patrick, what's at about 42.5 degrees declination that has a strong radio source?" he asked. Moore was reading his mail. Without looking up, he replied, "Cassiopeia."

"When?"

"Prominent from midnight till four A.M."

"OK, that's it then." The radio telescope had detected "Cas A," a supernova remnant 11,000 light-years from Earth. Wright wrote a note on the chart. He studied the chart further, then announced, "Between twenty-two thirty and twenty-three hundred hours last night one of two things happened: Either we had a supernova, or your neighbor was using his welder."

We went out onto the capacious lawn to inspect the radio telescope. It consisted of two wires, each 2.4 meters long, attached to an old receiver working at 60 megahertz. "The three hundred meters separating the two wires makes a three-hundred-meter-diameter dipole radio telescope," Wright announced. "Eventually it will be much larger. One wire will be installed at the South Downs Planetarium, while the other remains here in Patrick's yard. We obtain the direction of radio sources in the sky by measuring the slightly different time that a radio signal, traveling at the velocity of light, hits the two antennas. We also get a lot of interference, of course, including Spanish, Russian, and Italian TV signals coming in."

We inspected Moore's telescopes, each installed in its own observatory. A big green tub with a flat roof that resembled a petroleum storage tank housed a forty-year-old, 15-inch Newtonian reflector with a wooden tube and a gigantic mounting. "It looks quite old-fashioned, I'll admit," Moore said. "No electronics at all. Computerized astronomy is not my line. I'm an astronomical dinosaur—a visual observer of lunar and planetary phenomena. Anything beyond the orbit of Neptune is a bit far out for me."

Ten meters away stood a tall, rotund shed, about the dimensions of an ocean liner's smokestack, which held a five-inch refractor. Wright unceremoniously aimed it at the Sun and projected a solar image onto a weathered sheet of paper bearing a maintenance note. We were in a year of maximum solar activity, and sunspots dotted the disk. A comely octagonal hardwood observatory structure with windows in the sides contained an 8.5-inch reflector. Moore explained that he'd designed the building to look nice because it originally stood in the front yard of his previous house. We opened up yet another observatory, little more than a rude shed, to reveal a 12.5-inch reflector on a simple altazimuth mount.

"It's a good mirror," Moore said. "How many hours have I spent using this telescope? God, I don't know. Tens of thousands of hours, I suppose. I discovered Mare Orientale with this telescope." With a satisfied sigh, he leaned back against the shed door and stared up into the late-afternoon sky, where a bank of clouds was gathering. He seemed thoroughly at home, in his skin and at this house, but it required no great psychological acuity to perceive in him the solitary boy who was kept home from school and reached out to other worlds for companionship, or the man who fought in the war while keeping aspects of his identity secret and then become an author, itself one of the world's loneliest occupations, dwelling for half a century in celebrity's castle keep.

"I know Neil Armstrong very well," he remarked. "And I met Orville Wright. Do you know, the first man on the Moon and the first aviator could have met—could have, but they didn't. You have a wife and son?"

"Yes."

"I envy you that. I always wanted that, to have a wife and a son, but it was not to be. I grew up at an unusual time, through no fault of my own." His face brightened. "My twenty-year-old godson, Adam, who is getting his college degree, has decided to make his home with me," he said. "I'm quite happy about that." He leaned back against the observatory, relaxed and smiling, studying the darkening sky.

9.

Moon Dance

Summer Moon—
clapping hands,
I herald dawn.
—Basho

Soon as the evening shades prevail,
The Moon takes up the wondrous tale,
And nightly to the listening earth
Repeats the story of her birth.
—Joseph Addison

A T THE WINTER SOLSTICE of 1999, on the coast of Lana'i in the Hawaiian Islands, a few dozen tourists were sitting on lawn chairs and sipping from steaming mugs of coffee on a greensward overlooking the sea, watching a planetarium lecturer set up an 11-inch Schmidt-Cassegrain telescope on a massive German equatorial mount to view the full Moon. The press had been running front-page stories about this particular full Moon. It would occur on the longest Northern Hemisphere night of the year, and also when the Moon was near perigee, the point where it draws closest to Earth in its orbit. The Moon tonight would be 50,000 kilometers nearer, and hence would look 14 percent larger, than when at lunar apogee. The lecturer, a robust, middle-aged man in well-worn track shoes, was a veteran observer, but the sky was cloudy and he was getting flustered. He blundered, saying that the Moon was closer to Earth tonight than at any time in the past 130 years—actually, the Moon comes this close every perigee, and the near-perigee full Moon of January 15, 1930, was 416 miles closer than this one—but I'd be the last to correct him. He was trying to hold a restless audience, and so far had nothing to show them.

I'd been in his shoes. I recalled one night in the Yellowstone mountains when I was showing the night sky to a gaggle of billionaires who asked a lot of questions. I wasn't aware of being nervous, but when I looked up to aim my portable telescope I could recognize not a single constellation. They had vanished as completely as if we'd been transported to a distant star. Panic solves nothing, so I simply took a deep breath and waited a moment. The constellations reappeared, and I went to work, but the episode had since haunted me like a bad dream.

So tonight, under the Hawaiian clouds, I did what I could to avoid making things any harder for the planetarium man as he struggled through his impromptu lecture under cloudy skies. Then, just as a few members of the audience began edging toward the beckoning lights of their resort hotel, a hole in the clouds opened up to reveal the Moon, wonderfully bright though dulled a bit by intervening mists to the muted sterling hues of a storybook illustration. It was as if a silversmith had engraved the face of the Man in the Moon at the center of an oval serving platter, with coils of tropical clouds its peerless filigree. A line formed at the telescope, and the lecturer relaxed into practiced patter about lunar craters and seas.

THE MOON IS THE MOST familiar and yet in some ways the most unusual object in the night sky. It's uniquely big—at nearly a quarter of Earth's diameter the largest of all solar system satellites, relative to the planet it orbits. (The sole exception is Charon, which orbits Pluto, which probably isn't a planet.) So the sight of a fat, full Moon looming in the sky—blamed and praised for everything from brawls to love affairs—really is something special. Among the other terrestrial planets, Mercury and Venus have no known satellites at all, and Mars has but two tiny ones, Phobos and Deimos, which most likely are captured asteroids. Yet the Moon is also a lightweight, much less dense than the Earth is, a situation that raises intriguing questions about where the Moon came from. Some stargazers regard the Moon as a dull object, or assume that everything is already known about it. But the long, intertwined history of Earth and its Moon is far from being fully understood, and illustrates the dictum that we cannot expect to understand our home planet without looking and going beyond it.

To the unaided eye, the Moon presents a shimmering disk splattered with darker regions—the *maria*, or "seas." The skilled observer and astronomy popularizer Joseph Ashbrook noted that "a surprising amount of lunar surface features" can be seen with the naked eye if you view the Moon by twilight, when it produces less glare.[1] Since one can hardly ignore these light and dark patterns, you'd think that the ancients had all sorts of theories about them. And they did have a few. In the first century A.D., Plutarch, an acute astronomical observer, wrote a book, *On the Face in the Moon*, evaluating the more popular theories.[2] Among the explanations of lunar markings that Plutarch says were "current and on everyone's lips" in those days (along with claims that the Moon was a mirror reflecting back an image of Earth) was the seemingly sensible idea that the Moon was a ball of rock and that its markings were continents like the Earth's. But if so, what kept it from falling down on us? That was the nub of the problem. Common sense dictated—and many great thinkers, Aristotle among them, agreed—that the Earth was the only planet in the universe. Rocks fall, so if there were any rocks up in the sky, they would already have fallen to Earth. Hence everything up there must be made of some lighter substance. This "fifth element," called aether, was thought to be pure, unsullied stuff, so if the Moon seemed to have marks on it, they must be illusory.

The failure of ancient thinkers to better understand lunar features arose not from any want of reasoning ability—Aristotle and Plutarch were extraordinarily reasonable men—but from their lack of adequate technology: They could not properly investigate the markings on the Moon because they had no telescopes. The rise of science is often equated with the rise of "reason," and so it was, if we look no further than to the days when witches were being burned and flagellants literally beat themselves up over the Plague. But if we look back to classical times, and outward to the civilizations of Asia and the Arabic world, we find that reason, though seldom popular, had been around for a long time. What made science possible was the rise not of reason but of technology. Galileo, Herschel, and Hubble were no smarter than the great scientific thinkers of ancient times; they just had better equipment. Accounts of their exploits have tended to soft-pedal the technological side of the story, as if their authors were embarrassed to admit that eminent astronomers got results by using good hardware rather than through "pure" thinking. But they have nothing to be embarrassed about. Without the gear, astronomy wouldn't be here.

When telescopes were first trained on the Moon, in 1609, by Thomas Harriot in England, and then, more rigorously, by Galileo in Italy, it soon became evident that the Moon, like Earth, has mountains and valleys—and, perhaps, seas, although the term *maria*, coined in 1647 by the amateur astronomer Johannes Hevelius, arose more from poetic license than from any real conviction that they actually contained water. "The Moon is by no means endowed with a smooth and polished surface, but is rough and uneven and, just as the face of the Earth itself, crowded everywhere with vast prominences, deep chasms, and convolutions," Galileo reported, in his best-selling book *Sidereus Nuncius* (the "Starry Message," published in 1610).[3] Reasoning by way of analogy, he asked rhetorically, "On Earth, before sunrise, aren't the peaks of the highest mountains illuminated by the Sun's rays while shadows still cover the plain? Doesn't light grow, after a little while, until the middle and larger parts of the same mountains are illuminated, and finally, when the Sun has risen, aren't the illuminations of plains and hills joined together?"[4]

Well, yes: The Moon is like the Earth, in this regard at least, and Galileo's research established that it was a world in its own right—territory, in short, potentially subject to exploration in much the same way that the European powers in Galileo's day were exploring "new" worlds across the seas. The Jamestown settlement had just been established, the Dutch were opening up trade with Japan, and Galileo could purchase Chinese tea in the shops of Venice. In 1610, Sir William Lower, who had been studying the Moon with a telescope sent him by Harriot, wrote to his fellow lunar observer that in his opinion Galileo "hath done more . . . than Magellane in openinge the streights to the South Sea or the dutchmen that were eaten by the beares in Nova Zembla."[5] Kepler wrote to Galileo that space travel might prove to be easier than was currently thought—much as the Atlantic, on some crossings, was "calmer and safer" than the English Channel. He would go to the Moon, Kepler joked, and Galileo could go to Jupiter.

Comparisons of Earth and Moon can be pushed too far, however, and they soon were, by dreamers who leaped to the conclusion that, since the Moon resembles the Earth, it, too, must be inhabited. Galileo remained noncommittal on the subject, and Kepler's speculations about lunar life were mainly confined to his science fiction tale, *Somnium seu Astronomia Lunari* ("Dream or Astronomy of the Moon"), but many were less prudent. The German astronomer Johann Bode felt certain that intelligent beings inhabited the Moon, beneficiaries of "special arrangements of the wise Creator."[6] The shepherd-stargazer James Ferguson argued that the similarities among the Earth, Moon, and planets "leave us no room to doubt but that all the Planets and Moons in the System are designed as commodious habitations for creatures endowed with capacities of knowing and adoring their beneficent Creator."[7] Even the great William Herschel, normally a level-headed empiricist, went half-loony over lunar life. In a letter to the English astronomer Nevil Maskelyne (whom he implored to "promise not to call me a Lunatic") Herschel wrote, "I hope, and am convinced, that some time or other very evident signs of life will be discovered on the Moon." While conducting lunar observations with a new telescope on May 28, 1776, Herschel recorded seeing what he described in his journal as "a forest, this word being also taken in its proper extended significa-tion as consisting of such large *growing substances*," and he went on to identify lunar cities, canals, and turnpikes.[8]

David Rittenhouse, regarded as the greatest eighteenth-century American scientist other than Ben Franklin, took patriotic comfort in the prospect that the Moon lay beyond the stifling grasp of the imperial powers. "You, inhabi-tants of the Moon," he wrote, in his 1775 book *An Oration*, "are effectually secured, alike from the rapacious hand of the haughty Spaniard, and of the unfeeling British nabob."[9] In 1822 the mathematician Karl Gauss wrote to the astronomer Heinrich Olbers proposing that an array of "100 separate mirrors, each of 16 square feet, used conjointly . . . would be able to send good heliotrope-light to the Moon," creating a signal that he was quoted as saying could be employed to "get in touch with our neighbors on the Moon."[10] The Harvard astronomer William H. Pickering persuaded himself that he beheld vegetation, swarms of insects, and canals on the Moon: "I have seen every-thing practically except the selenites themselves running round with spades to turn off the water into other canals," he reported, in a 1912 letter to his brother written while he was observing with Harvard telescopes at Mande-ville, Jamaica.[11]

The apogee of such fancies was attained by the great Moon Hoax of 1835, when the *New York Sun*, a penny-press paper hungry for circulation, published a series of suspiciously exclusive stories reporting that the astronomer John Herschel (William's son), using a new super-telescope that it claimed he had built in South Africa, had observed exotic creatures on the Moon—among them horned bears, "a strange amphibious creature of a spherical form, which rolled with great velocity across the pebbly beach," and short, bearded "bat-men" given to lascivious liaisons with bat-women. Readership soared—the *Sun* briefly became the most popular paper in the world—and a booklet reprinting the articles sold sixty thousand copies. *The New York Times*, unable to contact Sir John, got on the bandwagon and pronounced the reports "both probable and plausible." Two Yale professors were dispatched to New York to examine the scientific papers cited in the articles (said to have been published in the nonexistent *Edinburgh Journal of Science*), but the hoax was not exposed until its author, *Sun* editor Richard Adams Locke, confessed to a colleague who was asking for reprint rights that he had made the whole thing up.

The Moon Hoax originated from a concatenation of literary imagination and scientific enthusiasm in the darkly fertile mind of Edgar Allan Poe—who, characteristically, never made a penny from it. Poe had been fascinated by astronomy from the age of sixteen, when he studied the Moon with a telescope that he set up on the second-story portico of his stepfather's house during a rare interlude of happiness in a notoriously doom-laden life, and he might have continued stargazing had he ever been able to afford as much as a spyglass of his own. But destitution clung to Poe like a second skin, inflamed by the fact that he drank "like a savage," as the poet Charles Baudelaire put it, and habitually forsook future royalties to raise ready cash. (Poe earned only one hundred dollars for his story "The Gold-Bug," which sold over three hundred thousand copies, and only nine dollars for "The Raven," an overnight success.) Unable to survive for long between paychecks, Poe produced mostly short works, and by October 1833 was, as an acquaintance put it, "engaged on a voyage to the Moon," for which he was reading books about astronomy, atmospherics, and ballooning.

Poe's Moon story, published in the *Southern Literary Messenger* three weeks before the *Sun*'s hoaxes appeared, was titled "The Unparalleled Adventure of One Hans Pfaall." Thinly disguised as nonfiction—Poe could write about science and technology so convincingly that bits and pieces of his fiction were quoted as if factual in newspaper stories, a Senate report, and an article on whirlpools in the ninth edition of the *Encyclopaedia Britannica*—it was full of broad, winking references designed to tip off the knowing reader. Hans Pfaall—the name is a joke on the sound a fizzled rocket makes as it leaps up, only to fall back to earth—is portrayed as a Dutch aviator who on April Fool's Day ascends to the heavens in a balloon "manufactured entirely of dirty newspapers" and decorated with "a circle of little instruments, resembling sheepbells, which kept up a continual tinkling to the tune of Betty Martin"—a reference to the English saying, "All my eye, and Betty Martin," meaning, "It's nonsense!" Hans arrives at the Moon to find himself in

> the middle of a vast crowd of ugly little people, who none of them uttered a single syllable, or gave themselves the least trouble to render me assistance, but stood, like a parcel of idiots, grinning in a ludicrous manner, and eyeing me and my balloon askant, with their arms set akimbo. I turned from them in contempt, and, gazing upwards at the Earth so lately left, and left perhaps for ever, beheld it like a huge, dull, copper shield, about two degrees in diameter, fixed immovably in the heavens overhead, and tipped on one of its edges with a crescent border of the most brilliant gold.[12]

Disappointed to see an inferior satire triumph while his original, and superior, version remained unknown, Poe chided the *Sun*'s readers for being taken in by its Moon Hoax articles: "That the public were misled, even for an instant, merely proves the gross ignorance which is so generally prevalent upon subjects of an astronomical nature," he wrote. Poe noted that even if Herschel's telescope were capable of magnifying an image forty-two thousand times, as the *Sun* had claimed, this would equal a naked-eye view of the Moon from 5.5 miles away, and "no animal at all could be seen so far."[13] But Poe also appreciated just how clever Locke had been in pegging his inventions on John Herschel, from whose popular *Treatise on Astronomy* Poe had lifted most of his astronomical information. Herschel himself had argued for the possibility of "animal or vegetable life" on the Moon. As Poe perceived, what Locke had written was really a brutal satire of all the astronomers who had tricked themselves into believing that they saw cattle, cities, and even chapels amid the lunar wastelands.[14]

When better telescopes were trained on the Moon, it became evident that there was no life, water, or air there, but that there clearly were many interesting things to study, and that the first step was to draw accurate lunar maps. This was promising work for amateur astronomers since it did not require particularly large telescopes. Among the amateurs who contributed to the effort were Johann Schröter, who had published two volumes of lunar drawings, in 1791 and 1802, before invading French troops looted and destroyed his observatory; Wilhelm Gotthelf Lohrmann, whose lunar studies provided guidance for Wilhelm Beer, who with Johann Mädler in 1837 published a map nearly a meter in diameter that remained the standard work for decades; Johann Nepomuk Krieger, who made beautiful drawings of lunar features in the late 1800s; Philipp Fauth, who produced Moon maps of unprecedented detail but besmirched his reputation by espousing the bizarre cosmological notions of Hans Hörbiger (whose claim that the stars are made of ice was later taken up by Nazi pseudoscientists); and Percival Wilkins, who with Patrick Moore compiled an intricate lunar map fully 7.6 meters wide. Thereafter, the Moon was charted mainly by photography—from Earth, by orbiting lunar probes, and by the Apollo astronauts.

Although the golden age of Moon mapping has passed, amateur astronomers today continue to draw and photograph lunar craters, mountains, and maria—attracted, in part, by the uniquely detailed views the Moon provides: Mars, the next-nearest object whose surface can be seen, never gets much closer to Earth than 150 times the Moon's distance, so Mars at 150x looks only about as distinct as does the Moon to the naked eye. The best lunar viewing is usually along the terminator—the dividing line between the light and dark sides of the Moon—where long, sharp shadows cast even small features into striking relief. Two weeks pass between lunar sunrise and sunset, making for a languid and captivating shadow play as, on successive nights, sunlight catches the peaks of crater walls like those of Plato and Archimedes in the northern highlands, or Walter and Purbach in the south, and then descends to flood the crater floors. Isolated mountain peaks anchor their lanky shadows like illustrations in a Euclidean geometry text, and the mysterious lunar domes—thought to have been formed volcanically, although scientists disagree on just how—are visible in the sidelight, fading from view as the Sun gets higher in the Moon's black sky. The full Moon is generally regarded as less interesting, although it can afford imposing, almost hypnotic, views of the brightest features, like the stark white rays emerging from the young craters Copernicus, Tycho, and Aristarchus, where relatively recent impacts excavated lighter material from beneath the old, gray surface.

It helps, in appreciating the view of the Moon through a telescope, to consider the dimension of time. The Moon is not just a lifeless, airless ball of rock but a rich tapestry of cosmic history. In the absence of wind and water, the lunar surface erodes very slowly—the main mechanism, micrometeorite "gardening," wears down only about a millimeter of lunar soil every 50 million years—so its appearance has changed little during the past 4 billion years. The bright lunar highlands, scarred with multiple craters and debris generated by the impacts that were frequent when the solar system was young, preserve a record of events early in the evolution of the solar system, when planets and satellites were being bombarded by pieces of rock similar to the ones from which they had formed in the first place. The maria are somewhat younger lava plains that evidently filled in the largest impact features and then froze over, once the bombardment ended. Sinuous rills are old lava pipes whose tops have caved in, while the straighter linear rills are fault lines. The more you know about the Moon, the more you will see there.

But is there any point in making such observations, now that orbiting spacecraft have extensively (although not exhaustively) mapped the Moon, astronauts have taken snapshots on the lunar surface, and Moon rocks have been brought to Earth? Well, if you enjoy it, why not? Some hardworking amateurs feel guilty about making observations that lack scientific value, but others regard the pleasure of observing as an end in itself. Amateur astronomy, like many other human activities, contains a certain tension between hedonism and the work ethic, to which the Moon can, perhaps, serve as a remedy. If you got into stargazing for fun but find that it's turning into work, you can always relax and train your telescope on the Moon. There, at least, it is unlikely that you will get entangled in doing real science.

Unlikely, but not impossible: Real discoveries may remain to be made on the Moon. Consider "lunar transient phenomena," or LTPs. These are flashes of light or patches of dark or glowing haze spotted on the Moon, mostly by amateur observers. Some LTPs may be nonlunar phenomena, such as sunlight flashing off the solar panels of an artificial satellite that happens to be passing in front of the Moon, and others may be caused by the detonation of meteoroids hitting the Moon, but not all of them can readily be explained. On January 24, 1956, an amateur and a professional astronomer, using two telescopes at two locations, both saw a flashing light in the crater Cavendish, near the terminator. On July 19, 1969, when the Apollo 11 command module was orbiting the Moon, Mission Control radioed the crew that amateur astronomers were reporting a transient phenomenon near the crater Aristarchus. Astronaut Neil Armstrong had a look and radioed back that he saw an "area that is considerably more illuminated than the surrounding area. It just has—seems to have a slight amount of fluorescence to it."[15] He was not sure of the location but believed it to be Aristarchus. Winifred Sawtell Cameron, a former NASA scientist who has made a study of LTPs, notes that three-quarters of all of them involve just a dozen lunar features, with one area, the Aristarchus-Herodotus-Schröter Valley, producing fully a third of such reports. "Most LTP activity occurs along the edges of the maria, near volcanic features, like domes, sinuous rilles, and craters with dark halos or floors," she writes, adding that "these regions, like the rest of the Moon, have long been considered geologically dead."[16] But if LTPs are real, there may be (geological) life in the old Moon yet.[17] As the amateur astronomer Alan MacRobert noted, in 1988, "Plenty of mysteries remain on this big, close world."[18]

One long-standing mystery has to do with the riddle of the Moon's birth. The Moon is much less dense than is Earth. The Earth has a dense, molten core—produced when iron and other heavy elements sank to the center, back when our home planet was still a molten ball—and a lighter, silicate crust, made of materials that floated on the surface and eventually solidified there. (Earth's core and mantle, heated by radioactive elements, remain molten to this day, which is why our planet has volcanoes and hence is said to be geologically active or "alive.") The Moon, however, has a global density close to that of Earth's crust. Theorists wondered why this should be, and whether it had something to do with the Moon's unusually large size. In recent years they have come up with a group of plausible explanations, known as the "giant impact" theories, that may explain the Moon's origin.

When the Sun formed, spun out of a dark cocoon of gas and dust in the heart of a nebula some 4.5 billion years ago, it was surrounded by a disk of rocks and ice fragments that stuck together, bit by bit, to form planets. It is unlikely that all these early planets survived. Instead, it is possible that some fell into the Sun, while others were whiplashed into the outer darkness by close mutual encounters, and still others collided, with destructive—and sometimes creative—results. Had the Earth and the Moon formed at the same time, they ought to have similar densities. Instead, it seems, the Moon emerged from a collision between the young Earth and a wayward, Mars-size planetesimal that sideswiped Earth, blasting loose a great swatch of its crust and upper mantle. Most of this stuff fell back on the afflicted planet, but much of the rest went into orbit, forming a short-lived ring. Computer simulations suggest that it took only about a year for this ring of debris to coalesce into the Moon, which originally hung only ten thousand miles high. Wider than an outstretched hand against the sky, this glowing new satellite looked down on a violent Earth awash in lava seas. Since the solar system was still full of leftover scrap that had not yet been swept up, meteoroids pelted Earth and Moon alike. The Moon has retained these old scars as craters, but volcanic activity, mountain-building, earthquakes, and erosion caused by wind and water wiped away Earth's ancient craters long ago.

Once the terrestrial oceans formed, the nearby Moon must have produced momentous tides. The tides, in turn, caused the Moon to move away from Earth, so that the lunar orbit has been growing larger ever since. This tidal dance between Earth and Moon is intriguing both for its effect on our home planet and for its sheer dynamic elegance.

Envision the Earth, a large ball of rock mostly covered by oceans, and the Moon, a smaller ball with no oceans. Each body is subject to the other's gravitational attraction at each point on its surface, and the force of this attraction is inversely proportional to the square of the distance. So if, say, the city of Bombay happens to be located directly under the Moon, it will experience the strongest lunar gravitational force, and hence the highest tides. Meanwhile, those lands that lie off at the limbs of the Earth as seen from the Moon—Japan and western Africa, located a quarter of the way around the globe from Bombay to the east and west—feel less lunar gravity, because they're thousands of miles farther from the Moon than Bombay is. The effect is to squeeze Earth, like stepping on a solid rubber ball, so that the planet is a bit wider along a line drawn from its center through Bombay to the Moon than along a perpendicular line. (Tidal forces deform both rock and water, but since water is much more pliant than rock the result is to draw the seas up onto the beaches at high tide. The Sun contributes to tides, too—which is why the highest tides occur at full and new Moon, when Sun, Moon, and Earth are lined up—but this effect is much smaller and I will ignore it here.) If this imagined Earth-Moon system were static, tides would be easy to understand—Bombay (and Baja California, on the other side of the globe) would always be at high tide, Japan and western Africa would have low tides, and the reasons for both sets of tides would be obvious. But the system is in motion: The Earth revolves on its axis once a day, spinning eastward, and the Moon also moves eastward, completing an orbit every 27.32 days. This makes the tides more complicated and produces a repelling effect on the Moon.

As the Earth spins under the Moon, it does not deform instantly but lags behind a bit, owing to the inertial resistance of the rocks and water. So high tide arrives at Bombay not at the moment that the Moon is overhead but a while later, when Bombay has already rotated on to the east. As a result, the point of Earth's maximum bulge always lies a bit east of the Moon. Since the Moon is headed east in its orbit, the bulge tugs the Moon onward—ceaselessly, day and night, with the Moon never catching up—and acts like a carrot on a stick, speeding up the Moon's orbital velocity.[19] Any orbiting object that is accelerated will climb to a higher orbit, and that's what has hoisted the Moon, from its literally lowly origins, to its present perch some 240,000 miles up. Laser beams fired from Earth and bounced off the reflectors placed there by Apollo astronauts confirm that the Moon is still climbing, at a rate of 3.82 centimeters (about 1.5 inches) per year. This means that the spectacle of total solar eclipses, with the Moon covering the Sun so neatly, is an accident of time as well as space. Millions of years ago the Moon looked so large that the solar corona was masked during totality. In the future, it will be too small in the sky to cover the entire solar disk.

The energy expended in jacking the Moon up to ever-higher orbits, which had to be paid for somehow, was purchased at the cost of slowing Earth's rotation rate: The Moon, while being accelerated by the eastward tidal bulge, also pulls back, and thus acts as a brake on the eastward-spinning Earth. This effect has been verified geologically, by examining tidal ridges deposited in geological strata at the Big Cottonwood Formation near Salt Lake City, Utah, the Elatina Formation near Adelaide, Australia, the Pottsville Formation in northern Alabama, and the Mansfield Formation in Indiana. These findings indicate that 900 million years ago, a day on the fast-spinning Earth lasted only about 18 hours.

The same braking force acts on the Moon as well, but has worked more quickly there, since the Moon has only one percent the mass of Earth. (It's easier to stop a spinning basketball than a spinning ball of marble.) As a result, the Moon "locked" long ago and has since kept one side always facing Earth. In theory, tidal braking should eventually stop Earth's spin, too, but since its force keeps diminishing with the Moon's growing distance, the Earth won't lock down for another 15 billion years—by which time the Sun will have expanded into a red giant the size of Earth's orbit and then shriveled up to become a white dwarf star.

Another old lunar mystery is known as the "Moon illusion." The rising full Moon seems enormous, yet a few hours later, when it has climbed high into the sky, its silvery disk appears to have shrunken dramatically. Quantitative studies show that people perceive the Moon when rising to be two and a half to three times larger than when overhead. But if measured objectively at these two positions—as by photographing it through a telephoto lens on both occasions, then superimposing the negatives—the Moon's diameter is unchanged.

The cause of the Moon illusion has long been debated among astronomers and students of perception. Claudius Ptolemy in the second century A.D. argued for what is now called the "refraction" theory—that the Moon looks larger near the horizon because in passing through a thick mass of dense, moist air its image is magnified, "just like the apparent enlargement of objects in water, which increases with the depth of immersion."[20] The refraction theory seems reasonable enough, and you still hear people explaining the Moon illusion in these terms today, but it's wrong. Had Ptolemy tested it—say, by measuring the apparent diameter of the Moon at low and high elevations through tubes of various sizes—he would have seen that refraction plays no role in creating the illusion. When this experiment eventually was conducted, in the seventeenth century, it showed that the diameter of the lunar disk against the sky is the same at all elevations. Benedetto Castelli, a student of Galileo's, reported after making similar measurements of the Big Dipper, "I always found that it subtended the same space and thus I felt sure that such a phenomenon . . . of necessity had to be a fallacy of judgment and of apprehension."[21]

Illusions can arise when the eye-brain system is called upon to deal with objects beyond the limited range of our binocular vision. Having no direct way to calculate their distances—or, therefore, how big they are—the brain makes educated guesses about the size of distant objects, based on various indirect clues. For some reason these guesses can assert themselves with great conviction, as students of perception learn from persuasive examples like the "Ponzo illusion," introduced by the Italian psychologist Mario Ponzo in 1913. In a drawing, two blocks of equal size are superimposed on a railway track that converges toward the horizon, as if the lower block was at the near end of the track and the upper block farther away. The brain guesses that the upper block must be larger than the lower one, in order to look as big as it does even though it's "farther away." So we mistakenly perceive the upper block as much larger than the lower block, even after we've learned that we're deluding ourselves. Evidently such perspective effects can contribute to the Moon illusion—as the eleventh-century Arab astronomer Abu Ali al-Hasan ibn al-Haytham noted and many investigators have since confirmed—but they aren't the whole story. The rising, or setting, Moon looks swollen even in the absence of obvious perspective clues—as, for instance, when seen over a featureless seascape from aboard ship. Why?

Two illuminating recent test results help answer that question. Lloyd Kaufman, a professor of psychology, and his son James, a physicist working in computer science, animated a stereo image of the full Moon on a laptop computer. By looking at the display and crossing your eyes you could see the Moon in three-D, floating in space. A computer animation program then moved one of the images so that the Moon appeared to be receding into the distance. You might expect, and many researchers had assumed, that people watching this animation would assign a smaller size to the receding Moon. Instead the brain does just the opposite: As the stereographic Moon on the computer screen moves away, it appears to grow larger. In the other experiment, the Kaufmans asked their subjects to identify a point in the sky halfway between the horizon and the zenith—that is, at an elevation of 45 degrees. Almost all of them put it much too close to the horizon. Evidently we perceive the sky not as a dome but as a lens-shaped ceiling, the top of which is much closer to us than is the horizon. The brain assumes that the Moon, when near the zenith, needn't be terribly big since it's not too far away. But when the Moon is near the horizon, the brain concludes that, if it's more distant than all those things, it must be *really* big.[22]

So the Moon boggles the mind, illustrating the dictum of Goethe that "thinking man has a strange trait: When faced with an unsolved problem he likes to concoct a fantastic mental image, one he can never escape even when the problem is solved and the truth revealed."[23] None of which is surprising, from an evolutionary standpoint. What is surprising is that, through science, humankind has been able to overcome its perceptual limitations and learn how far away the Moon really is, and how large it is, and what it is made of.

Yet many cling to the notion that scientific study of the Moon and its exploration by the Apollo astronauts has robbed it of its old romance. As Bob Dylan put it, "Man has invented his doom / First step was touching the Moon."[24] But it doesn't seem to me that heavenly bodies become less enticing once we learn about them or explore them. The real Mars and the real Sun are

more exciting now than when they were only lights in the sky with mythological pedigrees, and it does art little justice to imagine that romance requires
ignorance. As the poet James Dickey remarked, "Poetry occurs when the
utmost reality and the utmost strangeness coincide."[25]

Future Moon dreams belong to the young, heirs to fact and fancy alike.
One night at the observatory, I showed the Moon to a half-dozen eight-year-
olds. As they stepped in turn to the top of the little stepladder and peered
into the eyepiece, something remarkable happened. By the cold moonlight—a
sort of spotlight, painting an ill-focused portrait of the Moon on the eye, eye
socket, and a bit of brow and cheek—each child seemed transformed into an
adult. Nini, with her red hair and freckles, became a woman in her forties, the
prime of her considerable athleticism past but her effervescent spirit unsub-
dued. Nion, a shy and appealing boy, was suddenly an elegantly commanding
grown-up who might be director of a foundation or president of an airline.
Mischievous Kathryn became intent, capable, no-nonsense; in business for
herself, perhaps. My son looked only a bit under my own present age. Poised
and serious, he offered a vision of a time when I myself had become a memory.
Everything we see in the sky belongs to the past. The velocity of light is
fast—186,000 miles per second—but finite: We see the Moon as it was 1.3 seconds ago, the stars decades to centuries ago, the galaxies millions of years ago.
Children in their similarities to us and their differences from us embody our
concept of the future. We elders fall away into the past, like leaves from trees
in autumn, and the young fall away from us, too, as they dive toward the
depths of the future.

The Telescope and the Tomb:
A Visit with Percival Lowell

AT MIDNIGHT ON THE LAST winter solstice of the second millennium, while observing Jupiter and Saturn with the 24-inch Clark refractor at Lowell Observatory in Flagstaff, Arizona, I found myself becoming entangled in what might be called the spirit of its founder, Percival Lowell, whose blue-and-white marble tomb stood brooding in the starlight just down a footpath from the dome housing this classic telescope. Lowell died in 1916, but his presence is so strongly felt that staff members of the observatory still refer to him as "Percy," as if he might turn up at any moment. He chose Flagstaff to combine the clean, dark skies of a reasonably high-altitude (7,000 feet) western site with ready access to the transcontinental railroad. Even now, my observing session was accompanied by the night-bird hoot and call of freight trains being made up in the yards, and the fading rattle of big diesel lash-ups balling the jack down the Santa Fe line.

Lowell was a gentleman, his brother Abbott the president of Harvard University, his sister the poet Amy Lowell, his family immortalized in the witticism that "the Lowells talk only to the Cabots, and the Cabots talk only to God." The observatory evidences his wealth and good taste, along with his idiosyncratic flair for amateur experimentation. It cost him twenty thousand dollars in 1896, equivalent to about half a million dollars today, and he got good value. The vaulted dome, designed by an ex-cowboy machinist named Godfrey Sykes, was erected from hatchet-hewn ponderosa pine timbers by ten workmen in just ten days. Inside it resembles a country church, with creaky wooden floors and an elevated observer's chair, employed to reach the eyepiece when the telescope is pointed near the horizon, that looks like a pulpit awaiting the pastor's arrival. The telescope, one of the great nineteenth-century planetary refractors, was fashioned by Alvan Clark, a descendant of Cape Cod whalers who started as a portrait painter and was awakened to the splendors of astronomy by his future wife, Maria Pease, a boarder in the household of the minister and amateur astronomer Edward Hitchcock. The great comet of 1843 stirred Clark's interest in telescope making—as it did that of his son George, a student at Phillips Academy at Andover, who, when the school's dinner bell broke, melted it down and made a reflecting telescope out of the metal.

Clark refractors were the best in the world in their day, and Lowell put three of them on one mount—the 24-inch plus a 12-inch and a 6-inch, each in its own right capable of serving as the centerpiece of a small college observatory. Such auxiliary instruments are usually called "guide telescopes," since they can be used to refine the tracking when time-exposure photographs are being made. But, since seeing conditions often favor smaller telescopes—the important factor is the ratio of aperture to the size of the convective cells disturbing the air at a given time—I suspect that Lowell used his guide telescopes as alternative instruments for scrutinizing Mars, switching from one to the other to select the combination of magnification and aperture that performed best at each opportunity. In any event, that's how I employed them.

The common wisdom among stargazers holds that nothing beats a large, long-focus refractor when it comes to observing planets, but my first look through the Clark, at Jupiter, was disappointing. The planet was surrounded by red and violet halos, the signature of chromatic aberration. To minimize the aberration, observatory staffers had closed down the iris diaphragm on the big telescope's objective lens, turning it into something quite rare, a nine-inch refractor working at a focal ratio of F/44 (which is to say that it had a focal length 44 times the diameter of the light-gathering lens). Since stopping the lens down didn't eliminate the chromatic aberration anyway, I twisted the big brass knob attached to the 33-foot rod that controls the iris and opened it up to full aperture. This made the aberration worse, but aside from the spurious colors the view was pretty good. The disks of Jupiter's and Saturn's satellites were evident, and the cloud bands of the giant planets exhibited a wealth of detail during moments of steady seeing. I watched, fascinated, as Jupiter's satellite Ganymede emerged from behind the giant planet, vanished into its shadow nineteen minutes later, and then reappeared after another two hours and eight minutes had passed.

The regular motions of Jupiter's satellites and their resemblance to the spinning governors of clockworks were not lost on early observers such as Galileo, who proposed using Jupiter to adjust the clocks of ships at sea. Sighting on Jupiter with a telescope from the pitching decks of a square-rigger proved impractical, but the method was employed on land by French cartographers to obtain more accurate determinations of Earth's circumference. The Danish astronomer Olaus Rømer made ingenious use of the Jovian clockwork in 1676 when he inferred the velocity of light by timing Jovian eclipses. Rømer noted that eclipses of Ganymede and other Jovian satellites occur earlier than expected when the Earth is close to Jupiter, and later than normal when the two planets are farther apart. He realized that the difference could be explained by the finite velocity of light. When Earth is on the side of its orbit farthest from Jupiter the eclipses come "late" because their light has to cross almost the entire diameter of Earth's orbit before reaching our telescopes. So if you know the diameter of Earth's orbit you can calculate the speed of light from the difference in observed eclipse timings made when Earth is nearest to and farthest from Jupiter. Rømer's result, 140,000 miles per second, was admirably close to the correct value of 186,000 mps.

As the night wore on I got comfortable with the old Clark, which had become arthritic in over a century of use but still lived up to its pedigree. The clock-drive mechanism worked fairly well, although it had developed a few flat spots where it would stall momentarily, allowing the planet to drift out of the field of view. The mechanical slow-motion controls were shot: To correct for guiding errors or to move from Jupiter to Saturn I simply manhandled the big telescope, pushing it around as one would a Dobsonian. When the original dome-rotating mechanism moved unsatisfactorily it was replaced for a while by a strange system that involved pontoons floating in water. That failed, too, and for the past half-century the dome had been supported on a set of 1954-vintage automobile wheels and tires: When a tire goes flat, they jack up the dome and patch the tire. The three shutters that cover the dome slit originally were controlled by a rigging, like a sailboat's, but that approach remained in use only on the top slit, and the other two shutters now consisted of old electrically operated aircraft landing-gear doors. These alterations may have diminished the observatory's artistic integrity, but they did not detract from its atmosphere of living history. It was pleasant to sit there, taking in Saturn's wan sphere and icy rings, listening to the ticking of the clock drive and the mournful cry of the freight trains, and pondering the many nights that Lowell, the most prominent amateur astronomer of his time, had spent here, mapping the canals he thought he saw on Mars.

The persistent sense of Lowell's presence was enhanced when an old drawer turned out to contain one of his original eyepieces. Presumably made by Clark, it was huge—nearly as thick as the 2.4-inch refractor I used as a boy—and finished in an elegant ochre hue, the color of Mars itself. I blew the dust off it, admiring its clear, heavy glass, and found that when the modern assembly on the business end of the telescope was removed, it could be threaded into the original tube. Evidently, no one had used this eyepiece in years. What would I see?

It was spectacular. Saturn looked crisp and sharp, better with it than with the new and more technologically advanced eyepieces I'd been using. With

Saturn high in the sky near the meridian, the eyepiece position was profoundly uncomfortable—I had to lie on my side on the floor, with my neck turned to a painful attitude—but I stuck with it, gratified by the view and determined not to be put off by contortions that Lowell himself routinely endured.

I'd been observing in this dreamy state for a long time, alone, when a stray breeze stirred against the observatory door. It slowly opened, with a haunted-house creak. I turned my head toward the looming doorway in the dark—it was full of stars—and heard myself whisper, "Percy?"

10.

Mars

Year by year,
the monkey's mask
reveals the monkey
—Basho

Asking is half of knowing.
—Rumi

ON CALIFORNIA HIGHWAY 159 in the predawn darkness of July
20, 1976, I could spot the crescent Moon, hanging near the sparkling
Pleiades in the southeastern sky. The roadway tonight was host to
an exotic assortment of vehicles, from rental cars like mine to thundering
Lotuses and Ferraris, converging with abnormal urgency on NASA's Jet
Propulsion Laboratory, where the glass-walled buildings blazed with light
amid an otherwise benighted hillside above Pasadena. Mars, the object of our
attraction, was not in the sky—it was on the other side of Earth—yet we
were on our way to see it. This was the night when *Viking 1* was scheduled to
land there and send back the first photos ever taken from the Martian surface.

Inside the lab, nearly everyone was restless, tense, and overcaffeinated. By
now *Viking 1* had either landed safely or failed, but it would take nineteen minutes for its signals, traveling at the velocity of light, to reach Earth—so its
fate, already decided on Mars, remained in our local future. The event had the
post facto flavor of following election returns: There had been an outcome,
but none yet knew what it was. Owing to the lag time, flight controllers on
Earth could not direct the descent of the lander—a gray, insectlike robot
about the size of a touring motorcycle. On its own, it had to manage its fiery
entry into the Martian atmosphere, deploy and jettison its parachute, then
fire its three descent engines to touch down gently on the surface. If it happened to hit a boulder, or land on a steep crater wall, it would be doomed.

Gentry Lee, an intense young man who was losing his hair in front and had let the rest grow into a fan that hovered over his shoulders, called out interpretations of the raw data streaming onto the computer monitors: "Twenty-six hundred feet! Five hundred feet! Two hundred feet!" Then, touchdown. "We've got a good one!" cried Rex Sjostrom, chief of the Lander Performance and Analysis Group.

A few minutes later the first photo came in, in a series of vertical strips, a startlingly clear, black-and-white view of the rocky, rolling landscape of Chryse Planitia, the Plains of Gold.

"It's incredible," whispered Thomas "Tim" Mutch, the leader of the lander imaging team and one of the prime movers of the Viking project. (After Tim died—on October 7, 1980, while climbing in the Himalayas—the Viking 1 lander was renamed Mutch Station in his honor.) Television viewers in Asia had stayed up late to see the historic live broadcast, in Australia they got up early, and in Europe people interrupted work to witness it. Yet, although the image happened to come in just when the live morning shows were broadcasting on the three major commercial television networks in the United States, none carried it. The producers said the public wasn't interested.

MARS HAS FASCINATED PEOPLE since pretelescopic times, when little more was known about it than that it was a wandering ochre dot in the sky whose motions defied precise explanation. (It was by analyzing records of these motions that Johannes Kepler realized that they were produced by both Mars and Earth moving in elliptical orbits around the Sun.) The advent of the telescope inflamed the old romance. William Herschel, observing Mars during the opposition of 1783, identified the snowy south polar cap and realized, from its position, that Mars had a tilted polar axis and therefore must have seasons. "The analogy between Mars and the Earth is, perhaps, by far the greatest in the whole solar system," he wrote, adding that since "we find that the globe we inhabit has its polar regions frozen and covered with mountains of ice and snow, that only partly melt when alternately exposed to the Sun, I may well be permitted to surmise that the same causes may probably have the same effect on the globe of Mars; that the bright polar spots are owing to the vivid reflection of light from frozen regions; and that the reduction of those spots is to be ascribed to their being exposed to the Sun."[1] He was right, but the question of to what degree Mars resembles Earth—whether, specifically, there might be life on Mars—involved amateur and professional observers alike in controversies that persist today.

Observing Mars through a telescope is like watching the Dance of the Seven Veils: Seldom are you certain of exactly what you've seen, but it's enough to pique your interest. Several aspects of Mars's relation to Earth conspire to create this tantalizing situation.

For one thing, Mars comes close to Earth only once every couple of years. These approaches, called *oppositions* because they occur when the Sun is opposite Mars in the sky, happen once every 780 days, on average, when Earth, orbiting on the inside track, catches up to the slower-moving Mars.[2] Intervals between oppositions vary somewhat, owing to the changing velocities of the two planets: As Kepler discovered, each goes faster when close to the Sun and more slowly when far from it. Nor are all oppositions created equal. The orbit of Mars is more eccentric than that of Earth, so if opposition occurs at Martian perihelion (the point in its orbit when Mars is closest to the Sun) we can draw to within 35 million miles of the red planet, but when oppositions occur at Martian aphelion (when the Mars–Sun distance is greatest), Mars comes no closer than 61 million miles. Only during oppositions, when we haul abreast of the red planet, can much detail be seen on Mars through terrestrial telescopes. Most of the time the Mars–Earth distance runs to over a hundred million miles, and even the best telescopes show little but a splotchy blob with a splash of white at the pole.

Earth's atmosphere plays a particularly fretful role in Martian observations. The best part of the sky in which to observe is overhead, near the zenith, but at the latitudes where most observers live, Mars never gets that high. Like the other planets, it orbits in a plane, the ecliptic, which is overhead only for observers located within 23.5 degrees—the tilt of Earth's axis relative to its orbit—of the equator. So you have to be south of Miami and north of Rio if you are to see Mars at the zenith—and that, frustratingly, excludes most of the world's largest telescopes.

Mars has an atmosphere of its own. It's thin, and so doesn't much blur the view, but it is never entirely clear of wind-borne dust. A global "blue haze" embraces the planet (using a red filter helps cut through it), ice clouds form in the upper atmosphere, local ground fogs congeal in canyons and other low-lying areas, and gigantic Martian dust storms can obscure virtually the entire planet for months at a time. Martian days (known as *sols*) last 24 hours 37 minutes, so an observer training a telescope on Mars on successive nights sees a somewhat different part of the globe each time: Familiar features "set" 37 minutes later each night, making the Martian globe appear to rotate backward once every 40 days.

Early observers, confounded by such complexities and by the severe limitations of their telescopes, nonetheless managed to compile reliable, if brief, dossiers on Mars. They charted the major surface features, most of which are large dark areas readily seen against the reddish deserts. (Conspicuous among them were the sharp, broad arrowhead of Syrtis Major; circular Argyre and Solis Lacus; the long, jagged line of what is now known as Valles Marineris; and the bright spot that Schiaparelli, who discovered it in 1879, presciently named Nix Olympica, the "Snows of Olympus.") They noted that some of these features varied in prominence, standing out vividly on occasion, then fading away: Solis Lacus was so good at this disappearing act that Schiaparelli called it "the Land of Wonders." Some of these variations, they found, appeared to be seasonal. In particular there was a "wave of darkening" that swept across each hemisphere during local spring, in synch with the waxing and waning of the polar caps.

But as can happen when you look long and hard at something indistinct, observers also saw things that weren't there, notably the notorious Martian canals. The canals are an optical illusion created by the mind's tendency to connect splotchy dots to make lines. You may be able to see them for yourself. Put a good, clear, modern photo of Mars under a bright light and view it from across the room. If you are at the right distance and if your mind has a robust straight-line-making function—as most of us do, since the visual centers of the brain have programs devoted to identifying straight lines—canals will appear. A number of capable astronomers saw canals on Mars—among them Angelo Secchi and Giovanni Schiaparelli, who dubbed them *canali*, by which he meant natural features like rivers—but most remained skeptical as to their reality. Vincenzo Cerulli, a skilled scrutinizer of Mars, reported that while viewing Mars from his private observatory he had witnessed a canal called Lethes resolve into a "system of minute patches" in a rare moment of superb seeing—strong evidence that they were illusory. Schiaparelli noted in a letter to Cerulli that the eye can concoct geometrical patterns out of random dots, as when viewing the printed page of a book from a distance. "The naïve faith in the regularity of the lines [*canali*] is shaken," Schiaparelli wrote, predicting that "as optics continue to improve, the process will proceed to other stages of vision, or illusion."[3]

Such insights made little impression on Percival Lowell, who had a better telescope than Schiaparelli ever used but who kept seeing Martian canals anyway. Lowell was a persuasive writer with a splendid, if untamed, imagination, and his books and articles presented hyperbolic conjectures in a diffident, urbane style. ("That Mars is inhabited by beings of some sort or other we may consider as certain as it is uncertain what those beings may be.")[4] He believed that the canals had been constructed for irrigation purposes by an advanced Martian civilization, that the dark continents of Mars were covered with vegetation, and that the seasonal "wave of darkening" was the renewal of this plant life by a springtime return of meltwater from the poles. On the maps that Lowell issued from his private observatory in Flagstaff, the canals, which had been gracefully curved in Schiaparelli's drawings, became straight lines that met at what looked like pumping stations. Understanding that his telescope lacked the resolving power to discern an actual canal unless it was at least thirty miles wide, Lowell argued that the lines represented local vegetation nourished by water drawn off the canals to irrigate lands along their banks.

It all fit, and Lowell's gracefully written expositions of his theory excited millions with the heady vision of a Martian civilization. Lowell's soaring popularity embittered some professional astronomers, who viewed him as an egomaniac whose extravagant claims were poisoning the well of more legitimate Mars studies. The astronomer W. W. Campbell, of Lick Observatory, ostentatiously refused to read Lowell's published work on spectroscopic methods of looking for water vapor in the Martian atmosphere, even though Campbell himself was engaged in just such studies—and was keeping them a secret from the press, because "reporters are daft on [the] subject."[5] The astronomer and historian Donald Osterbrock writes that "in Campbell's mind Lowell was not only an amateur, but an unprincipled one at that, who tried to present his rhetorical arguments as if they were scientific research. As such, he was an outlaw. . . . Scientists did not leap to conclusions, and build observatories to prove them, or publish smoothly written articles that gave only one side of the story in glossy magazines written for nonspecialists."[6]

But, in fairness, it should be said that science has room for both Lowell's enthusiasms and Campbell's acidity.

Our eyes can indeed "play tricks on us," especially when we're trying to make sense of something as indistinct as a quivering telescopic view of Mars. The uncommonly keen-eyed astronomer Edward Emerson Barnard, whose observations of Mars with a 36-inch refractor at Lick Observatory during the 1894 opposition provided what were probably the clearest glimpses of the red planet attained up to that time—and proffered no canals—warned that "man is too quick at forming conclusions. Let him but indistinctly see a thing, or even be undecided as to whether he does actually see it and he will then and there set himself to theorizing, and build immense castles of conjecture on a foundation of whose existence he is by no means certain."[7] Had Lowell taken such advice to heart, he could have avoided misleading himself and his readers.

On the other hand, more discoveries have been made by scientists burning to prove a cherished theory than by disinterested observers who sat back and asked of the universe, "OK, what can you show me?" Most theories are flawed or false, but to scrutinize Mars in hopes of seeing cities or herds of buffalo is better than looking at it lackadaisically: A grouse hunter out at dawn will see more in the woods than will a casual tourist, even if he finds no grouse. Had Lowell not been obsessed with his vision of a civilized Mars he would not have built Lowell Observatory, where useful Martian observations were made and Pluto discovered.

When I was a boy, observing Mars from Key Biscayne, our information about the red planet was still mostly limited to what could be glimpsed during opposition through nineteenth-century–style refracting telescopes like those employed by Lowell and Barnard, and the existence of the canals remained a subject of active debate. There followed the period, unique in the history of our species, when spacecraft were dispatched to other planets and our remote views were replaced by much clearer images made from close range. The data collected by these robotic explorers taught us more about Mars than had been learned in all prior human history. At the time, I thought they spelled the end of the epoch when ground-based observers would contribute much more to our knowledge of Mars. I was wrong about that.

The first space probe to reach Mars was *Mariner 4*, which flew within 6,118 miles of the red planet on July 14, 1965. Its television camera transmitted twenty-two pictures back to Earth, at an agonizingly slow rate: Receiving a single image could take ten or twelve hours. With reporters clamoring outside the gates—for all anyone knew, the photos showed Martian cities, the existence of which the government was concealing—the scientists at the Jet Propulsion Laboratory worked through the night, trying to make sense of the first images, which looked almost blank. Fortunately the later photos in the series were clearer. Unfortunately, for those who had shared Lowell's imaginings, they showed large craters on Mars. "So much for Lowell's canals and seasonal plant life," declared the planetary scientist Bruce Murray.[8]

Big craters were bad news. They suggested that Mars was like the Moon. The prevalent syllogism was that the Moon has craters, and is both geologically and biologically dead; Mars has craters; therefore Mars is dead. That the logic was flawed failed to blunt its popularity. *The New York Times*, in a 1965 editorial, dubbed Mars "The Dead Planet."

A dissenter from this consensus was the planetary scientist Carl Sagan. *Mariner 4*, he noted, "got twenty pictures of the planet, with the finest detail one kilometer across. Now, if you take twenty pictures of the Earth at one-kilometer resolution, there's *no chance* of finding life here. And yet people say, 'Well, I didn't see anything alive on that planet, it must be a dead planet.' What *terrible* logic!"[9] Those who shared Sagan's views found little encouragement when two more probes, *Mariner 6* and *7*, flew by Mars in 1969, returning pictures that showed mostly more craters. In 1968 Sagan, Clark Chapman, and James Pollack had cautioned that we still knew too little about Mars to conclude that it was a dead world. "If substantial aqueous-erosion features—such as river valleys—were produced during earlier epochs on Mars, we should not expect any trace of them to be visible on the *Mariner 4* photographs unless they were of greater extent than typical features on the Earth," they wrote.[10] But theirs was a minority view.

And so the situation remained until 1971, when *Mariner 9* became the first probe to orbit another planet. As a Mars orbiter, it offered the prospect of a sustained reconnaissance—rather than just another quick flyby—of what was, as Sagan kept reminding everybody, a big planet. (Mars is only half the diameter of Earth, but as it has no oceans its land area is comparable to Earth's. The *Mariner 4*, *6*, and *7* missions had photographed only ten percent of it.) The first photos transmitted back by *Mariner 9* were featureless, except for the south polar cap and four mysterious dark spots near the equator. A global dust storm had covered the face of Mars.[11] Weeks passed. Then the ochre clouds began to dissipate, and the curtain went up on a new act in the old drama of Mars and the human mind.

As the atmosphere cleared, the four spots turned out to be craterlike rings, forty to fifty miles wide. But what were craters doing atop mountains—as these must be, since the dust was clearing from the top down? The answer was that they were the calderas of gigantic shield volcanoes. This was wholly unexpected, particularly to those who had thought that Mars was like the Moon, which has no evident volcanoes. Next to be unveiled by the descending dust was a gigantic canyon, three miles deep, up to one hundred miles wide and long enough to stretch from Los Angeles to New York. Named Valles Marineris in honor of *Mariner 9*, it is the largest canyon in the solar system. As the remaining clouds cleared, sinuous valleys could be glimpsed, some of which turned out to have been fed by fanlike tributaries. These ancient riverbeds or flood zones suggested that water once flowed there. If so, Mars must once have had a much denser atmosphere. Mars today is cold and its atmosphere is equivalent to what one encounters on Earth at an altitude of about 115,000 feet; in such conditions water goes directly from ice to vapor. "Our view of Mars had completely changed—once again," wrote Bruce Murray. "Lowell's Earth-like Mars was forever gone, but so was the Moon-like

Mars portrayed by our first three flyby missions."[12] Mars had emerged as a world all its own, with a unique history written across its vast and subtly beautiful landscapes. That history could not be deciphered from just a few brief glimpses.

Mariner 9 took more than seventy-three hundred images of Mars, at a best resolution of about one hundred meters (meaning that it could discern nothing much smaller than a ball park). The twin Viking landers and orbiters, which arrived at Mars in 1976, took forty-six thousand higher-resolution pictures from orbit, some at a resolution of better than ten meters, and thousands more from two landing sites, enough to begin bringing the full splendors of Mars within reach. In those pre-Internet days, the *Viking* images were printed out, bound in loose-leaf notebooks, and archived in thirteen little libraries, called Regional Planetary Image Facilities, at sites scattered around the country. I used to stop in at these centers and pore over the photos, amazed to find myself alive at this moment in history. One could see, now, that Argyre and Syrtis Major were crater fields, their dark coloration due not to vegetation but to subsurface sands wind-scattered from the craters; that Solis Lacus, which Lowell mapped as a canal junction station, was a region of lingering clouds and fogs; and that Schiaparelli's Nix Olympica was indeed a snowcapped mountain, the gigantic volcano Olympus Mons.

Mariner and *Viking* left many questions unanswered, as did subsequent missions such as *Mars Pathfinder* and *Mars Global Surveyor*. It remained unclear whether, if there was once substantial running water on Mars, it took the form of lakes, rivers, and streams, or instead was released in rare torrents during brief warming spells. Nor was it clear where the water went: Most of it should still be there, since no known mechanism could since have transported it all into space; perhaps it was locked into the soil, as permafrost. Nor was it known what caused Mars to freeze up, a billion years or more ago. But the likelihood that Mars once had liquid water sustained hope that there could have been life there, evidence of which might be found, someday, in the form of fossils or hibernating microorganisms in the soil.

Mars has seasons and constantly changing weather, a better understanding of which could teach us how our own planet works. To learn about weather systems, however, requires gathering data from as many places over as much time as possible. Space probes have contributed to this effort, but only in bits and pieces. The Viking mission imaged Mars for two years. The Pathfinder lander returned images and data from one site for a period of three months. *Mars Global Surveyor* operated for years, taking over seventy thousand images from orbit, but its high-resolution pictures of small areas were difficult to synthe-

size into an overall weather pattern. The Hubble Space Telescope, busy probing the depths of cosmological space, could spare time for only an occasional
glimpse of Mars.

Which brings us back to the amateur astronomers, and why I was wrong to
think that their planetary studies would be rendered obsolete by the space
probes.

Amateurs had, after all, been observing Mars for centuries, often with
spectacular results. Its two tiny satellites, Phobos and Deimos, were discovered in 1877 by Asaph Hall, an impoverished amateur who had been obliged to
drop his university astronomy studies because he couldn't afford the tuition,
but nevertheless had won appointment to the U.S. Naval Observatory (where
President Abraham Lincoln turned up one night, alone, to talk astronomy).
Amateurs observed Martian volcanoes and portions of the Valles Marineris
canyon system decades before their detection by professionals studying
Mariner photos, although they could not see them clearly enough to determine
precisely what they were. There is even a tantalizing report of observers having spotted craters on Mars in the nineteenth and early twentieth centuries.[13]
The coming of the space age opened new chapters in Mars exploration but did
not alter the fact that many of the best amateur observers had more experience in studying the red planet through a telescope than the professionals did.

And the amateurs had enough observers and telescopes to bring Mars
under wide and protracted scrutiny. Aided by a few sympathetic professional
astronomers, amateurs had already been organizing themselves into observing
teams like the International Mars Patrol, established in the 1960s by Charles
F. "Chick" Capen Jr. Observers contributing to these projects scrutinized
Mars through privately owned telescopes around the world, producing reports
that did more than anything else, before the Mariner and Viking missions, to
capture a sense of Mars as a sovereign world. Paging through Capen's reports
of the 1964–65 opposition of Mars as observed from Table Mountain Observatory in Big Pines, California, when nightly weather reports were made, in
his words, "to discern and predict atmospheric trends and surface conditions
during the *Mariner 4* encounter," one gets a sense of just how valuable it can
be to monitor the constantly changing Martian weather on a large scale.[14]
Among the amateur observations: "The North Polar Cap was seen clear and
bright and rapidly receding during the Martian spring season. The northern
peripheral gray melt-band was unusually weak." "An extensive morning limb
haze was recorded in blue light." "The Martian atmosphere contained much
visible moisture during this period of observation. A large cloud was recorded
in blue and violet light over the Elysium plateau. A thin, small morning haze

was noted over the eastern edge of the Chryse desert." "Frost was noted on the Libya, Crocea, Cenotria, and Aeria deserts bordering the southern end of Syrtis Major." "The North Cap was clear, sharp, and white. The Antarctic polar hood continued to fade."[15]

Rather than being discouraged by the close-up views provided by the space probes, the amateurs redoubled their efforts—if anything with fresh enthusiasm, since they now often had a better sense of what they were looking for. A major Martian dust storm in June 1988 was observed almost exclusively by amateurs. "It was a short-lived storm, so, though the professionals knew about it, either they couldn't get telescope time on short notice or they had bad weather or seeing," reported the physician and amateur astronomer Don Parker. "The amateurs really pulled us through, and the professionals are very happy about that."[16] Stephen James O'Meara noted that amateur astronomers "monitored when and where it started and stopped; they recorded how extensively the dust altered the appearance of the surface; and they documented its progress on videotapes, photographs, and drawings."[17] In 1990, three major Martian dust storms were observed by amateurs in seven nations, and by not a single professional astronomer. Amateurs equipped with CCD cameras established, during the 1992–93 opposition, that the north cap of Mars was shrinking more rapidly than the professionals had predicted. They also found abnormally abundant water vapor clouds on Mars, indicating that the Martian atmosphere was growing unusually damp as the polar cap shrank.

"Perhaps the most exciting clouds were equatorial band clouds composed of carbon dioxide and water ice," reported Don Parker and Richard Berry. "Before this apparition"—that is, the brightened, enlarged appearance of Mars during the months surrounding an opposition—"observers thought such clouds to be exceedingly rare, but during the latter half of this apparition equatorial band clouds were seen and imaged on numerous occasions. Many images taken during this apparition were made with CCD cameras. Because properly calibrated CCD images are very precise, features that are only one to two percent brighter than the planet itself can be enhanced. With electronic image processing, we can finally see these clouds clearly."[18] By the end of the century, the International Mars Patrol had collected over thirty thousand observations covering fifteen oppositions of Mars, and its data were being studied by professional researchers all over the world.

Not that the amateur reports were error-free. Dark Martian features like Syrtis Major were often reported to be "green," which would support Lowell's notion that they were covered with vegetation but which is now understood to be an optical illusion produced by the contrast between dark, neutrally colored regions and the reddish disk. As late as the 1960s some observers were still speaking blithely of "canals," although many used the term "canal-like features" and noted that they vanished under optimal seeing conditions. But through a combination of hard work, cooperation, high technology, and a rough-hewn semblance of internal peer review, amateur astronomers earned a significant role in the continuing human enterprise of trying to make sense of Mars.

Like the Moon, Mars is differentiated geologically into highlands, where many old craters survive, and lowlands, where more recent lava flows wiped out the older craters. Highlands predominate in the Southern Hemisphere and lowlands in the north, making Mars a surprisingly asymmetrical planet. The cause of its asymmetry is unknown. The big volcanoes dot a large, bulging area called Tharsis, wider than North America and about six miles high. (Valles Marineris starts as a crack on one edge of the Tharsis bulge.) The very enormity of the volcanoes indicates that even when young, Mars lacked enough internal heat to drive plate tectonics. Instead, it was from early times a "one-plate planet," where volcanoes grew very large because the plate they rested on never moved off the magma hot spot that created them. The implication is that the catastrophic event that spelled an end to the halcyon days of Martian warm weather had to do with a lack of sufficient internal heat to keep the volcanoes operating. But nobody yet knows how this worked. Solving the riddle may ultimately require on-site geological digs. I once asked the astronaut Kathryn Sullivan how long, if she went to Mars, she would be willing to stay there. "How long would you *let* me stay?" she replied, her eyes agleam. "Put me on Tharsis for a year, and I'll give you the keys to the planet."[19]

The polar caps constitute another mystery. Both poles are covered with dry ice during local winter, when temperatures drop below −125°C, but otherwise they are very different from each other. When the dry-ice cap at the south pole shrinks, it reveals a permanent cap below, thought to be made of mixed dry and water ice. At the north pole, though, the dry ice sublimes away entirely in late summer, exposing a water-ice cap larger than Ireland. The cause of these differences is unknown, but may be related to radical changes in Martian seasons over time.

Three aspects of Martian dynamics—the tilt of the red planet's axis, its precession, and the shape of its orbit—conspire to produce long-term variations in the seasons. At present, the polar axis of Mars is tilted 25.2 degrees, rather close to Earth's 23.5 degrees, but over time it is believed to have oscillated from about 15 to 35 degrees, owing to the gravitational force exerted on Mars by the giant planets Jupiter and Saturn.[20] (Mars is particularly susceptible to these influences because the concentration of mass bulging up at Tharsis makes it dynamically unstable, like a spinning top with a wad of gum stuck on it.) Since seasons are caused mostly by the tilt of the axis, with the hemisphere pointed at the Sun at a given time experiencing local summer, the seasons on Mars were more pronounced when the axis was extremely tilted than when the planet stood up straight. Complicating this effect is the precession—a slow, circular wobbling of the axis, like that of a spinning top slowing down—which on Mars takes 173,000 years per wobble, and on Earth 25,800 years. In addition, the orbit of Mars is much more elliptical than that of Earth, and the eccentricity of its orbit—the extent to which it deviates from circular—has changed much more radically over the course of its history.[21]

Mars evidently experienced particularly violent seasons, when the Northern Hemisphere was pointed sharply toward the Sun at just the time that its elliptical orbit also brought it closest to the Sun, interspersed with periods of weak seasons when the solar distance and the axial tilt worked at cross-purposes. Earth is thought to have suffered much less of this sort of thing, in part because its large Moon acts as a protective shield against gravitational "pumping" by the Jovian planets. Strong seasonal variations on Mars produced big dust storms, and dust deposited on a polar cap in local summer got trapped there when fresh ice froze over it. The south polar cap of Mars shows, in local summer, an intricate pattern consisting of long fingers of ice and dust. This laminated terrain, with further study, could yield a detailed history of how the Martian seasons have changed over time.

An accurate history of Martian water remains to be written. The geological record contains channels probably carved by floods, streams that may have been fed from beneath the ground, and tributaries that, as the planetary scientist David Morrison writes, "tell us that Mars once supported free-flowing rivers and experienced the miracle of rain."[22] Since these features are found only in the highlands, the Martian tropical epoch probably occurred during the first billion years of the planet's existence, before the lowlands were covered by lava flows. Perhaps the same events that bulged up Tharsis created the volcanoes, which in turn caused the flooding by releasing water from what had been frozen soil. "Volcanic heating might have released the water in dramatic, episodic floods," Morrison writes. "Perhaps each such flood lasted only a few days or weeks. Maybe the water quickly evaporated or refroze at higher latitudes. Or possibly temporary ice-covered lakes or seas formed in basins like Chryse. We do not know."[23]

As with Venus, finding the keys to planet Mars is a matter not just of scientific curiosity but of the strictest human self-interest. We are utterly dependent on Earth for our survival, yet we do not fully understand its origin, its present condition, or its future. We are ignorant of such rudimentary facts as what caused the ice ages, what produced the many reversals in Earth's magnetic polarity that have repeatedly swapped the north and south magnetic poles, and what impact global warming would have on human life and the ecosystem that supports it. Since science works best by comparing one thing to another, the best way to learn how Earth works is to study similar planets, and the best object for such studies is Mars. This effort requires all the help it can get, from amateurs and professionals alike, and it needs to proceed with dispatch, if we are not to reap the very nasty surprises threatened by tampering with a planet whose processes we have only begun to comprehend. It takes more than the work of a lifetime to come to grips with a world.

On the end of the long first day of the *Viking 1* sojourn on Mars, Carl Sagan and I sat side by side on a couch in the rented apartment where we were staying in Pasadena, our heads together, holding up a two-foot-long black-and-white panoramic lander photo and curving it so that it filled our field of view like a Cinerama movie screen. The print, less than two hours old, was still damp from its darkroom fixing bath.

"Bring your full concentration to it," Carl urged. "Try to imagine yourself there."

We studied the image in silence for a long while. In the foreground stood the top of the lander itself. In the middle distance, hundreds of scattered rocks and a few boulders cast the long shadows of a late afternoon in the Martian summer. At the horizon, perhaps a quarter-mile away, rose a chalk-white outcropping. Carl put his index finger at just that spot. "Look at this," he said. "What do you see?"

I stared at the outcropping until my mind, starved for information, wove it into an Arabian Nights fancy, which I then described to Carl: a miniature oasis, with a shimmering lake, mangroves rising from its banks, and a lone palm reared against the sky.

"I had," Carl said, "exactly the same hallucination."[24]

Light at the Edge of Darkness:
A Visit with James Turrell

As the American artist James Turrell tells the story, he was conceived on the night of February 25, 1942, when his parents celebrated completing a room they had added to their home, in Pasadena, California.[1] The room had three walls of windows, above a 42-inch wainscot, through which his father could observe birds and coax them inside by whistling variations on birdsongs. February 25 was the night of the "great Los Angeles air raid," when antiaircraft bursts were fired into the empty sky in response to spurious radar echoes. On that as on many other wartime nights, the windows had to be covered with blackout curtains. Turrell, the unplanned child of older parents, grew up in this windowed room. He soon began puncturing the blackout curtains to mark the positions of the stars.

"When I was six years old, in order to assert my own presence in the room, I took a pin or needle to these curtains and pierced them to make star patterns and the constellations," he recalled. "I would simply make bigger holes for stars of greater magnitude. Pulling down the curtains and darkening the room, you could see the stars in the middle of the day. These weren't just holes in the curtains, they were holes in reality. By changing the reality of the conscious-awake state of day, one could see further into imagined space to the stars, which were actually there but obscured by the light of the Sun."[2]

"I was intrigued with why you couldn't see the stars in the day—that it was light that blinded us," Turrell told me recently. "The idea that light does not necessarily illuminate, or that it is illuminating something that obscures what you want to see, was fascinating to me. It explained why I couldn't see the stars. I kept adding different portions of the sky, so it became a bit of a creative sky—but that's kind of how the real sky is anyway."[3]

Turrell manipulates light—without optics—in works intended, as he puts it, to "gently prod us into seeing ourselves see."[4] Many of his installations involve admitting carefully controlled amounts of light into a boxed enclosure, harkening back to the black-curtained room of his boyhood. Their abridging of a meaningful darkness evokes two of Turrell's main conceptual images, Plato's cave and the camera obscura (literally, "dark room," a chamber in which a pinhole projects an image of the outer world), and invokes the mysterious interior of the human skull, where, in darkness, appears all the light that each of us ever sees.

Turrell remodeled an abandoned gas station in Los Angeles, turning it into a chamber with an opening in the roof that, as he put it, "has the quality of bringing the sky right down to you."[5] He has lit long tunnels so that visitors cannot tell, while walking toward the source of light, how far they have to go—which, in a sense, is one's situation in going through life. Viewing such works can be disorienting.[6] Having fashioned what looked like solid walls out of nothing but a fog of light, Turrell was sued in 1982 by a woman who fell and injured her wrist when she tried to lean on one of his insubstantial walls while visiting the Whitney Museum of American Art in New York. Some Turrell installations are so dimly illuminated that it takes fifteen to twenty minutes before the eyes dark-adapt sufficiently to see anything at all. Even then, one is seldom certain how much of what is seen is actually out there. (Turrell's belief that low light acts "as a trigger that sets off the seeing from inside the head" was borne out when police in Orange County, California, closed a show of his featuring a dim, unadorned red lamp with a blue halo because of complaints that he was showing pornography.) Here, as when stargazing, one is made aware of being chained within the confines of the skull, trying to conjure up the objective world from flickering shadows.

A pilot since childhood—his father was an aeronautical engineer who built his own airplanes—Turrell was earning a living as an aerial cartographer when, in 1974, he first spotted Roden Crater, an extinct volcano on the rim of the Painted Desert outside Flagstaff, Arizona, and conceived of turning it into a gigantic artwork. Sculpting the crater's cone with earthmoving equipment would refine its contours to enhance its effect on "celestial vaulting," the flattened-bowl shape into which the mind shapes the sky. (The caldera, 814 feet in diameter, is almost perfectly circular, so only its interior slopes needed to be reworked.) Tunnels similar to those of the ancient pyramids—and, like them, aligned with the Sun, Moon, and stars—would connect "light observatories," where the sky intruded into the darkness.

It was an almost absurdly ambitious project for an individual artist to attempt. "I look at this today and I wonder, what was I thinking?" Turrell muses.[7] Yet he persuaded an art foundation to buy the land and he set to work. A few years later, the foundation ran into financial trouble. To persuade the bank to let him assume the mortgage on the land, Turrell had to purchase two adjacent ranches and go into the cattle-raising business on what was now a

155-square-mile spread with a $1.7-million mortgage. His wife left him, and skeptics wrote him off as one more tragic artist whose reach had exceeded his grasp. Turrell himself had doubts. "The megalomania of artists!" he told one interviewer. "You draw these things out on paper and then you do some chewing at the top and you realize you are moving eight hundred thousand cubic yards of earth just to make this thing the right shape. And you've hardly changed the damned thing at all!"[8] It took him twenty-five years, but Turrell paid off the mortgage—"thanks," he said, "to my beautiful cows"—and completed the Roden Crater project, one of the largest and potentially most lasting artworks to have been fashioned in modern times.

On the afternoon of the 2000 winter solstice, a few months before Roden Crater opened to the public, Turrell drove me out to see it. A big, bearded man, every inch the working artist and with a sense of humor about it, he threw my rented Jeep down the ruddy dirt roads with the easy familiarity of one who has been this way many times before. On the way, we talked flying. "A friend wanted me to take him skydiving, so I put him in a biplane without fastening his seat belts, and then I inverted the plane," Turrell shouted, over the roar of the engine. "I'll tell you, he just *dropped* out of that thing, fast as a Steinway. It was amazing to watch."

Roden Crater came into view, looming above a field of more than a hundred other volcanic craters and silhouetted against the shimmering pastels of the Painted Desert beyond. Aside from the road itself, there was nothing in sight that appeared to have changed much in the past ten thousand years. "It really looks like a planet, doesn't it?" Turrell said. "Of course, it *is* a planet."

We parked the car by an entryway on one side of the volcano and entered a chamber that opened onto an upward-sloping, keyhole-shaped tunnel, fourteen feet tall and a thousand feet long. At its far end was a perfectly circular aperture into which peeked the disk of the late afternoon Sun. The tunnel is aligned to the lunar "standstill"—the point at which the Moon stops creeping along the horizon and rises from the same point for a few days before turning back—and hence is also approximately aligned to the solar standstill.[9] Similar alignments are found in many ancient monuments, among them the pyramids of Egypt, Borobudur in Java, and Stonehenge in England. I tried to photograph the dark tunnel with its circle of stark white light at the far end, but the matrix metering system on the camera, which checks each pattern of light and darkness against a library of thousands of images stored in its memory, went haywire and froze for the first time in the eight years I'd been using it.

We trudged up the long tunnel toward the circle of light, only to find when we reached the chamber at the top that it wasn't a circle at all but an ellipse, tilted to look like a circle from the perspective of the tunnel. I laughed out loud, thinking of Kepler's joy at finding, after years of work, that the orbits of the planets weren't circular, as had long been assumed, but elliptical. At times like these, art, like science, seems less a sophisticated intellectual construct than a straightforward tool, a wrench to crank us out of conventional modes of perception.

Advancing past this room, we made our way by touch through ink-black tunnels—the apertures that will admit and sculpt light were boarded over during this last phase of construction—to the "eye" at the center of the crater. There we emerged into the light and lay on our backs, looking upward. The late-afternoon sky seemed to have been drawn down and attached to the crater's rim, like a lens. We walked backward up the smoothly sloping caldera walls, watching as the rim seemed to shrink in size. Nothing was as it seemed—or, rather, nothing would long conform to conventional models. Turrell's gigantic, optics-free observatory, though still unfinished, was already working.

The Sun had set and the western sky was a palette of fleeting reds and purples as we drove away from the crater, on our way to get a pitcher of cold beer at a hang-glider bar. "We define civilizations through their art and architecture, and from those leavings we try to infer their cosmology, which is a pretty big step—I mean, that's a leap!" Turrell said. "But remember, it's just in this last century that we came to understand the idea of a galaxy, and it wasn't all that long ago that we discovered the existence of the solar system. We don't celebrate those events, but we should. They are big moments, big leaps of thought, and I'm surprised that we don't celebrate them."

11.

Stones from the Sky

From following walls I never lift my eye,
Except at night to places in the sky
Where showers of charted meteors let fly.
—**Robert Frost**

On nights when meteors play
And light the breakers' dance . . .
—**Herman Melville**

*O*N THE NIGHT OF OCTOBER 9, 1993, *thousands of spectators at high school football games in the northeastern United States see a brilliant meteor—a "fireball"—streak overhead and break up in the air. Moments later, a football-size chunk of the meteor smashes the right rear fender of a 1980 Chevy Malibu parked outside the home of an eighteen-year-old high school senior, Michelle Knapp. She hears the crash, ventures out into the rain, and finds the meteorite resting in a washtub-size crater beneath the gaping hole it has punched in her car.*

August 31, 1992: In Noblesville, Indiana, at twilight, a small meteor whistles past Brodie Spaulding's right shoulder and plows into the grass a few yards from where Spaulding, age thirteen, is standing in his front yard, talking with a neighbor, Brian Kinzie, age nine. "When I sit down and think about it," Brodie reflects, "it was kind of scary."

November 30, 1954: Annie Hodges is napping on a couch in her home in Sylacauga, Alabama, when an eight-pound meteorite crashes through the roof and attic, punches a hole in the ceiling, smashes her console radio, bounces across the room, and hits her in the leg, raising a nasty bruise.

September 14, 1511: In Italy, a cluster of meteorites reportedly strikes and kills a monk and several barnyard animals.

January 14, 616: Meteorites fall on an army camp in China, killing "more than ten people."

Constantinople, 472: A meteor brighter than the Sun passes overhead and explodes,

139

knocking people down in the streets, capsizing sailboats in the harbor, blowing out shut-
tered windows, and covering the city with a cloud of black dust.

Yucatán Peninsula, Mexico, 65 million years ago: A comet or asteroid some ten kilo-
meters in diameter hits Earth, enshrouding the planet in clouds of dust mingled with the
smoke from millions of forest fires. The ecosystem collapses, resulting in extinction for the
dinosaurs and most other species of terrestrial life.[1]

ALMOST EVERYTHING THAT STARGAZERS observe is far away, and stays
far away—except for meteorites, stones that fall to Earth. Meteorites, by definition, are incoming. (They are called *meteoroids* while in space, *meteors* when
making their fiery plunge through the atmosphere, and *meteorites* if they hit
land or sea.) The Earth scoops up hundreds of tons of meteoric material every
day, most of it in the form of dust-grain–size particles that waft down to the
surface without anyone's noticing them. (Run your finger over a dusty mantel-
piece, and part of what you pick up comes from meteorites. There is an
astronomer in Pasadena who studies interplanetary dust by sifting it from random harvests of grit settling out of the air.) A hundred million of this daily
dose of particles—most of them about the size of sand grains or peas—are
large enough to produce the bright "shooting stars" we exclaim over at night.
Their glow comes from atmospheric friction, which heats them up, typically
when they are about 50 miles high, and slows them down, from an initial
velocity of about 24,000 miles per hour to a terminal velocity of only 300
mph. Meteors massive enough to survive their fiery rite of passage without
vaporizing will usually stop glowing once they have decelerated—atmospheric
friction doesn't generate all that much heat, now that they're moving slower
than a commercial jetliner—and then spend several minutes just tumbling
down through the air, a stage in their journey called "dark fall." Massive meteors decelerate less than small ones do, and so may explode in the air or hit the
ground without going through a dark fall phase.

Meteors that glow brighter than Venus are called fireballs or bolides. They
can cast shadows and make noises, ranging from sonic booms to distant rumbling or crackling sounds. One hears sounds from the lowest parts of the
meteor's fall path first; the rolling thunder effect as the sound fades away
occurs because, as seconds pass, the sound is coming from ever higher points
along the fall.

Nearly ten thousand meteorites are now in museums or in the hands of
private collectors. The great majority of them were found in places where
they stand out conspicuously—on Antarctic ice fields, the Sahara Desert in

North Africa, and the Nullarbor Plain in southwestern Australia. Scientists study them to learn about the composition of asteroids and comets, which is where nearly all meteorites came from, and to investigate such questions as whether the complex organic molecules found in them could have played a role in the origin of life on Earth. Terrestrial contamination is a problem, though. By the time it is discovered, a typical Antarctic meteorite, for example, has been lying in the snow for ten thousand to a million years, so it can be difficult to differentiate between the chemicals it brought with it from space and those it subsequently picked up here on Earth. Scientists prefer to get their hands on relatively pristine meteorites freshly collected immediately after a fall.

Fortunately a few have obligingly shown up in handy, if unsettling, locations. On September 30, 1984, moments after a brilliant fireball was seen arcing across the midmorning skies over Perth, Australia, two sunbathers on Binningup Beach, eighty miles to the south, heard a whistling sound followed by a loud thud, and sat up to find a one-pound meteorite embedded in the sand just twelve feet away. Ten days later, Don Richardson, a Vietnam veteran, was stepping out of his trailer near Claxton, Georgia, when he flinched at a whistling sound that reminded him of an incoming mortar round: A meteorite smacked into his neighbor's mailbox. Vital Lemay was feeding foxes on his farm northeast of Montreal on the evening of June 14, 1994, when the foxes all looked up. Following their gaze, he saw a "ball of smoke, like fireworks," then heard a hissing sound and a thump. A neighbor, Stéphane Forcier, went to investigate, and found a group of cows standing in a circle, staring at a foot-wide crater with a black, grapefruit-size meteorite in it.[2] A meteor that exploded above Uganda on the afternoon of August 14, 1992, rained stones down on Mbale, a city of fifty thousand inhabitants. One of them bounced off the top of a Shell Oil fuel tank; others hit a cotton factory, a sewage works, a railway station, and a prison. A boy from Doko was hit on the head by a four-gram fragment but lived to tell about it. Scientists who gathered dozens of the Mbale meteorites for study found that all had a "black fusion crust," indicating that they came from a meteor that exploded at high altitude, leaving its fragments to be individually toasted by atmospheric friction on the way down. (Reports of newly fallen meteorites being "warm to the touch" are thought to be ill-founded, though, since only a thin layer of meteor surfaces are heated, while their interiors remain cold.)

Such special deliveries are rare, so investigators looking for fresh meteorites more often collect reports of a recent fireball's path and try to triangulate on a possible fall zone small enough to be searched effectively. This can be done when amateur or professional observers at two or more locations simultaneously photograph or videotape a fireball, but fresh fragments have sometimes been found on the basis of nothing more than a few visual reports by eyewitnesses who knew the stars well enough to accurately report its trajectory. Jim Brook, an outdoorsman and pilot in the Yukon with an interest in geology, was driving on the ice of Tagish Lake on January 25, 2000, when he spotted fragments of a fireball that had burst overhead a week earlier. Having heard about it, he recalled, "I was watching closely for meteorites and suspected their identity as soon as I saw them, although I had been fooled several times by wolf droppings."[3] He bagged several dozen fragments, taking care not to touch them. Scientists found that they contained rudimentary organic molecules dating back to the origin of the solar system.

When the path of a bright meteor is accurately triangulated, it is sometimes possible for scientists to calculate the orbit that the meteoroid pursued out in space before it blundered into Earth. These data can help establish where the meteorite came from. The composition of a few meteorites indicates that they were knocked off the Moon or even Mars. (A meteorite that struck and killed a dog in Nakhla, Egypt, on June 28, 1911, came from the surface of Mars. Age-dating of the famous Mars meteorite in which scientists found tiny structures suggestive of microbial life, collected in an Antarctic ice field in 1984, suggests that it was blasted off the red planet by a meteoroid impact 16 million years ago, eventually falling to Earth in about the year 11,000 B.C.) Meteorites amount to free sample-return missions delivering specimens of asteroids, comets, the Moon, and at least one planet.

Asteroids, the progenitors of most meteorites, range in size from a few robust ones like Ceres and Pallas—which have diameters of 930 kilometers and 600 kilometers respectively, larger than the island of Jamaica—down to many more the size of office buildings and automobiles. Ninety percent of them are found in the asteroid belt, a flat band of debris located between the orbits of Mars and Jupiter. The composition of meteorites indicates that some asteroids are rich in iron, nickel, and other metals (these may be useful one day as mining sites), while others are made of lighter, stony materials or a mixture of stones and metals.

Comets, the other primary source of meteorites, contain metals and stone, but also lots of ice. These "dirty snowballs," as the American astronomer Fred Whipple memorably called them, are as black as soot and hence hard to observe—until they approach the Sun.[4] When that happens, sunlight warms the ice, which sublimes directly into gas in the cold vacuum. Gases trapped in the ice heat up and expand, break out of the comet, and spurt into space.

Their jets can blast hunks of ice and rock away from the nucleus, as the solid part of a comet is called, and push it around, altering the comet's trajectory in unpredictable ways. The vented gas produces a glowing spherical coma that surrounds the nucleus, as well as a streaming tail that can stretch for tens of millions of miles. Dust blown off the nucleus creates a second tail. Gas and dust particles react differently to the solar wind—charged particles emitted by the Sun—so the gas and dust tails often fan out in diverging directions. Much of the dirt and ice knocked off the comet litters its orbital path like a snail's trail. When Earth passes through such a trail, the result is an annual meteor shower.

Comets belong to two general families, short-period and long-period. Short-period comets orbit the Sun every two centuries or less. They reside along the ecliptic—the plane defined by the orbits of the planets—and move in the same direction as the planets and asteroids do. Long-period comets can take thousands or millions of years to make one circuit of the Sun. They come in from all sides, showing no preference for any particular orbital direction. This distinction has long suggested differing origins for the two families of comets, a hypothesis confirmed by recent astronomical findings.

It is thought that most short-period comets originate in the Kuiper belt, a larger and more distant version of the asteroid belt that lies out beyond the orbit of Neptune. The existence of such a belt was first postulated by K. E. Edgeworth in 1943 and was elaborated by Gerard Kuiper in 1951, but—since Kuiper belt objects are dark, are much smaller than planets, and shine only by the reflected light of the distant Sun—none was observed in the belt itself until the 1990s. Astronomers do not yet know how many Kuiper belt objects there are or how far out the belt extends; by some estimates it contains billions of cometlike objects, ten thousand of them larger than a hundred kilometers in diameter, making it much more massive than the asteroid belt. By other estimates, the belt is relatively narrow, stretching only from the orbit of Neptune to the outer limits of Pluto's eccentric orbit. (Pluto itself is probably a Kuiper belt object.) A Kuiper belt origin for short-period comets would explain why they orbit in the ecliptic and in the same direction as the planets, as this is the case for Kuiper belt objects generally.

The long-period comets, for their part, are thought to come from the Oort cloud—a vast, spherical congress of perhaps a trillion comets that may stretch halfway to the nearest star. The Oort cloud is named after the Dutch astronomer Jan H. Oort, who proposed its existence in 1950. (The Estonian astronomer Ernst Öpik had come up with much the same idea already, but astronomical nomenclature is an inexact art, and so we have these names rather than the Edgeworth belt and the Öpik cloud.) Oort studied the orbits of nineteen long-period comets and calculated that the inner edge of the cloud lies perhaps six hundred times farther from the Sun than Neptune is—a third of a light-year from Earth—while its outer limits are perhaps two light-years out. If Earth's orbit were the rim of an espresso cup set at the center of a circular dining table, the asteroid belt would be like the rim of a salad plate placed under the cup, the Kuiper belt would start at the edge of the table and extend at least to the backs of chairs drawn up to the table, and the inner fringes of the Oort cloud would lie out at the city limits.

Kuiper belt objects are generally too dim to be studied by amateurs, but one dramatic exception to this rule testifies to the stargazer maxim that you never know what you can observe until you try. Hughes Pack, a science teacher at Northfield Mount Hermon School in Northfield, Massachusetts, was overseeing an asteroid project in October 1998 with his students Heather McCurdy, Miriam Gustafson, and George Peterson, when, against all odds, they discovered a "trans-Neptunian object," a denizen of the Kuiper belt. Remotely observing through a University of California, Berkeley, telescope as part of an educational "Hands-On Universe" project, the students inspected pairs of CCD images taken in the same square of sky at different times to see if anything had moved against the stars, as asteroids do. In this way they had already found two new main-belt asteroids, or "minor planets," and the business of asteroid discovery was beginning to seem "almost mundane," Pack recalled. "In reality it was anything but," he added. "Just imagine being the only human being on the planet Earth to observe [an] undiscovered minor planet.

"About halfway through the class one of the search teams called out that they had found another candidate. With an almost bored tone, I urged them to go ahead and follow through with the process and gather whatever information they could about their new object. As I look back on this scene, I am a little embarrassed that I took this new potential discovery so casually." Pack watched over their shoulders as they examined a pair of white and black dots on the computer monitor that they concluded were spurious—probably caused by cosmic-ray impacts on the CCD chip. Then Pack saw another pair of dots that the students hadn't yet noticed. "The hackles on the back of my neck rose in shock and excitement," he recalled. "This pair of dots showed the exact signature that I was expecting to see if we ever found a trans-Neptunian object! I had been dreaming of this moment for several years and was well aware of what to look for. We have known from the very start that such objects should be in some of our images—that if we were careful with our image processing and inspection, we should be able to find them in their hiding places out beyond Neptune and Pluto.

"With all the self-discipline I could muster, I held myself in check and maintained my role as teacher, clasping my hands behind my back to avoid pointing at the screen. I forced myself to back up and walk away, telling them to keep looking. To my great glee, I did not get more than three or four steps before they called me back. It is hard to put into words the thoughts and feelings of elation that were coursing through my mind and body at that moment. I spun on my feet and returned with a great big smile, saying with suppressed excitement, 'Yeeess!' This was simply an exquisite moment in my life as a teacher."[5] Pack asked the students if they could explain the significance of the fact that the two dots—two images of a moving object taken at different times—were so close together, compared to the asteroids they had imaged earlier. They answered, correctly, that it meant the object was moving slower than asteroids do, and so must be much farther from the Sun. As Pack told me the story, it put tears in his eyes. The icy object they discovered, a hundred miles in diameter and now designated 1998 FS144, has since been studied extensively by astronomers investigating the anatomy of the solar system.

Comets in the Oort cloud are as dark as Kuiper belt objects but much more distant. At present they can be detected only when one happens to pass directly in front of a star, causing the star to wink out momentarily. But when a comet comes chuffing and fuming into the inner solar system, constantly brightening as it nears the Sun, it can be seen directly. Once in a long while, a massive comet or asteroid actually hits Earth—a sample-return mission with a vengeance. So the back-yard stargazer who searches with a telescope for previously undiscovered asteroids and comets is simultaneously engaged in two missions—a study of our origins and a reconnaissance that just might bear on our survival.

Observing meteors is one of the easiest and most pleasant ways of stargazing, and requires no equipment whatever. One simply reclines on a blanket or a lawn chair, enjoys the starry sky, and waits for a meteor to streak by. The human eye can see only 10 percent or 12 percent of the sky, so the more stargazers the merrier. In the old days virtually all organized meteor observations were made by amateurs, and some convened sober sessions aimed at making accurate counts of meteor rates. Observers were assigned square pieces of sky, sometimes delineated by big wire frames placed on stands, and each was given strict instructions to keep looking at his square and not under any circumstance to turn away from it when others cried out "Oooo!" and "Ah!" at the sight of a particularly beautiful meteor, lest he miss one passing through

his particular square at the same time. But now that meteor counts can be garnered by radar reflections and time-exposure photography, the pressure is off
and meteor observers can relax and enjoy themselves. If they want to make a
scientific contribution they can set up cameras. When a fireball explodes in the
sky such images are useful in locating the fall zone where fresh meteorites may
be found.

A stray meteor or two might show up on any dark night, but the best time
to see them is during one of the periodic showers that occur when Earth passes
through the trail of an old comet. Meteor showers are named after the constellations from which the meteors seem to "radiate," meaning that if one
drew the tracks of all the meteors backward, they would cross at a point in
that constellation. It isn't necessary to look at the radiant point—the meteors
are coming *from* there, and most don't fire up until they've reached another
part of the sky—but knowing where the radiant is enables you to distinguish
between interlopers and meteors that belong to the shower. Moonless nights
are best, since moonlight obscures the dimmer meteors. Many more meteors
are likely to be seen after midnight than before, because it is then that Earth's
rotation—counterclockwise, as seen from above the North Pole—carries us
onto the side of the planet that faces in the direction of Earth's orbital
motion. The Earth picks up more meteors on its leading than on its trailing
side, for the same reason that a car speeding through the rain gets wetter on
its windshield than on its rear window.

The Perseid meteor shower peaks around August 12, ideal for pleasant
summer viewing in the Northern Hemisphere. On many a summer night I've
slept in the open during the Perseids, awakening in my sleeping bag every hour
or so—one can tell the time pretty well from the shifting positions of the
stars, which move fifteen degrees per hour—to watch a few meteors fall before
dozing off again. The Perseids occur when Earth encounters the rubble trail
left behind by comet 109P/Swift-Tuttle, discovered in 1862 by Lewis Swift, an
amateur astronomer who ran a hardware store by day, and independently
observed three days later by Horace Tuttle at Harvard College Observatory.
The comet has an orbital period of about 120 years, but its debris is spread out
fairly evenly along its orbital path, making the Perseids a relatively reliable
shower.

Most other annual showers come from less homogeneous comet trails, and
hence are less predictable. The Quadrantids (January 3–4) can be spectacular,
but the debris trail is so narrow that the peak of the shower lasts only a few
hours. The Orionids (October 16–27) come from the trail of Halley's comet
and have produced showers ranging from fewer than ten meteors per hour per

observer, in 1900, to a high of thirty-five per hour in 1922. (A decent shower should produce twenty or more per hour.) The Leonids, associated with comet 55P/Tempel-Tuttle, are often disappointing, with individual observers spotting only eight or ten meteors an hour. But there's at least one big clump in the Leonid stream, and Earth hit it on the night of November 12, 1833. As the astronomy writer Agnes Clerke described the scene, "The sky was scored in every direction with shining tracks and illuminated with majestic fireballs. At Boston, the frequency of meteors was estimated to be about half that of flakes of snow in an average snowstorm." By dawn, an estimated ten thousand bright meteors per hour were blazing across the sky. Their vast numbers made it clear that there was a radiant point, in Leo, and that the radiant moved westward with the stars in the course of the evening, a finding that helped lay to rest any lingering Aristotelian claims that meteors originate in the atmosphere (owing to which venerable belief the study of weather is still called "meteorology").

The historical records suggest that Earth passes through the rich part of the Leonid stream every thirty-two to thirty-three years, but eager observers who stayed up all night on those once-in-a-generation occasions were often let down. When astronomers used radar to map the debris of the Leonid trail they learned why the shower is so erratic: The rich part of the stream is only 22,000 miles thick, so Earth, clipping along at some 66,000 miles per hour in its orbit around the Sun, spends less than an hour in the most densely populated part of it. If that hour happens to be when the leading night side of Earth is located, say, at the Pacific Ocean, few will see it.

By 1999, meteor shower experts equipped with radar were able to predict that the Leonids that year would be spectacular over eastern Europe and fairly good over western Europe, but would peter out by the time the United States rolled around to Earth's bow side. This prediction proved to be extremely accurate. Leonid fireworks blazed over eastern Europe, where observers thrilled to the spectacle of as many as three thousand meteors per hour, but from my location in California only a few stragglers could be seen, skating in low over the eastern horizon, on long, flat trajectories, like artillery rounds from a distant battle.

Unseen meteors are detected, even in the daytime, by amateur and professional radio astronomers who listen for radio broadcasts from distant stations. When commercial FM radio broadcasts bounce off meteors' ionized trails, stations normally out of range can be picked up, if only for a few moments, on an ordinary receiver. One simply tunes the radio to a place on the dial where no station is heard, and waits. Three amateur radio astronomers—Moh'd Alawneh,

Moh'd Odeh, and Tareq Katbeh of the Jordanian Astronomical Society—report detecting FM broadcasts bounced off meteor trails in the Al-Azraq Desert. "To distinguish the signals reflected from a meteor [from] others such as airplanes, the signals reflected from a meteor [are] very sudden, mostly loud and clear, and end gradually," they write.[6]

For centuries, people occasionally have seen what looked like the flashes of meteoroids hitting the Moon. A monk in Canterbury named Gervase reported that on June 25, 1178, at twilight, "some five or more men who were sitting facing the Moon" saw one horn of its crescent "split in two" and that "from the midpoint of the division a flaming torch sprung up, spewing out over a considerable distance, fire, hot coals, and sparks," after which "the Moon from horn to horn, along its whole length, took on a blackish appearance."[7] Telescopic observers have seen flashes on the dark part of the Moon, and a photograph taken in 1953 shows what appears to be one. But professional astronomers tended to be skeptical about such reports, pointing out that what looked like a lunar flash could have been the blinking light on an aircraft or a flaw on a piece of film.

Not until the 1999 Leonids—the shower that lit up the skies of eastern Europe—was the issue resolved. On the night of November 17, an amateur astronomer in Houston, Brian Cudnik, saw an orange flash on the dark side of the Moon with his 14-inch telescope. He reported it to David Dunham, a stargazer whose daytime job is calculating spacecraft trajectories. Dunham checked a videotape he had made of the Moon during the shower through his five-inch telescope, in Mount Airy, Maryland, and found that it had recorded a flash at just the place and time reported by Cudnik.[8] In all, five lunar impacts by Leonid meteorites were confirmed on two videotapes made that night. This pioneering amateur effort established that flashes of light on the Moon are indeed produced by meteor impacts and can be observed from Earth. Ultimately it may be possible, by taking spectra of such flashes, to study the composition of lunar soil and search for signs of water on the Moon.

Dunham was one of the founders of IOTA, the International Occultation Timing Association, an amateur group. Occultations occur when a foreground object, such as the Moon or an asteroid, passes in front of a more distant object, usually a star. Occultations of stars by the Moon occur instantaneously, since the Moon has virtually no atmosphere, so timing them can provide accurate data on the Moon's location. "Grazing" lunar occultations, in which stars disappear and then reappear from behind the mountains of the Moon, have contributed to charting the generally ill-mapped contours of the lunar polar regions.

Forecasting the occultations of stars by asteroids is a complicated matter. An occultation occurs when the long, thin shadow of an asteroid, cast by a distant star, traces its narrow course across Earth's surface. Hundreds of occultations are observable every year, but owing to errors in the cataloged positions of asteroids and stars it is easy to make mistakes in predicting the exact path the asteroid shadow will pursue. Occultation observers can be almost fanatically devoted, so much so that one astronomy handbook warns them not to set up their telescopes on train tracks for fear they will be run over while concentrating on a cherished pinpoint of starlight, and you risk incurring their wrath if you send them off to observe from a remote location and they then see nothing.

Data obtained by timing occultations can provide information on the asteroids' sizes, shapes, and compositions. Asteroids "take on a carnival of forms, resembling lizard heads, kidney beans, molars, peanuts and skulls," notes the planetary scientist Erik Asphaug.[9] It is theorized that many asteroids are not solid objects but rubble piles. Such asteroids cannot be rotating very rapidly—otherwise, centrifugal force would have torn them apart—and initial indications are that there is indeed a sharply defined upper limit to asteroid rotation rates, suggesting that at least some asteroids are slag heaps, not boulders. But only a handful have yet been imaged with sufficient accuracy to make this determination, so amateurs can improve the database by using occultations to map the shapes of additional asteroids. Dunham derived the silhouette of Kleopatra, a main-belt asteroid, by videotaping its occultation of a 9th-magnitude star on January 19, 1991, and combining his data with those obtained by other observers at various locations in the northeastern United States. The profile showed that Kleopatra is shaped like a peanut or a dog bone, roughly 240 by 70 kilometers in size. This unusual finding was confirmed when astronomers at the giant Arecibo radio telescope bounced radar waves off Kleopatra, and others took images of it with the European Southern Observatory's 3.6-meter telescope at La Silla, Chile, that showed the asteroid's outlines.[10] They derived a slow rotation rate for Kleopatra, consistent with a rubble-pile structure.

The best way to discover previously uncharted asteroids is to look for dim dots of light along the ecliptic that move against the background stars. Amateur observers had been finding asteroids photographically in this way for decades—Joel H. Metcalf, a New England minister, had discovered 150 of them by the time of his death, in 1925—but the discovery rate accelerated once CCDs put increased power in amateur and professional hands. Takao Kobayashi, of Oizumi, Japan, discovered a hundred asteroids in a single month, using a CCD camera and a ten-inch reflecting telescope. This was only a tenth of the discovery rate of the largest professional asteroid-locating project operating at the time, but it demonstrated that useful asteroid work could still be done by amateurs with modest equipment.

The amateur astronomer and science writer Dennis di Cicco accidentally discovered eight asteroids on CCD images taken for other reasons, and wondered what he could do if he set his mind to finding more of them. One October night in 1995 when the Moon was too bright to do much other work di Cicco took CCD images of five overlapping areas of the ecliptic in Pisces, and found three asteroids in the first field alone. "One turned out to be a known asteroid somewhat off its predicted position," he reported, "but the other two were new and I received credit for their discovery. Lucky night? Not really. By the close of the year I had searched on another eight evenings [and] on all but one night I was successful, chalking up an additional 21 confirmed discoveries."[11] Di Cicco's published account of his work encouraged other amateurs to try their hand at the field—among them Jeff Medkeff and David Healy, who wrote a computer program that helped them discover three asteroids in their first three nights of operation, on CCD images taken through a telescope while they were in bed asleep.

Astronomers lose track of many of the asteroids they have discovered, for want of enough follow-up observations to get an accurate orbit. Asteroids aren't supposed to be named and numbered until their orbits have been established sufficiently well so that they can be "reacquired" years later—but, embarrassingly, even some named and numbered asteroids have been misplaced. Asteroid 719 Albert, discovered in 1911, was lost for eighty-nine years before being reacquired by members of the Spacewatch team at Kitt Peak, and asteroid 878 Mildred was missing for seventy-five years before it was reacquired in 1991. Amateurs can help with this problem, both by reacquiring previously detected asteroids to refine their orbits and by following up on the ones they've discovered themselves. "Unlike professionals, who can rarely afford to follow new but otherwise typical-looking asteroids, it's easy for amateurs to take a personal interest in their newfound celestial real estate," di Cicco notes. "By observing a new object on several pairs of nights spread over a month or so there's always a chance the resulting orbit will be sufficiently accurate to link with previous sightings."[12]

Comets can be discovered with nothing more elaborate than large binoculars or a small wide-field telescope. The recommended technique is to sweep the sky near dusk and dawn, looking for comets that have brightened sufficiently to become visible but have not yet been detected because they were previously lost in the Sun's glare. What counts most is that the observer have a dark site, familiarity with the night sky, and persistence.

If you don't observe from a dark-sky location, the odds are that a comet you might otherwise have discovered will first be found by someone else who does. William Liller, a veteran professional astronomer who retired to devote himself to amateur astronomy, advises that to get seriously involved in comet hunting you need to be able to see objects at least as dim as 12th magnitude. Stars that dim can be seen with a telescope of as little as three inches aperture, provided that you have a dark sky and clean optics, but the fuzzy blur of an undiscovered comet may require a bit more light-gathering power, especially since you don't know where to look for it.

A familiarity with the night sky speeds up the process, since you won't need to check the charts every time you see a nebulous glow. The Messier catalog—a list of a hundred fuzzy-looking objects in northern skies, compiled by Charles Messier in the eighteenth century to aid fellow comet hunters who might otherwise mistake them for comets—covers most of the brightest nebulae, star clusters, and galaxies, but if you are observing things as dim as 12th magnitude you will encounter ten times that many such objects. Most carry NGC numbers, meaning that they are listed in the New General Catalog, created by Johann Dreyer in 1888 from John Herschel's data. Familiarizing yourself with them isn't terribly difficult—you need learn only their faces, so to speak, not their names—and is a pleasure in itself, like learning your way around the streets of London.

It takes an average of three hundred hours of observing to discover a comet, so patience is a requisite virtue. Lewis Swift averaged a comet discovery per year from 1877 to 1881. Although he limped badly from a fractured hip, he regularly rose before dawn and made his way to Duffy's cider mill in Rochester, carrying the optical parts of his telescope in a shopping basket. There he climbed three ladders to the roof and swept the skies before going on to work. "One cannot discover comets lying in bed," he said.[13] George E. D. Alcock, a retired English schoolteacher plagued by ill health, was observing through the closed, double-glazed windows of his bedroom when he codiscovered Comet IRAS-Araki-Alcock 1983d.[14] It was the elderly Alcock's fifth comet discovery, but his first in nearly eighteen consecutive years of fruitless searching.

David Levy, sweeping the sky at dawn on March 19, 1988, saw "what appeared to be an edge-on spiral galaxy . . . elongated but more diffuse than I would expect for a spiral." When he returned to the spot the next night, "The galaxy was gone! But a short distance to the north there was another spiral galaxy, this one having a sharply defined central core. Since it didn't resemble the first object, I drew its position, accused myself of completely misplotting the other galaxy, and again went to bed." On the third night, Levy checked again. He found nothing nebulous in the first position, while in the second, what had appeared to be a "galaxy's central core" was now clearly a star, with no surrounding nebulosity. "'There must have been a comet superimposed on this star,' I thought. Holding on to the telescope, I realized this was a special moment. If my theory was correct, simply moving the telescope less than a degree to the north would produce a comet. Holding my breath and trembling, I nudged the 16-inch reflector northward. There it was, a playful new comet that had just lost its game of hide and seek. It was duly reported and announced as comet Levy 1988c."[15]

On occasion, comets have been discovered by observers who were simply looking at another comet they already knew was there. Yuji Hyakutake, a courtly former newspaper photoengraver whose demeanor suited his last name—it means "the chivalry of a hundred samurai"—was fondly observing comet C/1995 Y1, which he had discovered five weeks earlier, when he spotted a new comet nearby, through a hole in the clouds. "I said to myself, 'I must be dreaming,'" he recalled. "I left my binoculars"—he was using a pair of giant, pedestal-mounted 25x150s—"for a while to calm myself down, and then I started drawing the cometlike object relative to the background." *Hyakutake C/1996 B2* eventually became a spectacular naked-eye comet that incited scientific study and untutored awe around the world, and it changed its discoverer's life. "I wonder why I could find the comet, because I have not been a person who devotes all himself to the hobby," said Hyakutake, who observed only a few hours a week out of deference to his family. "My wife can't make phone calls because the phone is always ringing. I'm a bit perplexed by all the attention paid to me, when it is the comet that deserves the credit."[16]

Alan Hale, a professional astronomer who studies extrasolar planets, discovered comet Hale-Bopp almost simultaneously with Tom Bopp, an amateur astronomer who worked in the parts department of a Phoenix construction materials company. The two were observing from sites ninety miles apart, in Arizona and New Mexico, and had never met. Hale had observed comet Clark shortly before midnight. While waiting for a second comet, *d'Arrest*, to rise, he recalled:

> I decided to pass the time by observing some deep-sky objects in
> Sagittarius, and when I turned my [16-inch] telescope to [the globular
> cluster] M70, I immediately noticed a fuzzy object in the field that

hadn't been there when I had looked at M70 two weeks earlier. After verifying that I was indeed looking at M70, and not one of the many other globular clusters in that part of the sky, I checked the various deep-sky catalogues, then ran the comet-identification program at the IAU Central Bureau's computer in Cambridge, Massachusetts. I sent an email to Brian Marsden and Dan Green at the Central Bureau at that time informing them of a possible comet. . . . I continued to follow the comet for a total of about three hours, until it set behind trees in the southwest, and then was able to email a detailed report, complete with two positions.[17]

At the same time, Bopp was observing M70 through a friend's homemade 17.5-inch Dobsonian reflector. Like most Dobsonians, the telescope had no clock-drive mechanism. While Bopp examined the globular cluster it drifted out of the field of view and the comet drifted in.

The fame bestowed on comet discoverers can turn sour when a comet that astronomers predicted would be "great"—meaning easily visible to the unaided eye—fizzles instead. The much-ballyhooed 1973 comet Kohoutek literally ran out of gas and never amounted to much, turning the surname of its discoverer, the Czech astronomer Lubos Kohoutek, into a synonym for inflated expectations. I can fearlessly predict that *Kohoutek* will be more prominent the next time it comes around—fearlessly, since it won't happen for another seventy thousand years.

Although photographs tend to make many comets look alike, when viewed through a telescope each is unique, and can change in appearance within hours. One uses low-power eyepieces to behold comets' ethereal tails, higher powers to study their fulminating nuclei. Observing *Hale-Bopp* from Rocky Hill on the night of April 4, 1997, I was startled to see, at high magnification, a distinct spiral pattern in the dust surrounding the nucleus. (It had a big nucleus, more than forty kilometers in diameter, that spewed huge clouds of dust when it approached the Sun.) Having never seen anything like it, I at first suspected that the spiral was an illusion of some kind—due perhaps to an optical defect. But the telescope performed normally when aimed at nearby stars, and the spiral in the comet's nucleus held up when viewed through various high-power eyepieces. Images taken subsequently by the Hubble Space Telescope showed that the spiral was the result of a single, robust jet shooting out dust from the spinning nucleus, like water from a lawn sprinkler.

Comets don't streak across the sky as meteors do, but they typically move against the background stars rapidly enough that taking a time-exposure image of one requires tracking the telescope on the comet and not the stars, and the closer they come to Earth the more conspicuous their proper motion becomes. Comet IRAS-Araki-Alcoçk 1983d passed within three million miles of Earth, the closest comet encounter since 1770. Observing it with a portable telescope from amid the bright lights of Hollywood, California, I could see it moving against the background stars, in a matter of minutes: It was *trucking.*

Short-period comets—the ones that have relatively small orbits, and so return close to the Sun at intervals of years or decades rather than centuries— offer observers an opportunity to compare their appearance and behavior on successive apparitions, rather like that of old schoolmates whom one sees only at class reunions. The prototypical short-period comet is *Halley.* It was named after Edmond Halley not because he discovered it (which he didn't) but because, by calculating its orbit and comparing the result with historical records, Halley correctly inferred that what had been taken for three different comets—those of 1531, 1607, and 1682—were actually recurring visits of the same one. (Thanks to the commendable record-keeping habits of early stargazers in many parts of the world, observing records have since been found for every prior appearance of Halley's comet dating back to 239 B.C.) The seventy-six-year orbital period of Halley's comet falls within the reach of a human lifetime, so that those born at the right time can hope to see it twice. My son is one such person. Born on March 23, 1986, he was only a few weeks old when his mother and I took him to the dark deserts of Joshua Tree National Monument and held him up so he could see *Halley* at its closest approach to Earth, outward bound. (He does not remember this, of course, but I keep reminding him of it in the expectation that by the time he is seventy-six years old he will at least remember having been told that he saw the comet.) Mark Twain, born when *Halley* was visible in the sky, predicted in 1909 that his life would end upon its return. "I came in with Halley's Comet in 1835. It is coming again next year, and I expect to go out with it," he wrote. "The Almighty has said, no doubt: 'Now here are these two unaccountable freaks; they came in together, they must go out together.'"[18] Twain's prophecy came true. He died on April 21, 1910, one day after *Halley* passed perihelion.

Comets have long been feared as harbingers of doom—and, as we shall see in the next chapter, in some ways deserve their baleful reputation, since comet and asteroid impacts evidently have wreaked devastation on Earth many times over the eons—and meteorites traditionally have been revered as emissaries from the great beyond. The Egyptian word for iron means "thunderbolt of heaven," while the Hittites and Sumerians similarly called iron "fire from heaven," and the Assyrians, "metal of heaven." A dagger found in the tomb of Pharaoh Tutankhamen evidently was forged from an iron-rich meteorite, and some scholars maintain that early metalsmiths, under orders to make ceremonial weapons from meteorites, were thus motivated to push their forges past the melting point of copper, 1,980°F, all the way to 2,795°F, the melting point of iron. "Meteorites jump-started the Iron Age," concludes the astronomer Bradley E. Schaefer.[19] If so, the foundations of human civilization were built in part with stones from the sky.

Comet Trails:
A Visit with David Levy

WHEN I FIRST WENT to see David Levy, in November 1991, he was known to the inner epicycles of the amateur astronomy community as a skilled comet hunter, but was otherwise obscure. Three years later his name had become a household word. What happened in between was his discovery, with the astronomers Gene and Carolyn Shoemaker, of Comet Shoemaker-Levy 9, and its gaudy demise in a series of impacts that Levy described as "the most dramatic ever seen on another world."

Levy's small home in an outermost suburb of Tucson, Arizona, was a modest affair, inconspicuous and remote as a comet at aphelion, with rude wooden screens posted here and there to block out local lights and a rooftop observing platform from which he'd discovered a comet the first night after a local handyman finished building it. "I told him, 'If I find a comet it will be worth the cost,'" Levy recalled. "Well, it was worth it."

His telescopes, clustered together in a cramped shed like barn animals in winter, looked unimposing. A white reflector was mounted on a tiny ten-dollar bench where Levy would sit, like a shy suitor in a park, controlling the telescope's motion by manipulating a fishing reel. A 16-inch Dobsonian reflector that Levy named "Miranda" sported a blue cardboard tube and a second-hand three-dollar finder. A sleeker-looking pair of Schmidt cameras completed the array. The sole luxury was an elaborate motorized chair, which enabled Levy to keep his seat while sweeping wide swaths of sky; he traded an antique telescope for it. With this modest equipment, Levy by the year 2001 discovered twenty-one comets, making him the third most prolific comet-sweeper in history. Brass plaques mounted on the Dobsonian testified to his comet discoveries, among them "1984t, November 13," "1989r, August 25," and "1990c, May 20," referring to comets listed in the record books as Levy-Rudenko 1984t, Okazaki-Levy-Rudenko 1989r, and Levy 1990c.

Born in Montreal, Levy suffered from severe asthma and was sent, at age fourteen, to the Jewish National Home for Asthmatic Children in Denver, Colorado. He took with him a small telescope he'd been given as a bar mitzvah present and started sneaking out at nights to observe with it. His nocturnal escapades came to the attention of his doctor, who asked, "Why are you waking up at night?"

"I'm not waking up," the young Levy replied. "I'm going outside to observe Neptune with my telescope." The physician thought for a moment, then said, "As your doctor, I am ordering you to *keep observing Neptune*. You're not to let your asthma stop you from doing what you want to do."

Levy continued observing, and on November 13, 1984, after fifteen years of searching, discovered his first comet. He was having dinner that night with a friend, Lonny Baker. She grew impatient when he kept gazing over her shoulder out the window, where the sky was finally clearing after days of clouds. Unable to endure the suspense any longer—dusk and dawn are prime time for comet hunting, since that is when a previously undetected comet may be caught emerging from the Sun's glare—Levy cut the evening short and bolted.

"OK, stand me up," Lonny called after him, "but you'd better find a comet for me tonight!"

He did. After only an hour and seven minutes of sweeping the sky, he spotted a fuzzy patch in the sky near the star cluster NGC 6709 in Aquila. "The contrast between cluster and fuzz was so beautiful I knew that something was out of place, for such beauty would have appeared in all the astronomical picture books," he recalled. "Within ten minutes I had my answer, for the fuzzy patch had moved!" He reported the comet (thereafter designated comet Levy-Rudenko 1984t) to the Harvard authorities, then called Lonny. "Well, did you find a comet for me tonight?" she asked. She laughed when he answered "Yes," Levy recalled. "When I told her the magnitude she laughed again. It was only after I provided the position and direction and rate of motion that she stopped laughing. 'My God, you're serious!'"[1]

By the time I met Levy he had discovered over a dozen comets, some with his back-yard telescopes and others—more recently and at an accelerating rate—on photographs he'd taken with the Shoemakers using the 0.46-meter Schmidt telescope at Palomar. A typical observing run in this project, called the Palomar Asteroid and Comet Survey, consumed seven nights and produced more than three hundred photographs. Levy would pack up his battered old car, drive to Palomar, work all night for a week, then drive home again. He was reimbursed for his travel expenses but was paid nothing.

Although he'd never taken an astronomy course or held a job in science, Levy took his observing seriously. "While amateurs are not doing astronomy for a living, it's certainly not just a hobby for most of us," he told one interviewer. "It is part of our nature. . . . If you are a professional astronomer, you are doing astronomy as a daily activity to earn money. There's nothing wrong with this, and the fact that you're a professional astronomer doesn't stop you from being an amateur, too."[2]

Levy's bible was *Starlight Nights: The Adventures of a Star-Gazer*, a rhapsodic memoir written by the late Leslie Peltier, whom Harlow Shapley of the Harvard College Observatory called "the world's greatest nonprofessional astronomer." In the course of an amateur career that started in his boyhood, Peltier discovered twelve comets and six novae and recorded 132,000 variable-star observations. Levy had been carrying around his copy of *Starlight Nights* for so many years that he'd had to have it rebound, in handsome midnight-blue leather, adding blank pages in which he listed every talk he'd given in which he quoted from it.[3] This turned out to be almost all his talks. During my stay in Tucson, Levy and I attended an amateur astronomy meeting for which the scheduled speaker failed to appear, so Levy gave an impromptu lecture. He produced his ever-present copy of *Starlight Nights* and read a passage in which Peltier, in his mid-twenties and still living on his parents' Ohio farm, describes spotting a comet through his six-inch telescope on Friday, November 13, 1925, riding his old bicycle through the night to a railroad signal tower in town, and dispatching a telegram describing the comet to Harvard. Heading home, Peltier wonders, "What would happen to my message? Would Harvard relay it on so that the big scopes in California could pick it up that night? Or would it arrive at Cambridge only to hear in cultured accents: 'I say, here's a good one, some chap out in Ohio has just found that comet that was reported about six weeks ago!'"[4]

The discovery was genuine—Peltier's first—and he carved the new comet's name, "Peltier," along with the year, 1925, into the wooden pier of his telescope. Levy read this passage in rhapsodic tones, and once his talk was over dutifully recorded its place and date in blue ink in his talismanic copy of *Starlight Nights*.

"Comet hunters are like watchmen," Levy told me, as we opened up his observatory. "You have to be out there almost every night. You can look for comets any time you want, but if you want to find them, you have to keep looking all the time. I usually look for about an hour at a time, examining each field of view for a second or two. My record is nine hours and forty minutes without stopping."

He turned on an old shortwave radio, a gift from his grandfather. Its tubes glowed amber, warmed up, and after a minute or two it began crooning vintage rock. The sky turned dark and we peered through the big blue Dobsonian at the Andromeda galaxy, the Crab nebula, and Stephan's Quintet. A couple of meteors streaked through the sky, emissaries of the sparse but persistent Taurid shower. The desert skies were coal black and we could readily see the pale dagger of the zodiacal light. I thought about all that stuff out there, orbiting the Sun—asteroids, comets, dust grains, countless rocks and snowballs. . . .

"There's something about actually looking at these things—something magical about it," Levy mused. "Amateur astronomy means you do it from the heart—that you *have* to do it. It connects you, heart and soul, to the sky."

The last thing we observed was Jupiter—which was also the first thing David Levy had viewed through a telescope, at age twelve. Although nobody yet knew it, there was a comet in orbit around Jupiter that night, too dim for us to see but destined for big things. When Levy and the Shoemakers discovered it sixteen months later, the world turned its attention to Jupiter, and our sense of cosmic security—our assumption that all that stuff out there could be pretty much relied upon to *stay* out there—was shattered forever.

12.

Vermin of the Skies

Thee fully forth emerging, silent, gazing, pondering the
    themes thou lovest best,
Night, sleep, death and the stars.
   —Walt Whitman

There are as many comets in the sky as fishes in the sea.
   —Johannes Kepler

A T NASA HEADQUARTERS, in Washington, D.C., on a late afternoon at the end of the millennium, eight of us were seated around a
table in a glass-walled conference room that looked out on rush-hour
traffic clotting on an elevated freeway. This was the annual meeting of the
NASA Steering Committee on "Near Earth Objects," asteroids and comets
that threaten to hit our planet and are large enough to do real damage when
they do. (What we were steering wasn't comets and asteroids—nobody has
tried that yet—but the efforts of other relevant committees.) We'd been here
for most of the past eight hours, during which time the coffee had congealed to
a dangerous density in its simmering pot; nobody was drinking any more of it,
although a Swedish astrophysicist had retrieved an aging breakfast roll and
was nibbling at it dubiously. Our current speaker, nearly the last of the day,
was discussing asteroid 1998 OX4—a "short-arc" asteroid, meaning that its
position on the sky was recorded only a few times before it was lost. Short-arc
asteroids that have been lost—owing to bad weather, a full Moon, or their
moving too close to the Sun in the sky, among other reasons—often cannot
thereafter be reacquired, because the scant orbital information adduced about
them in the first place is insufficient to determine where they were going.

Asteroid 1998 OX4 had been imaged for a period of only nine days, and
what little we knew about its orbit was slightly unsettling. "It still has a
nonzero impact probability," the speaker reminded us, meaning that the possi-

bility of its hitting Earth sooner or later could not yet be ruled out. Nor did anyone have much idea where to look for it: By now, two years after its loss, it could be anywhere in nearly half the sky. One might, the speaker noted, calculate what the asteroid's orbit would be if it *was* going to hit us, project that orbit backward in time to a place in the sky when it would be bright enough to observe, then look for it and breathe easier if it wasn't there. But no such steps had yet been taken. We discussed the value of having amateur astronomers search for lost asteroids like this one, noting that it would help if more amateurs had "meter-class" telescopes—that is, telescopes of around forty inches aperture. But only a few amateurs have instruments that large, mounted well enough to reliably search for asteroids in selected areas of the sky. We moved on to the next agenda item.

ASTEROIDS AND EVEN COMETS used to be dismissed as "vermin of the skies" by astronomers who regarded them as unwelcome distractions, like streakers cavorting on center court at Wimbledon. Asteroids made unsightly trails on long-exposure photographs taken to map deep space. Comets were showy but of limited interest, like photogenic movie stars who can't act. The public had few opinions about asteroids, since few had ever seen one, but the consensus regarding comets was that they were bad news—"a common death of man and beast," as the English mathematician Leonard Digges put it in the sixteenth century.[1] Such claims, based on little more than the superficial resemblance of a comet's tail to a fearsome sword, were rightly dismissed by scientists as mere superstition. But in one of history's little ironies, recent scientific findings show that in one respect comets and asteroids have lived up— or down—to their reputations: They can be killers.

No roof protects the Earth. Over the eons our home planet has been hit by a great many asteroids and comets large enough to do significant damage, and there are still plenty of big rocks out there now. An asteroid one hundred meters in diameter can flatten a city; a kilometer-class one could wind back the clock of human civilization to the time of Vlad the Impaler; and a ten-kilometer comet would exterminate most terrestrial life. The good news is that big impacts are much rarer than small ones: Ten-kilometer impacts like the one that wiped out the dinosaurs occur only about once in a hundred million years, while kilometer-class ones occur perhaps once in a million years. Smaller but still potent potential impactors are, however, pretty commonplace. Supertanker-size chunks of comets and asteroids, packing the punch of strategic nuclear warheads, hit once or twice per century. One blew up over a

Siberian forest in 1908, knocking down trees across two thousand square kilometers of forest; had it arrived four hours later, it could have destroyed St. Petersburg. Every day a dozen or so bungalow-size rocks zip by, missing us by less than the distance to the Moon, and a couple of them hit per year, usually impacting in an ocean or exploding above it. An airburst over the Pacific in 1994 was witnessed by no one but triggered military sensors and aroused enough attention that the president and vice president of the United States were awakened at four in the morning to be briefed on the blast. An object of comparable size hit western Honduras on November 22, 1996, excavating a crater 165 feet wide and setting fire to acres of coffee plants. It missed taking out downtown Bangkok or Manila by about ten hours.

The fearsome impact of a ten-kilometer–class object can substantially alter the course of biological evolution here on Earth—and evidently has done so in the past. Most of the damage is collateral: The impactor vaporizes when it hits, as does a comparable amount of the dry land or ocean floor in the area, and the resulting explosion hurls molten boulders all over the place—some of them flung in suborbital arcs, like those of intercontinental ballistic missiles, to land on the other side of the planet. These torches set the world afire. Soot from the fires and dust from the impact shroud the atmosphere in blackness for months. Deprived of sunlight, the ecosystem collapses, dooming most of the creatures on land and in the normally sunlit shallows of the sea. The impact at Yucatán that ended the reign of the dinosaurs 65 million years ago drove the majority of all living species into extinction. In the Permian-Triassic die-out of 250 million years ago, which also may have resulted from an impact, 85 percent of marine species were exterminated and the death rate on land was even higher.

During the waning decades of the twentieth century, as scientific research brought the brutal reality of impacts to light, the amateur and professional astronomers who study asteroids and comets were transformed, like Superman changing clothes in a phone booth, from pedestrian pursuers of celestial vermin to heroic lookouts who might one day save the world from the peril of Death From Above. The story of how this came about aids in understanding the interplay among amateur science, professional science, and superstition.

For centuries, reports of meteorite falls were discounted by learned professors confident that stones cannot drop from the sky. When residents of Barbotan, France, reported that a shower of meteorites had pelted their farmlands on July 24, 1790, they were jeered at in the *Journal des Sciences Utiles*. Eyewitness accounts of the fall of a large meteorite in Weston, Connecticut, on December 14, 1807, although vouched for by a pair of investigating Yale professors, were dismissed so summarily that popular books to this day repeat the apocryphal story of Thomas Jefferson saying loftily, "It is easier to believe that two professors could lie than admit that stones could fall from heaven."[2] Tales of earnest peasants, meteorites in their callused hands, being turned away in scorn by the authorities are recounted with some relish by those who claim that science is just another religion, led by a priesthood hostile toward evidence that confounds the prevailing dogma. But the truth is more subtle than that.

Eighteenth-century scientists did dismiss reports of meteorite falls, but their problem wasn't insular dogmatism. They had an alternate theory—that the scorched stones being brought in by village folk had been struck by lightning—that answered well to the preoccupations of researchers in an age hallmarked by Ben Franklin's 1752 experiments with kite and key. As for meteors seen streaking across the night sky, scholastic professors relied on the authority of Aristotle, whose *Meteorology* described meteors as jets of fire in the atmosphere.[3] By the late eighteenth and early nineteenth centuries, scientists began to doubt these theories as further reports were made connecting meteor flashes in the night with stones being found on the ground in the morning. Goethe, who wrote more scientific studies than he did poems and plays, examined the evidence and by 1801 was convinced that meteorites originate from beyond the atmosphere. Other converts included the influential French scientists Jean-Baptiste Biot, Pierre-Simon de Laplace, and Siméon-Denis Poisson, who on behalf of the French Academy of Sciences investigated the reported fall of hundreds of stones at L'Aigle, Normandy, on April 23, 1803, and pronounced it valid.

Scientists did, of course, exercise a requisite professional skepticism, and it was this skepticism, not a close-minded adherence to dogma, that caused many to have reservations about meteorite stories. Thomas Jefferson's actual opinions about meteorites were similarly nuanced. Concerning the Weston, Connecticut, meteorite, said to be of extraterrestrial origin by the two Yale professors whom he is supposed to have accused of lying, Jefferson wrote to Daniel Salmon, who had a Weston fragment in his possession, "We certainly are not to deny whatever we cannot account for. A thousand phenomena present themselves daily which we cannot explain, but where facts are suggested, bearing no analogy with the laws of nature as yet known to us, their verity needs proof proportioned to their difficulty."[4] Intrigued by a reported meteor fall in England and France on the night of November 12, 1799, Jefferson wrote to the surveyor and amateur astronomer Andrew Ellicott:

I do not know that this would be against the laws of nature, & therefore I do not say that it is impossible but as it is so much unlike any

operation of nature we have ever seen, it requires testimony proportionately strong. . . . A most respectable sensible and truth speaking
friend of mine gave me a circumstantial account of a rain of fish to
which he was an eye witness. I knew him to be incapable of speaking an
untruth. How he could be deceived in such a fact was as difficult for me
to account for, as how the fact should happen. I therefore prevailed on
my own mind to sojourn the decision of the question till new rains of
fish should take place to confirm it.[5]

Stones—and fish, too, evidently, when sucked into the air by water-
spouts—do fall from the sky, and eventually the extraterrestrial origin of
meteorites came to be accepted by scientists. But many rejected the implication that Earth has ever been hit by rocks big enough to alter the geological
record and drive species to extinction. Underlying their conservatism on this
score was a professional distaste for catastrophism—the belief, voiced by religious opponents to Darwin, that the Earth is only a few thousand years old
and that the evidence of great changes found in the geological record was created during that brief span of time by various catastrophes, such as the flood
survived by Noah and his zoological crew. Talk of Death from Above sat
poorly with scientific gradualists, who associated it with ignorant opposition
to Darwinism.

Comets in particular were distasteful to scientists, having come into
astronomy like Marley's ghost, dragging the chains of superstitious lore.
Comets were widely believed to be the literally ominous heralds of such dire
events as the London plague of 1665 and the deaths of luminaries ranging
from Julius Caesar in 44 B.C. to Charles the Bald in A.D. 877. "Comets are vile
stars," wrote Li Ch'un Feng, in the seventh century A.D. "Every time they
appear . . . something happens to wipe out the old and establish the new." A
verse known to every German schoolchild went:

> Eight things there be a comet brings
> When it on high doth horrid rage:
> Wind, Famine, Plague, and Death of Kings
> War, Earthquake, Floods, and Doleful Change.[6]

Churchgoers were told that these heaven-sent misfortunes amounted to
apt punishment for their many sins. The Lutheran bishop Andreas Celichius
in 1578 concocted a kind of moral astrophysics from this claim, asserting that
comets are made of "the thick smoke of human sins, rising every day, every

hour, every moment full of stench and horror, before the face of God, and becoming gradually so thick as to form a comet."[7] When the great comet of 1680 was moving like a ghost ship across the night sky, the theologian Christopher Ness warned that it was "a Sign from Heaven [that] signifies Drought, and portends War."[8] So many of the faithful believed him that the philosopher Pierre Bayle complained of his work constantly being interrupted by the anxious inquiries of terrified sinners.

Meanwhile, the amateur astronomer Edmond Halley saw the great comet, too, from a ship on the English Channel. Upon reaching France he met with the astronomer Jean-Dominique Cassini, who discussed with him the theory that comets were not missiles hurled down by God but interplanetary objects following predictable orbits, and that some of the ones recorded in the annals of history might be the recurring visits of the same comet. Back in England, Halley managed to get an interview with the generally unapproachable Isaac Newton. When Halley asked him about comet orbits, Newton said that he had solved the problem long ago. Halley urged Newton to write up his findings, and arranged for publication of the resulting book—the *Philosophiae Naturalis Principia Mathematica*, which galvanized the literate world and kindled the Enlightenment. In the *Principia*, Newton proved that comets are indeed interplanetary voyagers pursuing determined orbits, a finding that Halley then confirmed in grand style by computing the orbit of the comet that now bears his name and predicting that it would return in 1758—as it did, sixteen years after his death.[9]

Thanks to the efforts of Newton and Halley, comets became a focus of the conflict between science and superstition—a serious matter in those days, when devotion to science was limited to such a small minority that members of the Royal Society, in London, felt obliged to cast magic spells to demonstrate that they didn't work. As with one's opinion of astrology today, what you thought of comets in the eighteenth century was emblematic of your seriousness of mind. If you were ignorant and superstitious, you regarded comets as dangerous harbingers of doom; if scientifically literate, you understood that they were predictable parts of a clockwork universe and nothing to be afraid of. The culture wars came to a boil with each return of Halley's comet, almost as if the Scopes trial were re-enacted every seventy-six years, with science eventually gaining the upper hand. *Halley*'s 1910 apparition generated more science than silliness, although an Oklahoma sheriff had to intervene to prevent a group of fanatics from sacrificing a virgin to appease it. When *Halley* returned in 1982 no virgins were imperiled, and the comet was interrogated by a flotilla of six space probes, one of which, *Giotto*, launched jointly by the

European Space Agency and NASA, imaged its porous, percolating nucleus accurately enough to measure its dimensions, some sixteen kilometers long and eight kilometers wide.

As eighteenth-century astronomers plotted the paths of comets, hoping to identify other periodic ones, they confirmed that some do come uncomfortably close to Earth. Between 1702 and 1797, eight comets were observed passing within 20 million miles of Earth, and one, comet 1770 I Lexell, skimmed by at a distance of only 1.3 million miles, its coma swelling in the sky to five times the size of the full Moon. Clearly an outright collision was, as Halley himself noted, "by no means impossible."[10] Voltaire picked up on this prospect in his popular book *Elements of Newton's Philosophy*, asking, "What a disaster would it be for our Earth, if unhappily she should find herself in the same point [as a comet]? The idea of two bombs, which burst on clashing together in the air, is infinitely below what we ought to have of such an encounter as this."[11] The mathematician Pierre-Simon de Laplace noted that a cometary impact could mean "whole species destroyed [and] all the monuments of human industry reversed."[12] An 1857 French cartoon showed a fiendish-looking comet tearing the world into pieces while a grinning Moon looks on like a spectator at a demolition derby.

But astronomers generally played down the notion of comets hitting Earth. Textbooks stressed the low odds of an impact in any given year, adding—mistakenly—that since comets contain ice, getting hit by one wouldn't be much worse than being struck by a snowball. (Comets more nearly resemble frozen snowballs with rocks in them, and big ones like *Halley* can pack a mighty punch.) Based on close calls like that of *1770 I Lexell*, scientists could calculate that a large comet ought to have hit Earth every 10 million years or so. But back then few thought that Earth had been in existence for that long.

All that changed in the twentieth century, when improved astronomical and geological age-dating techniques made it clear that Earth is billions of years old: On that time scale, even world-shaking, once-in-a-hundred-million-year impacts like the Yucatán dinosaur doomsday event must have happened repeatedly. Studying photographs taken from the air and from orbit, scientists found that the Earth is scarred with many impact craters, most of them so eroded that they had previously been overlooked. More than 150 terrestrial impact craters eventually were identified. In Manicouagan, Quebec, a 100-kilometer ring made of two lakes back-flooded by a hydroelectric dam turned out to be a 212-million-year-old crater that had remained undefined from the ground but leaped out to the eye when viewed from orbit. In the Vienne Valley of central France, Rochechouart and two neighboring towns were found to lie inside a 23-kilometer-wide crater, and the stones composing an elegant manor house at Rochechouart proved to be made of rock fragments that were fused together in the impact fireball, 186 million years ago. The world was starting to look less safe, comets and asteroids less benign.

The copestone on the arch of potential catastrophe was laid on the night of March 23, 1993, during an observing run at Palomar by the astronomers Eugene and Carolyn Shoemaker and the amateur astronomer David Levy. The three had been surveying the sky for ten years, concentrating on comets and asteroids, but this night the sky was too cloudy to risk film in making observations. "We spend almost four dollars each time we put film into the telescope," Gene reminded Carolyn and David. "We simply can't afford to waste film on a bad night like this one." But at Levy's instigation they made a few exposures anyway, using partly fogged plates drawn from the bowels of a box that had inadvertently been opened in sunlight.

When Carolyn Shoemaker examined the plates two days later, she saw a strange smear near Jupiter. "I don't know what this is," she said at the time, "but it looks like a squashed comet."[13] Improbably, it turned out to be a comet indeed—one that had been captured by Jupiter back around the year 1929 and had been orbiting it undetected ever since. Now tidal forces from the giant planet were tearing the comet into fragments, producing what looked like a string of pearls. The pearls bloomed into miniature comets—not all that small, really; the most massive ones were probably over five kilometers in diameter—arrayed in a straight line, each with its own glowing tail, a sight unique in stargazing history.[14]

A Japanese amateur astronomer, Shuichi Nakano, calculated that Shoemaker-Levy was destined to hit Jupiter, and when Paul Chodas of NASA's Jet Propulsion Laboratory ran an impact-prediction program for the fragmented comet, the computer output made him sit up straight. "I'd always seen zeros before," he recalled. "Suddenly a fifty-percent number came up."[15] Chodas went home at dinnertime to help his wife care for their four-month-old baby, while his colleague Donald Yeomans stayed behind and kept working the numbers. Yeomans confirmed that the pearls of Shoemaker-Levy were going to hit Jupiter, and soon.

And that is just what happened, sixteen months later. Starting on the night of July 16, 1994, the comet fragments trundled in, one after another, like trucks skidding into a pileup on an icy highway. They exploded in Jupiter's upper atmosphere in extravagant, rising fireballs that left the giant planet's salmon- and sand-colored atmospheric bands scarred by a chain of lurid black splotches that endured for weeks. Few who witnessed this unprecedented sight could fail to take the threat of terrestrial comet impacts to heart. Just before the great comet crash, the astronomer Kevin Zahnle, of NASA's Ames Research Center in Mountain View, California, had ventured that "the removal of this comet leaves the solar system a slightly safer place for us all."

"But it doesn't feel that way, does it?" Zahnle reflected, after the crash. "The solar system no longer seems quite so far away as it did before. Here we are, close to the edge, protected from the true enormity of the universe by a thin blue line. A day will surely come when the sheltering sky is torn apart with a power that beggars the imagination. It has happened before. Ask any dinosaur, if you can find one. This is a dangerous place."[16]

Before the impact of *Shoemaker-Levy 9*, the total number of professional astronomers working on scanning the sky for comets and asteroids "was less than the staff of one McDonald's," recalled David Morrison, a planetary scientist at NASA.[17] Thereafter, prompted by what amounts to enlightened self-interest, more got involved. Most of their searches concentrated on asteroids, since there are a great many uncharted asteroids within reach of existing telescopes. The U.S. Air Force, invigorated by the prospect of defending the entire planet, assigned time for asteroid searching on a previously top-secret telescope at its Ground-Based Electro-Optical Deep Space Surveillance observatory at the White Sands Missile Range. The observatory was built to keep track of some ten thousand man-made objects that orbit Earth—from spy satellites to cameras and tools fumbled away by spacewalking astronauts. In one ten-month period, astronomers using this instrument discovered nineteen thousand asteroids, of which twenty-six were in the one-kilometer, nation-busting class.

By the end of the millennium, astronomers estimated that they had identified about half of the Near Earth Objects (NEOs) with diameters of one kilometer or larger. (NEOs are defined as asteroids and comets with orbits that can bring them within 28 million miles of Earth.) Locating the other, less conspicuous half was expected to be more difficult. Hunting for asteroids is like hunting for Easter eggs: The biggest and brightest are found first. Among the remaining asteroids were many that were dim, either because they had darker surfaces or were near the outer extremes of relatively long, elliptical orbits. At annual meetings of NASA's NEO Steering Committee, we discussed the fact that larger telescopes would help but also considered that the time spent constructing them might better be allocated to making longer exposures with the telescopes already available.

This meant that if amateur astronomers were to continue contributing to the effort they would have to have large telescopes, or be particularly persistent, or just lucky. One of the persistent ones was Len Amburgey, a teacher of astronomy and environmental science at Leominster High School in Fitchburg, Massachusetts. Amburgey normally used his telescope for the unglamorous but essential work of following up on recently discovered asteroids, obtaining additional images so their orbits could be refined. One night, distracted by the conversation of two other amateurs who were visiting his observatory, and fatigued—he had four-year-old twins who woke him most mornings at 6:30, when he'd often had only an hour or two of sleep after finishing an observing run—Amburgey made an error while typing in the coordinates for a follow-up asteroid observation. The resulting CCD image showed an unexpected streak that he at first attributed to a cosmic ray hit, but when he imaged the area a second time the streak turned up again, now almost at the edge of the field. Amburgey reported the find to Brian Marsden at the Minor Planet Center of the Harvard-Smithsonian Center for Astrophysics, in Cambridge, who confirmed that he had discovered a near-Earth asteroid. Further studies by Amburgey and professional astronomers determined that the new asteroid, 2000 NM, posed no threat to Earth for another million years, so on this occasion everyone could breathe easier. Marsden suggested that the reason the professionals had missed it was that it comes into opposition— asteroids appear brightest when they are opposite the Sun in the sky—in July and August, the months when Arizona and New Mexico, where most of the professional search telescopes are located, experience so much rain that local residents call it the "monsoon season."

"What an adventure!" Amburgey reflected, once his discovery had been posted, complete with animations and orbital plots, on professional Web sites around the world. "It appears that the real factor here is patience and persistence. If you want to contribute to astronomy keep observing, as often as you can. If you do it enough, something will come your way."[18]

Another amateur astronomer, Roy A. Tucker, discovered a rare Aten asteroid—Atens spend most of their time inside Earth's orbit—using a 14-inch telescope and a CCD at his home in Tucson. Tucker was searching the skies twenty to forty degrees off the ecliptic, in order to avoid "the clutter of common main-belt asteroids." When he first spotted the streak of the Aten on a CCD image, he could see from its length that it was moving fast, indicating that it was nearby. He was so startled that he had to talk himself down. "OK, what do I do now?" he recalls thinking. "Pause. Relax a moment. Take a deep breath. Suppress your excitement. You have work to do. There's only an hour and a half before morning twilight and additional observations must be obtained or this thing will get away." He managed to keep tracking the aster-oid, and within a few nights observers in Japan, Australia, the Czech Republic, and Italy had locked on to it, too. The following year, Tucker discovered two more asteroids with Earth-crossing orbits. "The big professional search opera-tions are proving to be tough competition," Tucker noted. But, he added, "amateurs have a long tradition of astronomical discovery, and I hate to totally surrender that to the pros."[19]

Frank Zoltowski of Woomera, South Australia, discovered a potentially dangerous asteroid, since designated 1999 AN10, using a 12-inch telescope in his front yard. "This is the first known object which is on a potential collision course with Earth which is large enough to cause a disaster on a global scale," he told a reporter for the *Sydney Morning Herald*. "If *AN10* hit it would lift enough dirt into the atmosphere to cause an impact winter."[20] Fortunately, subsequent observations showed that its orbit misses Earth's.

The risk posed by a comet or asteroid is the product of three elements— its orbit, its impact velocity, and its mass. Its orbit determines whether it can hit Earth in the first place. Its impact velocity depends on its trajectory: A comet or asteroid falling through Earth's orbit typically is moving at a rate of about 42 kilometers per second relative to the Sun, while Earth's orbital veloc-ity is 30 kps. So a NEO involved in a head-on collision slams into Earth at a speed of about 72 kps, while one coming up on us from behind impacts at a relatively mild 15 kps. The mass of the NEO determines the potential damage it can cause at a given impact velocity: Asteroids consist almost entirely of rock, while comets also incorporate ice, which is less dense than rock, so a comet has to be roughly twice the size of a given asteroid in order to equal the asteroid's mass. Otherwise, size matters. A one-hundred-meter NEO of either variety can take out a city; a kilometer-size one can destroy a nation; and a ten-kilometer object like the one that hit Yucatán pretty much means the end of the world, as far as most of its current inhabitants are concerned.

To calculate the risk, scientists multiply the odds of an impact by the mass of the impactor, producing what is called the Torino scale. Named for the Italian town where it was adopted, it quantifies the NEO risk in a single num-ber, rather like the Richter scale for earthquakes. An object with a Torino number of zero is either too small to do any damage or in an orbit that never brings it close to Earth. (The sand grains sizzling up in the atmosphere while you read this have Torino numbers of zero. So does the massive planet Uranus, which would be disastrous if it came our way, but never does.) Torino number one is assigned to objects "meriting careful monitoring." Quite a few Near Earth Objects belong to this category: They aren't going to hit us any time soon but we'd like to improve our knowledge of their orbits just to make sure, because they are big enough to do damage. Torino classes two to four— "events meriting concern"—designate objects big enough to cause "regional devastation" that have at least a one percent chance of impact. "Threatening events" of Torino classes five to seven would be invoked if an object were found that was both on a dangerous orbit and big enough to cause a global catastrophe. Torino classes eight to ten involve the "certain collision" of massive NEOs. The 1908 airburst over Tunguska, Siberia, was a Torino eight event ("localized destruction"), and so, probably, were the explosions reported in 1930 in a Brazilian jungle, in 1947 at Sikhote-Alin in Kamchatka, and in 1972 in the southwest Pacific. Class nine impacts ("regional devastation") happen on time scales of a thousand to ten thousand years. Torino ten impacts are rare but calamitous. The dinosaurs' dominion ended in a Torino ten event.

Reassuringly, no comet or asteroid has yet been found that has a Torino number higher than one. (Asteroid 1999 AN10 was rated Torino one when Frank Zoltowski discovered it, then demoted to zero once its orbit had been calculated more accurately.) But it's only a matter of time until an amateur or professional observer spots an asteroid or comet with a high Torino number— one that's big and deadly and is almost certainly going to hit us. What will we do about it?

The answer depends, first of all, on whether it's an asteroid or a comet.

Asteroids are like your neighbor down the street: If you see him pass by and wave, you can safely assume that he's been by on prior occasions when you didn't see him. An asteroid that threatens to hit Earth presumably has come close to us many times before, affording astronomers ample opportunity—if they know it's there—to calculate its orbit, recognize the threat, and send a mission to put a rocket engine on it and slowly, carefully move it to a safe vector. (Such missions could be profitable, too: John S. Lewis of the NASA/University of Arizona Space Engineering Research Center estimates that the smallest known metal-rich asteroid—*3554 Amun*, a kilometer in diameter— contains \$3.5 trillion dollars' worth of cobalt, nickel, iron, and platinum.)[21]

Comets, however, are like a mysterious uncle you haven't seen in years who suddenly turns up, looking impressive, telling exotic tales of faraway places, and bearing a veiled air of menace. A long-period comet falling out of the Oort cloud for the first time, or returning after an absence of thousands of years, would not become visible in most amateur or professional telescopes until it reached the vicinity of Jupiter's orbit. If it was going to hit Earth, we might have only a few months in which to mount a defense. A short-period comet that was pursuing a safe orbit when previously observed could turn lethal on its next visit, owing to an orbit-altering gravitational interaction with Jupiter or another giant planet. Jupiter generally protects us by hurling comets out of the inner solar system, but sometimes it tosses a curve ball our way.[22] Comet 1770 I Lexell, the one that came within 1.3 million miles of Earth, originally had a safer orbit, but sometime after 1767 it interacted with Jupiter and was dispatched on a new and more dangerous trajectory. A subsequent encounter with Jupiter perturbed it again, presumably to our benefit, but we cannot be sure because the current location of *1770 I Lexell* is unknown.

The clockwork model of the solar system promulgated by Newton's *Principia* depicted planets, asteroids, and comets as paradigms of predictability. But with the discovery of the Kuiper belt, the amassing of growing databases on asteroids and comets, and the aid of computers in re-creating the evolution of orbits, it has become clear that chaotic events, too, have shaped the solar system. There are, for instance, gaps in the Kuiper and asteroid belts. Studies show that these gaps represent zones where the gravitational environment discourages stable orbits, so that objects that once were there have since been ejected into higher or lower realms. Neptune, it seems, has long patrolled the inner edge of the Kuiper belt, nibbling at it in much the same way that Jupiter does the outer edge of the asteroid belt. In one scenario based on computer simulations, Jupiter-family comets start as Kuiper objects that, as a result of instabilities in the belt, eventually encounter Neptune and have their orbits altered. The energy they pick up from the transfer can loft them all the way to the Oort cloud. But their perihelion—the point in their orbits closest to the Sun—remains near Neptune's orbit, so that sooner or later they encounter Neptune again, whereupon their orbits are randomized anew. Some are passed down to Uranus, then Saturn, and then Jupiter, which may fling them back into the Oort cloud or settle them down into inner-solar-system orbits as Jupiter-family comets. That so many handoffs are required to sculpt a short-period comet's orbit helps explain why predictable periodic comets like *Halley* are so rare.

These departures from clockwork regularity—the solar system's helter-skelter aspects, with asteroids and comets and even, in the earliest days, hefty planets being flung hither and yon—remind us that the overarching laws of nature, grandly reassuring as they may be, are not the whole story. Nature has a creative side—dangerously but also delightfully so—owing to which an observer toting a telescope into the yard never knows whether a previously undiscovered comet will swim into view that night, or just how its tail will fan and filigree across the sky.

The Camera's Eye:
A Visit with Don Parker

FOR THE FIRST HALF of the twentieth century, planetary photography was dominated by professional astronomers, whose expertise, experience, and access to custom-made cameras and photographic emulsions made them hard to beat when it came to obtaining crisp, high-definition images of Mars, Jupiter, and Saturn. Then the amateurs started catching up. Hobbyists mounted war-surplus cameras on home-made telescopes, loaded them with the new, faster films that were becoming available, and took good pictures. Others experimented with video, and with light-amplifying devices taken from military night-vision goggles. CCDs completed the transformation. CCD chips are sensitive enough to image a bright planet with a short exposure that can capture the fleeting, clear glimpses afforded by moments of optimal seeing. Observers making drawings had relied on such moments but were obliged to spend hours at the telescope accumulating details piecemeal before a drawing was complete. With a CCD, you might get only one clear image out of dozens, but when you got it you could often see detail across the entire disk of the planet. It was only a matter of time before an amateur put optics, observing conditions, and CCDs together to produce results rivaling those obtained by the planetary-science professionals.

Even so, it happened more suddenly than anyone expected, when the amateur astronomer Don Parker started shooting whole stacks of planetary images that equaled or surpassed those taken with large observatory telescopes. The Hubble Space Telescope could do better, as could some of the best mountaintop telescopes, but having limited time for such things they could produce only occasional snapshots, while Parker was covering the planets night after night. When I first met Don, at a star party in 1992, he struck me as a big, hardy man with a laugh like a clap of thunder who endlessly made fun of himself. (Of his career as an anesthesiologist at Mercy Hospital in Miami, he jokes, "Why did I go into a field I can't spell?") But behind his bluster resided an astute, watchful intellect and a persistent streak of perfectionism. Curious to see how he took his extraordinary photos, I went to see Don eight years later at his home in Gables By The Sea, a waterway community in South Miami. The curving driveway in front of his big two-story house was crowded with cars and sports-utility vehicles, and a 29-foot racing sloop rocked gently in its moorings at a dock out back.

Don lurched out to greet me, supporting himself on a cane, his knees having succumbed to arthritis on top of injuries suffered while playing football for Loyola Academy in Chicago. (He'd been offered football scholarships to Notre Dame and the Air Force Academy, but his family physician convinced him that an aspiring doctor should stick to safer recreations.) As we went inside, Don told me that he'd been attracted to astronomy in the fifties by reading popular books and seeing science fiction movies. His youthful observations of the planets with a simple "stovepipe telescope" were interrupted by medical school, residency, internship, marriage, and a tour of duty in the Navy, but when he settled down in South Florida, attracted by its opportunities for year-round diving and sailing, he reclaimed his old telescope, trained it on Mars, and was astonished by what he saw. "I couldn't believe it," he told me. "Beautiful, steady images! I had assumed, because everyone told me so, that the skies here were terrible, that the atmosphere was too damp. But we get world-class seeing here. People used to say, 'Oh, it gets cloudy,' but they didn't realize how clear it is *between* the clouds. We get good laminar flow here, nice smooth air, and that's the most important thing—good laminar flow, and good optics."

We climbed to the master bedroom, on the second floor, and proceeded out to Don's observatory—a closet converted to a control room, opening onto a narrow deck where the telescope was mounted. Knowing that expert observers don't necessarily use shiny new equipment, I wasn't expecting much, but the reality was still a bit startling. The telescope, its black tube scarred like the flanks of an old freighter, was bedecked with oddities—a set of jury-rigged turnbuckles that held the CCD camera square to the focuser, a handmade wood truss for the mirror-cooling fans, and a stepper motor that Don described as "a one-thousand-to-one gear reducer from an analog computer in an old Eastern Air Lines flight simulator." Its mounting, a clutch of steel axles and pig-iron counterweights, looked like something dredged out of a maritime dump. The computer, a 486 PC running a DOS program, belonged in a Silicon Valley museum, and the imaging monitor resembled the black-and-white TVs found in motel rooms in the fifties. I studied this setup as intently as would a music student confronted by Yo-Yo Ma's 1712 Davidoff Stradivarius.

"Telescopes are for looking *through*, not *at*," Don said cheerfully. "This one's a sixteen-inch F/6 Newtonian with a real small—two-inch—secondary mirror. There are pieces on here dating from nineteen fifty-six."

The house blocked out the entire northern sky but there was a sweeping view of the ecliptic, where Saturn, Jupiter, and Mars were emerging in the twilight sky. Low in the west, a day-old crescent Moon lay on its back like a ranch brand—the Lazy Moon—with glittering Mercury immediately to its north. Don rummaged around in a drawer full of eyepieces and emerged with one in hand, announcing exuberantly, "I got a clean eyepiece here!" We trained the telescope on Mars, which was far away—194 million miles—and boiling in the still-warm western sky. It displayed little detail. "I always look, though," Don said. "I've been looking at Mars since 1954. It's like drinking Scotch—an acquired taste that requires practice. At the start of an apparition, I can hardly see anything on the planet. It takes three or four nights before I learn to see it again."

Don moved the telescope to Jupiter, and the view through the eyepiece was so clear that I nearly fell off the stepladder. "Let's try a higher power—it doesn't cost any more," he said. "I've had this scope up to sixteen hundred power. People say, 'You can't do that.' Well, OK, so call the police. I wasn't keeping track of the magnification at the time, but eventually I ran out of eyepieces."

Saturn was next. "I never get tired of Saturn," Don said, as he adjusted the focus. "I think you'll like this." I saw an orange-sherbet-colored globe wrapped in pale atmospheric bands, a set of sharp, finely etched rings, and a little blizzard of moons. "This," I said, "is a good telescope."

I asked Don to demonstrate his CCD imaging technique and he obliged, hooking up cables and tightening turnbuckles to align the camera and his handmade filter wheel. "I make everything adjustable, so I can fix things when I drop them," he said. "There—good enough for amateurs!" He slipped a salvaged rubber belt onto his hand-rigged electric focuser, and, to counterweight the camera and filter wheel up top, slapped a bright green orthopedic belt, the kind used by patients rehabilitating from injuries, onto a Velcro strip near the base of the telescope. "Whatever works," he said. "I think if you can do something simply, just do so. I take the same approach in medicine. If you hear hoofbeats, don't think zebras."

With the eyepiece holder now occupied by the CCD camera, Don looked up the position of Jupiter on an antiquated DOS program and pushed the telescope into position, consulting the two old-style, leaden-looking setting circles on the telescope mount. ("I made them myself, in the garage!") Groaning and muttering to himself ("Come on, Parker"), he lurched back to the control shed and centered Jupiter in the camera's field, squinting at the motel TV. The CCD camera, he informed me, was one of the first ever made. Its chip was only a tenth of an inch square, "but that's all I need for the planets. OK, where are we? There you are, you devil! Let's see, Jupiter, what do I give it—about a three-second exposure, I guess."

Seated at the cramped console with Jupiter in position and everything ready, Don danced his fingers over the keyboard in a seamless sequence of motions. He took three images, one each in red, green, and blue, to combine into a color photo. This had to be accomplished rapidly, before features on Jupiter rotated out of position, and he worked with astonishing speed, apologizing all the while for being so slow. I remarked that the red image of Jupiter looked quite different from the blue and green ones. "The deeper parts of the atmosphere have a markedly blue color sometimes," Don explained.

"With planets, it's quantity, not quality. I've probably taken twenty thousand pictures of the planets. I can show them to you if you like," he said, with a wicked grin. I asked if these images constituted the largest continual record of the appearances of Mars and Jupiter anywhere, as I suspected they might. "There's a fellow in Japan, Isao Miyazaki, who probably has as much as I have," he replied. "We're archrivals. He's head of the Jupiter section of the Oriental Astronomical Society, a really nice fellow, and a fine photographer. We have almost identical telescopes and cameras, and we're coauthors on some papers. His big thing is Jupiter; mine is Mars. I use his data and he uses mine. That's what science is all about.

"I made some digital images of Jupiter twelve degrees from the Sun. The people running the Galileo mission needed them. *Galileo* dropped a probe into the atmosphere of Jupiter, but the main spacecraft had a jammed tape recorder and they didn't want to run it to play back images of Jupiter, for fear it would stall and they'd lose the probe data. So they needed images at that precise time, in order to know what was visible on Jupiter when the probe went in. We got them, exactly at the time of insertion.

"I do all the planets, but Mars is really my passion. In ALPO, we've assembled more than seven thousand Mars observations from all over the world. Our primary objective is to look at the atmosphere of Mars—its clouds—and my specialty is the north polar cap. We managed to convince the professionals that the appearance of the north polar cap of Mars changes from one apparition to the next. It had been written in stone that it does not. I've had papers published here and there—about a hundred fifty papers with my name on them, in fifteen or twenty professional journals."

Big white clouds were ambling across the sky like grazing cattle, and while we waited for Jupiter to reappear Don told me how he and other amateurs in Europe and Asia had helped the professionals target the Hubble Space Telescope to catch Mars at the right times, and how they'd imaged a massive dust storm on Mars that the professionals mostly overlooked: "They sent me the paper—it was in French—and it said, 'This is the greatest picture of Mars ever taken by the greatest astronomers with the greatest telescope, and amateurs can't even see this,' and all their CCD image showed were a few little clouds: I mean, they just missed the whole damn thing." He laughed. "Since then we've helped them revise their data."

Jupiter flashed in and out of the parading clouds. "I don't feel that well," Don said; he was recovering from a cold. "I don't think I'm going to do any imaging tonight." Then the clouds parted and Jupiter reappeared. "Gee, it doesn't look so bad," he said. "Maybe I'll try a few more." He turned back to the keyboard. His fingers flew.

13.

Jupiter

Up through the darkness,
While ravening clouds, the burial clouds, in black masses
  spreading,
Lower sullen and fast athwart and down the sky,
Amid a transparent clear belt of ether yet left in the east,
Ascends large and calm the lord-star Jupiter.
  —**Walt Whitman**

Under the heaven of our holy ruler,
All things turn to spring.
  —**Su Tung-P'o**

HAVING ONLY A FEW MINUTES to spare for casual stargazing on a warm spring evening in 1993, I took a portable telescope into the garden and focused it on Jupiter. Bumblebee-fat, the giant planet was low in the southeastern sky, where heat from the cooling hillsides roiled the air. At low power in the little telescope, the four Galilean satellites sprang instantly into view—Callisto, Ganymede, and Io to the left, Europa to the right. Switching to higher power, I inspected Jupiter's oblate disk. Few features were visible at first, but when the swimming air fell still for a few moments I could see at once that something was wrong, or, more properly, that something unexpected was there—a dark blob, with a curving line extending from it like a hook, adjacent to the south equatorial belt. What could it be? Lacking a camera or even a drawing pad, I contented myself with memorizing the feature as best I could.

The next day I checked a few astronomical Web sites and confirmed that something had happened—an eruption in Jupiter's opaque atmosphere, heralding a "revival," or increase in brightness, of the South Equatorial Belt. It had been discovered by amateur astronomers in Spain and the United

States, and then photographed by professionals at the Pic du Midi Observatory in the French Pyrenees. The hook was a protuberance from the original, oval eruption, drawn out by the jet stream winds adjoining the belt. As José Olivarez of the Association of Lunar and Planetary Observers' Jupiter section reported, "The initial eruption was very bluish and so were the dark slanting festoons that developed later. . . . Over time, these features lost their blueness.

"It is my view that nearly all the blue features in Jupiter's upper cloud deck are fresh from below and warmer than their surroundings—at least at first," Olivarez theorized. "As support for this view, I offer the case that temperatures of the spots at the bases of the blue festoons found on the southern edge of the North Equatorial Belt are the warmest measured on Jupiter. So I am inclined to believe that the initial eruption of the [South Equatorial Belt] revival that I observed was due to warm material upwelling from below."[1]

I read the news with a renewed sense of the enchantment that had got me into stargazing in the first place. Is this not the wonderful thing about astronomy, and science generally—that we can learn about nature directly, without awaiting the verdict of priests or princes or any other authority beyond our own careful and informed observations? Jupiter is named for the king of the gods, but speaks to all alike, his pronouncements available to all who turn a telescope his way. Nothing is more egalitarian than the night sky.

JUPITER PATROLS THE FRONTIER between the warm, inner solar system and the icy realms beyond. It's cold out there: Jupiter is five times farther from the Sun than Earth is, and since starlight, like gravitational force, decreases by the square of the distance, only one twenty-fifth—four percent—as much sunlight shines on each square meter of Jupiter as Earth. This is thought to explain why the outer planets are giants. When the solar system formed, theorists reckon, the disk from which the planets condensed was made of much the same mix of elements found throughout the universe, which is to say that it was mostly hydrogen and helium. But these light gases could not linger in the vicinity of the newborn Sun: Sunlight and solar wind blew them away, so the inner planets formed primarily from the residual heavier materials—the rocky, silicate stuff—that was left behind. In the outer solar system, sunlight was too weak to sweep away much of the hydrogen and helium, so the planets that formed there incorporated these lighter materials as well. The result was a set of large planets with rocky cores surrounded by big, deep envelopes of liquid hydrogen and helium, with gaseous atmospheres up top. Hence what we see of Jupiter through a telescope is not a solid surface but the upper level of its deep and complex atmosphere. The disk is etched with a series of parallel belts and zones, bounded by darker north and south polar regions. Intricate patterns—called garlands, festoons, rifts, dents, knots, spots, and rafts—may be seen within and between the belts and zones. Close inspection by space probes such as *Cassini*, which passed near Jupiter while performing a momentum-transfer maneuver on its way to Saturn, reveals that the belts and even the zones are full of manic activity. As with most astronomical objects, the better you can see Jupiter, the prettier it looks.

Because it is so large—all the Sun's other planets could be put inside it with room left to rattle around in—Jupiter is easy to observe. Brighter than any star in the sky, its fat, oblate disk is more than forty seconds of arc in diameter, compared to only fifteen seconds for Mars at opposition. Jupiter's mass, more than double that of the other planets combined, exerts a crushing weight at the core. This heats the core—for roughly the same reason that a hammered nail gets hot—and as a result, most of the thermal energy reaching Jupiter's surface comes not from the Sun but from inside the planet. Were Jupiter a hundred times more massive than it is, the core would be hot enough to sustain nuclear fusion, and the Sun would belong to a double-star system. As it is, Jupiter belongs to a class of objects—rather less massive than "brown dwarf" stars—that glow with gravitationally induced rather than thermonuclear heat. As the astronomer Eugene Antoniadi put it, in 1926, "Jupiter is a chilled Sun."[2]

If we could ride a balloon down through Jupiter's colorful cloudtops and somehow steer it all the way to the center of the planet—which would require a capsule that, among other things, could endure crushing pressures while somehow making us comfortable at more than two and a half times our normal weight—here's what astronomers think would happen:

First we pass through the gigantic yellow, salmon, purple, brown, and gray cloud banks of the upper atmosphere, reared against blue skies stretching to horizons thousands of miles away. The light drops toward a dull red as we fall deeper, passing through ammonia ice clouds, then a combination of ammonia and hydrogen sulfide, then water ice and rain. As we sink, the temperature approaches that of a spring day on Earth, while the atmospheric pressure rises to about that of a terrestrial ocean at a depth of 100 meters. Darkness falls, and at a depth of less than 1,000 kilometers—a fraction of the 71,000-kilometer trip to the center—the density of the atmosphere increases to the point that it turns from gas to liquid. If our craft is capable of descending through the resulting ocean—it's 15,000 kilometers deep—we should find below it a second sea, composed of metallic hydrogen and helium, that constitutes the bulk of

Jupiter's interior. Ultimately we arrive at a rocky core, about the size of Earth but ten times more massive.

The probe launched from the Galileo spacecraft in December 1995 entered Jupiter's atmosphere at a velocity of 106,000 mph. It decelerated at an alarming 230 Gs but managed to keep transmitting for nearly an hour, by which time it was 600 kilometers deep, more than halfway to the liquid-hydrogen transition level. Before falling silent it measured temperatures of over 300°F and atmospheric pressures 23 times sea level on Earth. Aided by amateur astronomers observing Jupiter from Earth, scientists were able to pinpoint the probe's entry point. It lay near an infrared hot spot—a dry Jovian "desert" and hence probably not typical of the planet as a whole—but the data nevertheless contained information valuable in comprehending the planet. Among the surprises were less powerful lightning bolts than expected, higher-velocity winds, and, most remarkably, traces of argon, krypton, and xenon. Jupiter could only have captured these noble gases in extreme cold—at temperatures colder than Pluto and characteristic of the remote Kuiper belt. Either the protosolar nebula was a lot colder when Jupiter formed than had been thought, or else Jupiter originally was much farther from the Sun than it is today and has since descended to its present orbit, perhaps in part owing to a loss of momentum occasioned by its having ejected billions of Kuiper belt and other objects into deep space.

Jupiter rotates rapidly—a day lasts only 9.8 hours at the equator—so visual observers making detailed drawings of its features are obliged to work quickly, before features disappear over one limb and new ones appear on the other.[3] Observers can avoid this problem by making strip maps, which depict the belts and zones all the way around the planet. Compiling a full strip map normally takes several nights of part-time observations, although when Jupiter is properly placed in the sky it can be followed through a complete rotation in the course of a single terrestrial night. Its rapid rotation, plus the fact that it is made mostly of liquid and gas, has flattened Jupiter considerably: The ratio between the length of its polar and equatorial diameters is 14:15, so observers who sketch Jupiter start by tracing a suitably oblate outline for its disk.

As Galileo noted when he first studied it through a telescope, in 1610, Jupiter resembles a miniature solar system. The four large "Galilean" satellites, Callisto, Ganymede, Europa, and Io, could be seen with the unaided eye from Earth were they not lost in Jupiter's glare.[4] As it is, a few sharp-eyed individuals can spot at least one of the Galilean satellites with the naked eye, and all four are readily observable with binoculars. In medium-size telescopes they display disks, and with large telescopes traces of surface markings can be seen on the disks—although the planetary scientist John B. Murray, who managed to photograph features on all four in September 1973 with the 105-centimeter reflector at Pic du Midi Observatory in France, allowed that "the difficulty of observing markings on these satellites cannot be overemphasized."[5]

Each Galilean satellite is a distinct world, with its own tale to tell.

Callisto, the outermost of the four and the size of the planet Mercury, is geologically dead. Like Mercury, it is heavily pockmarked with craters, whose presence indicates that there has been little geological activity on Callisto since the bombardment days of the early solar system. Asgard, a bright crater that superficially resembles Copernicus on our Moon, is prominent enough to be discerned from Earth under favorable conditions.

Ganymede is the largest satellite in the solar system—larger, indeed, than Pluto and Mercury. Its surface is a beautiful array of glassy ice interwoven with older, darker materials. The replenishing of surface ice suggests that geological activity of some sort has altered this world over time, but nobody yet knows what it was, or when it happened, or why.

Europa, slightly smaller than Earth's Moon but only two-thirds as massive, has an icy surface besmirched by few craters and retaining no trace of primordial materials like those that dominate the face of our Moon. This indicates that the surface is geologically young. Several lines of theory and evidence, including measurements of Europa's magnetic field by the *Galileo* Jupiter orbiter, suggest that there is an ocean of liquid water beneath the ice. What kept the ocean from freezing? Presumably Europa has a molten core, which has been kneaded and kept hot over the eons by Europa's tidal interactions with bulging Jupiter and neighboring satellites. Since undersea thermal vents on Earth support abundant life and have been proposed as sites where terrestrial life began, Europa conceivably could support some type of marine life.

The ability of the Jovian system to keep a small world geologically active by tidally palpitating its core is dramatically demonstrated by Io, the innermost of the Galilean satellites. Io has an orbit larger than that of Earth's Moon, but Jupiter's intense gravitational field flails it through that orbit every 1.8 days. Tidal forces compress and release Io's core like a strong man squeezing a rubber ball, making it the most volcanic body in the solar system: Dozens of active plumes poke through its thin, leaky crust. Slower-moving volcanic ejecta splash back onto the sulfur-yellow and granite-gray Ionian landscape, while ejecta moving faster than Io's escape velocity replenish the Io Torus, a band of plasma that circles Jupiter along Io's orbit. On the surface, granite barges float on molten lava lakes. As the molecular biologist and amateur astronomer John H. Rogers writes, "Here is a world that surpasses science fiction."[6]

Insights into Io's fiery nature were presented in a paper by three physicists—S. J. Peale of the University of California, Santa Barbara, and P. Cassen and R. T. Reynolds of the NASA-Ames Research Center—who with exquisite timing managed to get it published only three days before *Voyager 1* arrived at Jupiter and got the first close look at Io. They calculated that owing to orbital eccentricities imposed by gravitational interactions of the Jovian satellites and to "the enormous tides induced by Jupiter," Io might be "the most intensely heated terrestrial-type body in the solar system," replete with "widespread and recurrent surface volcanism . . . leading to extensive differentiation and outgassing."[7]

James Secosky, a high school science teacher and amateur astronomer from Bloomfield, New York, was granted time to study Io with the Hubble Space Telescope in the early 1990s. Secosky's proposal was to image the fuming satellite during the first fifteen minutes after it emerged from Jupiter's shadow. Since the 1960s, terrestrial observers had reported that Io gets ten to fifteen percent brighter during those first few minutes, perhaps because sulfur dioxide there turns to frost when the satellite is plunged into Jupiter's shadow and then sublimes to form an expanding vapor when back in sunlight. On the eve of his observing run, Secosky spent an all-but-sleepless night in a Baltimore hotel room near the Space Telescope Science Institute. "I'm kind of fearful," he told a reporter at the time. "Maybe somebody made a mistake, maybe me. It doesn't take much to mess it up. . . . It's like the night before an athlete competes in the Olympics or a pilot flies an F-16. It's the biggest day in your life. For me, you don't get any higher than this. I'm using a $1.5-billion instrument. It's beyond my wildest dreams."[8]

At the Institute, Secosky could see his observing run, "SECOSKY/SO2 CONCENTRATION AND BRIGHTENING FOLLOW IO," listed on the computer screens among with those of eminent astronomers and astrophysicists like James Westphal of Caltech, Reta Beebe of New Mexico State, Sandra Faber of the University of California, and John Bahcall of the Institute for Advanced Study in Princeton. When his time came, there was nothing for Secosky to do—the space telescope operates itself, driven by computer instructions preloaded into its memory—and he twirled his fingers nervously. Then, when the first image of Io appeared, he lost his quasi-professional reserve and said, *"Hello!"* The data, disappointingly, were negative. No significant brightening of Io was recorded, suggesting that the phenomenon is either illusory or the intermittent result, say, of volcanic eruptions. Still, Secosky's work aided in astronomers' studies of Io's atmosphere. "Amateur astronomer Jim Secosky made near-infrared images (7100 Angstroms) of Io which [provide] new constraints on Io's surface composition," read the NASA announcement, and he wound up as the coauthor of a paper in *Icarus.*[9]

"It was really a great experience to be there and have this information come in," Secosky said afterward. "I was very excited to know something nobody else has known, or see something that nobody else has seen, [advancing] knowledge a little bit."[10]

Riccardo Giacconi, the director of the Space Telescope Science Institute and the man responsible for the amateur program—he'd donated time from his director's discretionary allotment to make it possible—admitted that there had been "some criticism about this amateur program. The thought was that it cost a *zillion* dollars to put this thing up there, and how dare you use it for an amateur. But amateurs are just as bright as anybody, in fact sometimes brighter. . . . To see the face of this guy—I mean, he was here in this office and he was flying two feet off the ground. Science would be terrible if you couldn't share it and make it become part of the culture. It should make you wonder and dream. Otherwise it's just numbers."[11]

In addition to the Galilean satellites, Jupiter has three other "families" of moons. The innermost family consists of four small bodies—Metis, Adrastea, Amalthea, and Thebe. Three of them were discovered by *Voyager,* while the fourth—Amalthea, an irregularly shaped object nearly 150 kilometers long—was first spotted in 1892 by Edward Emerson Barnard, using the 36-inch refracting telescope at Lick Observatory. Out past the Galilean satellites orbits another family of four, consisting of Leda, Himalia, Lysithea, and Elara, all well under 100 kilometers in diameter. Himalia, at magnitude 14.8, strays as far as one degree from Jupiter, where it can be seen by keen-eyed stargazers armed with powerful telescopes. Finally, there is an outermost family, made up of Ananke, Carme, Pasiphae, and Sinope. These four tiny moons pursue retrograde orbits, meaning that they go around Jupiter opposite the normal way; presumably they are captured asteroids or pieces of comets.

Jupiter also has a dim ring—a smoke ring, one might say, in that the particles composing it, which may consist of dust blown off the inner moons by micrometeorite bombardment, are typically as small as those of cigar smoke. It has three components—an inner halo that reaches halfway to Jupiter's cloud tops, a main ring entangled in the orbits of Adrastea and Metis, and a pair of extremely dim outer rings, patrolled respectively by Amalthea and Thebe. Discovered by *Voyager,* the ring subsequently was imaged from Earth by pushing large professional telescopes to their limits, but to my knowledge no amateur has observed it—yet.

To get a sense of the scale of the Jovian system, consider that if the Earth were placed at the center of Jupiter, our Moon would lie inside the orbit of Io, while distant Sinope would be a third of the way to Mars. Jupiter's magnetosphere, the wasp-shaped zone within which its magnetic field takes precedence over the wash of charged particles constituting the solar wind, extends more than seven million miles ahead of the planet in the direction of its orbital motion, where it stacks up against the solar wind at what is called the magnetopause, and trails so far behind that it sometimes impinges on Saturn. Were it visible to the eye, it would loom four times the size of the full Moon in our skies here on Earth. Auroras larger than the surface of Earth dance near Jupiter's poles, where its magnetic field lines converge. Gigantic lightning storms stitch the upper atmosphere. Meteors plunge into the Jovian atmosphere at a rate that makes Earth's 400 tons per day look paltry, and in the course of its long tenure Jupiter has ingested millions of comets.

The thousands of close-up photographs returned from Jupiter by the Voyager and Galileo missions might well have discouraged amateur astronomers from continuing to study the giant planet. How could they hope to compete? With regard to Jupiter's satellites this attitude was to some extent justified. Given the astonishingly detailed images returned by the space probes of the cracked surface of Europa and the volcanoes of Io, little motive remained for earthly observers to try drawing the surface features of these satellites, as had been attempted by venturesome observers such as Edward Holden in the 1800s, Percy Molesworth in the 1920s, Eugene Antoniadi in the 1930s, Bernard Lyot in the 1950s, and Audouin Dollfus in the 1980s. Still, some of these terrestrial observers did remarkably good work: They seem to have glimpsed signs of volcanic eruptions on Io, for instance, and Lyot's 1953 map of Ganymede, made at Pic du Midi, comports well in some respects with the far more detailed *Voyager* mapping of that icy world. In the long run, the spacecraft results spurred amateurs to undertake more extensive and better-informed Jupiter observing projects. Even when there is a probe in orbit there, collaboration between astronomers making use of it and amateurs on the ground can result in a more complete picture of Jupiter than would otherwise be attainable.

Besides which, Jupiter is great fun to look at. What one beholds there is weather made visible—a phenomenon experienced on Earth when hurricanes and thunderheads are viewed from orbit, but larger, longer lasting, and more colorful. The overall color is a pale yellow. Observers differ on the specifics, but to my eye the belts are brown or tan, with accents of red and salmon, while the zones are mostly pale yellow, off-white, or a stark, ammonia white. The atmospheric colors are thought to be due, at least in part, to the presence of phosphorus compounds in the clouds, excited by radiation, lightning, rapid vertical wind currents, and the other violent mechanisms that make the planet such a tumultuous place. Infrared studies indicate that the colors of the clouds betray their altitudes: Blue clouds are the deepest (and hence can be seen only when the skies above them are clear), while brown clouds are the next highest, with white above them and the red clouds up top. Since Jupiter gets warmer the deeper one probes, the highest clouds are also the coldest. The belts, for instance, are lower and warmer than are the zones.

The mechanisms driving the circulation of Jupiter's atmosphere are not fully understood, but one certainly is the Coriolis force, in which a planet's rotation sets up circulatory cells that spawn cyclones and anticyclones.[12] On Earth, where the equator is moving at 1,000 mph, the circulation cells created by the Coriolis force are fat ovals, taller than they are wide, shaped rather like aircraft windows but tilted eastward at their higher-latitude ends. On Jupiter, the equator races along at some 27,000 mph, creating a more dramatic Coriolis force that stretches the cells longitudinally into thin, flattened ovals, like a rubber band stretched between two fingers. High-velocity, high-altitude winds—jet streams—move in opposite directions along either edge of each of the Jovian belts and zones. The spots seen on Jupiter through a telescope are vortexes, created like water eddies along the boundaries between opposing winds. Because the atmosphere is so vast and sluggish, vortexes and other Jovian weather patterns persist a lot longer than terrestrial ones do. The lifetimes of big terrestrial hurricanes are measured in days or weeks. The largest and longest-lasting vortex on Jupiter, the Great Red Spot, an oval feature larger than Earth, has been raging for centuries.

The Great Red Spot rides along a ragged line that divides the broad, dusky south equatorial belt from the lighter South Tropical Zone, taking a bite out of both. Thermal measurements made by *Voyager* suggest that the roots of the Red Spot reach down only about 200 kilometers into the atmosphere, which would make it a hundred times wider than it is tall. (Terrestrial hurricanes are typically ten or twenty times as wide as they are tall.) Like nearly all the spots and ovals on Jupiter, the Great Red Spot is an anticyclone, a storm centered on a high-pressure zone. Whereas hurricanes are low-pressure cyclones, high-pressure anticyclones ride a bit above the surrounding clouds, as their white or red colors reveal.

Among the first observers to identify spots on Jupiter were Robert Hooke in England in 1664, Giuseppe Campani in Rome in 1665, and Jean-Dominique Cassini of Bologna and Paris. From their drawings and descriptions it cannot be ascertained whether any of the objects they saw was the Great Red Spot, although a lovely painting done by Donato Creti in about 1700—depicting observers with a small telescope under a sky in which Creti has obligingly rendered Jupiter as an enormous disk, to show how it looked to them through the telescope—displays what appears to be the Great Red Spot, in which case the storm is at least three hundred years old. The first unambiguous scientific records of the spot date from the 1830s, and it was recognized as a semipermanent feature by 1879. The historical record makes it clear that the spot's intensity waxes and wanes, so that it is sometimes quite red, then pallid, and during some periods amounts to little more than a colorless bay cutting into the South Equatorial Belt. For reasons not well understood, these changes evidently are synchronized with the appearance of the South Equatorial Belt itself, which grows faint when the spot is red and "revives" when the spot gets pallid.

The Great Red Spot is best seen at transit, when it crosses a line drawn between Jupiter's poles and so is nearest to the center of the disk. By timing transits of this and lesser spots, generations of amateur astronomers have accumulated records useful in charting the course of Jovian weather. Their logs record that the vortexes float semi-independently in the atmosphere, like bearings in oil, and that they sometimes cannibalize one another. This information has helped scientists construct fluid-dynamic models of the Jovian atmosphere, in which the emergence and survival of vortexes may be viewed as examples of the spontaneous appearance of islands of order in an ocean of chaos. In the long run, studying the visible weather of Jupiter and the other giant planets will aid in modeling and predicting the weather on Earth.

With the advent of CCD cameras, clear images of Jupiter could be made that documented its changes in detail over sustained periods of time. Although space probes and orbiting telescopes like *Hubble* made the sharpest images, ground-based amateur and professional observatories supplemented these snapshots with their continual observing projects. As Don Parker demonstrated with the telescope off his bedroom, a skilled amateur photographer can make CCD images of Jupiter that compete with those of the *Hubble* Wide-Field Planetary Camera in terms of the information available in a single image—and, since *Hubble* has many other tasks to perform, amateurs can do better in terms of the amount of data collected on Jupiter over the years. In December 2000, images of Jupiter were made only thirty hours apart by the *Cassini* space probe and by the Portuguese amateur astronomer António

Cidadão. The *Cassini* image did reveal somewhat more detail—as you might expect, given that the spacecraft was ten times closer to Jupiter than the Earth was—but otherwise the two pictures looked almost identical. "Considering that one of these images was taken with a mass-produced ten-inch Schmidt-Cassegrain telescope from an apartment balcony in Portugal, and the other from a billion-dollar spacecraft closing in on its subject, the differences seem almost trivial," remarked the Canadian amateur astronomer and telescope maker Gary Seronik. "Image after image reinforces the fact that amateurs are reaching heights of resolution and depths of faintness as never before."[13]

The four Galilean satellites sometimes pass in front of Jupiter. In small telescopes they can be difficult to see against the Jovian disk, especially when they are transiting across a light zone rather than a dark belt, but their shadows are more conspicuous.[14] Before opposition the shadows precede the satellites' motion, and so appear on the disk before the transit begins; after opposition the shadows trail the satellites. The shadow of an outer satellite can fall on an inner one, eclipsing it, and sometimes a satellite passes directly in front of another, occulting it. Amateurs making precise timings of these "mutual" satellite events, as they are called, can contribute data useful in refining scientific understanding of the moons' orbits, which are complicated by tidal effects. This information is valuable in targeting space probes.

Amateur radio astronomers observe Jupiter, too. The planet emits radio bursts in several frequency bands. Of these, amateurs tune in mostly to the continuous decimeter-wavelength radio noise produced by charged particles caught in Jupiter's magnetic field, and to the intermittent decameter noise produced by interactions of Io with the Io torus. Millimeter-length radio waves are emitted as part of the thermal energy of Jupiter's atmosphere. A simple receiver purchased at a ham flea market, connected to an antenna consisting of a single wire strung between three poles, is sufficient to make radio observations of Jupiter. There's something eerily stirring about listening to radio noise from Jupiter. "L-bursts," long-duration emissions from the Io torus, resemble the sound of huge, distant sheets of metal being flapped in the air, like the offstage effect in a Mahler symphony. "S-bursts," much more rapid, call to mind old recordings of freight trains rattling through switch yards. When an S-burst tape is slowed to the point that a half-second of data takes a minute to play back, the result is a series of weird, descending whistles, like the calls of alien birds.

Aesthetics aside, such data can be scientifically useful. When the fragmented comet Shoemaker-Levy 9 struck Jupiter, students at Tri-County Technical College in Pendleton, South Carolina, recorded radio noise evidently generated by at least two of the impacts. They sent the data to astronomers at Berkeley and the University of Florida and presented them at meetings of the Association of Lunar and Planetary Observers and the Society of Amateur Radio Astronomers. Their antenna consisted of an 18-inch loop inside a 10-foot parabolic dish, the design of which was obtained from a magazine article. "We had a very high quality RG-8 cable run until my dog Jupiter ate the 16-foot [cable] coming out of the ground," reported project supervisor John D. Bernard. "After they stopped laughing, the Northland Cable Company of Liberty, simply by being asked, replaced the 16-foot out-of-the-ground piece of cable with connectors for free."

The amateur team also "collected unique, unexplained signals from the SL9/Jupiter collision," Bernard reported:

> Astronomers were suggesting that we might hear something in the afternoon if Jupiter's magnetosphere was dramatically impacted [by comet fragments], especially in the higher frequencies. We happened to be sitting around the receivers, locked on fixed frequencies, when the spectrum analyzer lit up like a Christmas tree and a low rumbling came over the receivers for forty minutes. It startled us enough that we went outside looking for the blimp overhead. . . . We will not speculate on the origin of this signal (leaving it to professionals) but . . . one can only ask the question, "Is there seismic information embedded in these waves?" We have the data [but] will leave the theoretical explanations to the professionals. . . . As hunters we have brought home some interesting data; now all we have to do is explain it.[15]

Which is often the case with amateurs. Having made competent observations, they may require professional help in interpreting their data. But while amateur radio astronomy emails are fraught with naive questions, they can also be refreshingly unpretentious. "Don't forget, tune in to WJUP—the big red spot on your dial," advertises one email spam, while another offers "fresh Jupiter/Io data, served up on the Internet."

Observing Jupiter's tantalizing and rewarding array of violent yet austerely beautiful weather patterns and the endless dance of its satellites is reason enough to get a telescope. But, as it happens, Jupiter is not alone in the ice palace of the outer solar system. There are three more giants beyond it.

Storms on Saturn:
A Visit with Stuart Wilber

GUILELESS AS BILLY BUDD, Stuart Wilber is a gentle guy who says what he feels. As we toted his telescope out behind his home in Las Cruces, New Mexico, on a warm evening in 1992—the ruddy distant hills reminding me of Mars—his four-year-old daughter, Rigel, named for the brightest star in Orion, was playing in the grass. Watching her, Stuart whispered, "Isn't life amazing?"

Our target was Saturn. The telescope was a ten-inch F/7 Newtonian that Stuart had made by hand, writing a computer program to test the figure of the mirror as he ground it. Its high-gloss white tube was mounted on a streamlined Dobsonian-like box mount made of padouk, a fine African hardwood, with black walnut trim. "Let us glean from the night whatever we can," Stuart said quietly. "It's remarkable what you can do in your back yard if you have a little optical aid. I just love to look at the sky."

At Stuart's invitation I peered through the eyepiece and watched Saturn drift rapidly across the field of view. (The telescope, built on a tight budget, had no clock drive to compensate for Earth's rotation.) The rings were razor-sharp, the ball of the planet alive with festoons of sand, plum, and auburn, streaming like women's hair in a windstorm. In ordinary telescopes, planets look like pictures. In a superior telescope like this one, they look like worlds.

Stuart, a part-time math teacher at a local community college, was barred from teaching full time because he lacked a master's degree, and could not afford the tuition for more than about two courses a year in pursuit of it. "I'm studying vector equations," he said, with the enthusiasm of a hungry man ordering a steak. "Pretty soon I'll be able to tackle the Bessel functions."

At 8:30 P.M. Mountain Time on September 24, 1991, Stuart was scrutinizing Saturn through his telescope at a magnification of 300x when he noticed "a white pinprick of light" near the center of the planet's disk. He called his wife out to the yard and asked if she saw anything unusual on the planet. "I see a white spot," she said, and she identified the location. Stuart, satisfied that "I wasn't just seeing things," phoned his neighbor Clyde Tombaugh, the discoverer of

Pluto. Tombaugh suggested that Stuart contact Reta Beebe, an astronomer at New Mexico State who specialized in the giant planets. At her instigation Scott Murrell, an observer at the university's 24-inch telescope at Tortugas Mountain Observatory, had a look.

"Scott got there a little late," Stuart recalled. "By that time the spot had rotated out of view." A day on Saturn lasts 10.2 hours, a spot that rotates to the other side of Saturn on a given night usually won't be back till after sunup the next morning, and by sundown the next day will have gone round to the far side again. The upshot is that Saturn observers live on three-day cycles. "Three days later Scott called and said, 'Yes, your spot's coming around,'" Stuart told me. "I must have caught it just when it was beginning to emerge."

Saturn is a ball of gas and liquids, massive as ninety-five Earths but so low in density that it could float in water. It is the second largest of the Sun's planets, after Jupiter, with a diameter of 75,000 miles, more than nine times Earth's. Vast turbulences roil in the dark depths that extend from its small, rocky core to the surface. Once in a long while gigantic bubbles form in the depths—held there, perhaps, by an increase in the opacity of the cloud tops—then burst through into view, rising like thunderheads, capped with gleaming white ices of ammonia and water. One such great white spot was observed in 1933 by Will Hay, the English film, radio, and music-hall comedian. Another was spotted in 1960 by the South African amateur astronomer J. H. Botham. Stuart Wilber had discovered the third.

The spot, less than ten thousand miles in diameter when Stuart saw it, soon hit the high winds in Saturn's upper atmosphere and blossomed into a long, conspicuous oval. Astronomers all over the world began training their telescopes on it. Reta Beebe contacted colleagues working with the Hubble Space Telescope, whose director immediately freed time for *Hubble* to make images of what was being called the Wilber White Spot.

"That data set is still being used for analysis of the equatorial region of Saturn," Reta Beebe told me nearly nine years later. "We're now working on how the atmosphere recovers from the storm. The combination of amateur with *Hubble* observations enabled us to get significant data. Since there have been three storms of this magnitude on Saturn in its recorded history, occurring at intervals of roughly fifty-seven years, it appears that the period of buildup, storm, and recovery is about that long. Saturn's orbital period is thirty years, so it may be that such storms occur about every two Saturnian years. . . . We don't really know. After all, we've only been watching Saturn for about four Saturnian years.

"I've grown quite dependent on amateurs to monitor major storms on the giant planets, because such storms are readily observable in small telescopes," she added. "Amateurs typically own telescopes in the range of six to twelve inches, and they have a high probability of being able to see the planet through a turbulence bubble in Earth's atmosphere. Given the rapid response of the human eye, it's hard to beat what an amateur observer may see in an instant, and an instant is all that an observer may need to discern that there's something strange going on.

"When Stuart discovered the white spot, Saturn was only fifteen degrees above the western horizon. That's not a great place for professionals to be making observations, because there's so much of Earth's atmosphere in the way for objects so low in the sky, but it's a great place for the amateurs, who enjoy the beauty of the planet emerging just after sunset. I consider theirs to be very valid observations. When I'm working on a particular problem, I calmly assume that the amateur crowd is out there, watching the other planets for me."[1]

14.

The Outer Giants

Spring night—one hour worth a thousand gold coins . . .
—Su Tung-P'o

The hidden harmony is better than the obvious one.
—Heraclitus

O N A WET DECEMBER EVENING in 1997, I set up a small telescope on the deck of my house in San Francisco to watch the Moon pass in front of Saturn. This is an old pastime for human stargazers, dating back to at least 650 B.C., when a cuneiform inscription by Babylonian astronomers reports that "Saturn entered the Moon."[1] Tonight's occultation was predicted for 11:18 local time. It looked like Saturn would disappear behind the house before then, but I couldn't climb the vertical steel ladder to the roof while carrying the telescope, and it was too slippery to ask my wife or son to help, so I left the telescope where it was, showed my son Saturn through it, and described the occultation that I hoped to observe after he'd gone to bed. His response was pleasingly undoctrinaire: "What would we think if Saturn passed in front of the Moon instead of the other way round?"

As the appointed time drew near, the Moon and Saturn had indeed disappeared behind the house, but by jamming the telescope's tripod into the northeast corner of the deck and extending its legs to their maximum height, I managed to sight through a tunnel between the roof and the windblown branches of a tree. Although a nearby streetlamp flooded the scene, by pressing close to the eyepiece I could shut out most of the stray light.

There, sure enough, was Saturn with its incomparable rings, the moist air turning its color to that of an old doubloon seen through brown river water, with the stately cold white gibbous Moon almost on top of it. Moments later, the western end of the rings started to be eaten away by the total blackness of

the lunar dark side. Soon the disk, too, was being bit into, like a vanilla tea bis-
cuit, and was gone. Then the last of the rings was swallowed up, and there was
nothing to be seen but the Moon. I looked at the watch: It had happened right
on time. How Kepler would have envied modern astronomical computation.

AS I WRITE THIS, on a chill November evening in the year 2000, the giants
are in the sky. Out the window I can see Saturn, glowing a dull tarnished
bronze, with brighter Jupiter climbing in the east just twenty minutes behind
it. (Such a close conjunction of the Sun's two largest planets occurs only about
once every twenty years. I was a high school student when I first saw one, a
thirty-seven-year-old bachelor when the next one rolled around, and now I'm
an emeritus professor with a wife and son. If I survive to see another conjunction of Jupiter and Saturn I'll be pushing eighty.) Uranus and Neptune are
sinking in the west tonight, their slow-grinding clockwork reminding me that
Uranus, the god of heaven, was the father of time. Saturn takes almost thirty
years to complete one orbit of the Sun, Uranus eighty-four years, and Neptune 165 years.[2] Things are starting to get genuinely far away, now, as we enter
realms where the little compass of human life is boxed within ever more commodious frameworks of time and space.

The spectacle of Saturn seen through a telescope has turned more people
into stargazers than any other—or such is my impression, from asking amateur and professional astronomers over the years about how they got interested in the night sky. Reactions to seeing Saturn's rings for the first time tend
to run along the lines of "Oh, my God!" "Amazing!" and "Is that *real*?" Some
call Saturn too good to be true, whatever that may mean.

Like Jupiter, Saturn is a liquid planet, but somewhat smaller—its mass is a
third of Jupiter's—and only half as dense. Hence its rotation rate, although
comparable to Jupiter's, produces an even more oblate shape. The inference for
Saturn's internal structure is that it has proportionately less metallic hydrogen and more molecular hydrogen than Jupiter does. Like Jupiter, Saturn generates more heat internally than it receives from the Sun, but because it is less
massive and almost twice as far from the Sun, Saturn is a colder planet.

Seen through a telescope, Saturn exhibits belts, zones, and anticyclonic
storms, as Jupiter does, but these features are more muted in appearance.
They are sunk in a colder upper atmosphere, where frozen ammonia crystals in
the main cloud layer obscure our view of clouds below, and Saturn's greater
distance also degrades our view of them. Were it not for the rings, Saturn
would be just a smaller, dimmer, and more subtle version of Jupiter, a connois-

seur object left principally to the professionals and to advanced amateurs eager for a challenge.

But, of course, Saturn does have rings—a spectacular golden array of rings that stretch as far from edge to edge as the distance from the Earth to the Moon, yet may be no thicker than a coconut palm is tall.[3] Earthbound observers see three main ones, designated—from the outside in—as A, B, and C. The bright A and B rings are separated by a sharp black line, the Cassini division. These features are readily visible in small telescopes, but the C ring, dubbed the "crepe" ring by William Lassell when he observed it with William Dawes's refracting telescope in England on December 3, 1850, is more elusive: To see it usually requires steady air, a practiced eye, and a telescope of at least six inches aperture.[4] In addition to the Cassini division, observers over the centuries reported discerning various other gaps in the ring system, and occultation observations of the winking and blinking of starlight when the rings moved across a star confirmed that the rings were complicated. As the Voyager 1 space probe approached Saturn in late 1980, some thought there might be as many as several dozen undiscovered rings and divisions. Instead, the *Voyager* pictures revealed thousands of them. There are rings outside the A ring, hundreds of rings and gaps within those visible from Earth, eccentric rings, and even a "braided" ring that is twisted upon itself.

A multitude of studies, ranging from space probe images to the bouncing of radar off the rings, indicates that they are composed of ice, rock, and dirt—that is, that they are made of "dirty snowballs" rather like tiny comets, with diameters typically of only about ten centimeters and generally no more than five or ten meters. Early observers had theorized that the rings were vaporous or liquid. The idea that they are instead composed of many small orbiting bodies was first proposed by the French poet Jean Chapelain, in 1660. Cassini's discovery of the gap in the rings that now bears his name added weight to Chapelain's hypothesis, and the architect Christopher Wren eventually abandoned his own corona theory in favor of it, declaring that "the Ring of Saturn is a number of small moons like a swarm of Bees."[5] The Scottish physicist James Clerk Maxwell laid to rest the alternative theories in 1857 by demonstrating that the rings would be unstable if made of anything other than many small bodies. James Keeler at the Allegheny Observatory in Pittsburgh provided observational evidence in support of Maxwell's calculations in 1895 by taking spectra that showed the rings rotating differentially, the outer ones moving more slowly than the inner ones, as one would expect of millions of moonlets.

The intricate structure of the ring system is thought to be generated by gravitational resonances set up between the ring particles, nearby satellites, and Saturn's bulging equator. Resonances occur when the orbital periods of two bodies form an integral ratio, like 2:1 or 5:2. In such cases, repeated close encounters between the orbiting bodies function rather like pushing a child's swing at the apex of each arc to pump it into a wider arc, and the effect is to carve gaps in Saturn's ring system. The Cassini division is one such gap, generated by a 1:2 resonance with Saturn's moon Mimas. Other gaps are generated by resonances with the small moons Janus, Epimetheus, Pandora, and Prometheus.

The rings lie within the Roche limit—the distance from a given planet inside of which tidal forces will overcome a moon's internal binding energy and tear it apart.[6] This fact leads to two general theories of how Saturn's rings originated. In one, the rings—which have a total mass approximately equal to that of a medium-size moon—formed when Saturn itself did, out of material that was unable to congeal into a satellite because it lay inside the Roche limit. If so, the rings are ancient. In the other scenario, the rings formed later on, when an existing satellite wandered too close to Saturn and was dismantled. In that case, the rings could be much younger than Saturn is. We do not yet know which, if either, of these theories is correct.

Saturn's polar axis is tilted 26.7 degrees relative to the plane of its orbit, as is the ring plane. Combined with the fact that Saturn's orbit is itself inclined by 2.5 degrees relative to Earth's orbit, this means that over the long course of a Saturnian year the rings present themselves to us at a variety of angles, ranging from an open view, inclined at over 26 degrees, to an edge-on view when the rings disappear entirely through all but the largest telescopes. The last such "ring-plane crossing" occurred in 1996, and the next one is due in the year 2011. Galileo had the misfortune to first observe Saturn telescopically near a ring-plane crossing, in July 1610, and was confounded by what looked like a pair of handles. He diverted his attention to Jupiter's moons, and upon returning to Saturn in autumn 1612 was startled to find that the "handles" had disappeared. Ring-plane crossings have offered an excellent opportunity to search for previously undetected, dim satellites of Saturn, absent the glare of sunlight reflecting from the rings—which is how Cassini discovered four satellites of Saturn, and William Herschel two more. In all, thirteen of Saturn's moons were first spotted during ring-plane crossings.[7]

The dusky, hourglass-shaped "spokes" that sometimes appear on Saturn's rings are often said to have been discovered by *Voyager*, but as we have noted they were observed from Earth long before that. Stephen James O'Meara's independent discovery of the spokes in the mid–twentieth century was preceded by the nineteenth-century observations of Eugene Antoniadi and E. L. Trouvelot. The spokes, which probably consist of meteoric dust particles levitated by the planet's magnetic field, are suitably Saturnian in scale: The large ones are as wide as Brazil and long enough to go halfway around the Earth. Saturn emits bursts of kilometer-wavelength radio energy in cycles that last 10.66 hours, approximately the same period during which the spokes make an appearance.

Saturn's eight "classical" satellites, the ones discovered before the age of spaceflight, pursue orbits aligned along the ring plane and with radii of up to nearly five times the rings' diameter, so that when the rings are visible the outer satellites are often so far-flung that inexperienced observers mistake them for background stars.[8] The outermost three, Iapetus, Hyperion, and Titan, have orbits with more than twice the radii of their sister satellite Rhea, inside of which orbit Dione, Tethys, Enceladus, and Mimas.[9]

Titan, nearly as large as Ganymede and almost twice the mass of Earth's Moon, is the only satellite in the solar system that has a substantial atmosphere. Titan was discovered by the astronomer and mathematician Christiaan Huygens in 1655, when he was twenty-six years old. Huygens was a skilled inventor, who with the help of his brother Constantyn built a refracting telescope twelve feet long and capable of magnifying objects fifty times. (Telescopes in those days typically had rather small objective lenses and were designated by their tube length or focal length rather than aperture.) With this telescope, Huygens on March 25 observed two "stars" near Saturn and noted their position. On the next night he found that one of them had moved, and subsequent observations confirmed that it was orbiting around Saturn. "Its greatest digression [i.e., distance from Saturn] was seen to be a little less than three minutes," Huygens recorded in his journal, later recalculating the distance as 3 minutes 16 arc seconds.[10] I checked Huygens's observation on a computer program and was able to confirm that his positions for both Titan and the background star (SAO 99279, evidently) were admirably accurate.

Although Titan looks tiny in a telescope, acute observers were able to glimpse signs of its atmosphere long before the Voyager mission verified its existence. Arthur Stanley Williams, a British amateur astronomer who conducted many valuable Saturn observations, scrutinized Titan through a 6.5-inch reflecting telescope on April 12, 1892, during one of the approximately four-year periods when Saturn's orientation permits us to see its moons silhouetted against its disk, and reported that Titan "was darkest in the middle part, fading off somewhat on the edges, where it was nebulous looking, not sharp and clean-cut."[11] Sixteen years later, in 1908, the Spanish astronomer

Jose Comas Sola similarly reported having observed limb-darkening on Titan, and suggested that it was due to an atmosphere. These intimations were confirmed in 1944, when Gerard Kuiper took spectra of Titan and discovered methane gas there. In November 1980, *Voyager 1* took a close look at Titan (which the space scientists regarded as so interesting that they sent the space probe out of the ecliptic, forsaking its opportunity of going on to Uranus and Neptune, just to fly it over Titan) and found that its reddish-orange atmosphere is so opaque that the surface is seldom if ever visible.[12]

Titan is cold—something like −290°F at the surface—but in other respects it resembles the infant Earth, and consequently is of great interest to scientists investigating the origin of life. The Earth when life began was no vacation spot; were humans around back then, they would have found its poisonous atmosphere of hydrogen, ammonia, and methane—all present on Titan, by the way—to be extremely hostile. Christopher Wills and Jeffrey Bada describe such a visit in their book *The Spark of Life:*

> You must wear a space suit, for there is no oxygen in the atmosphere. And the suit will have to be Teflon coated, because nasty gases present in the thick air, such as hydrogen sulfide and the vapors of sulphuric and hydrochloric acids, would certainly kill you very quickly by eating through any protective suit made of ordinary chemicals. Can you see your surroundings? Not very well—the atmosphere is so smoggy that little or no light can penetrate to the surface, even at high noon. But lightning flashes do give you brief, lurid glimpses of your immediate neighborhood. And you might be able to see the red glow of a nearby active volcano.[13]

As is said of legislation and sausage-making, the outcome of the creation of life may be more appealing to contemplate than the process that goes into it.

The only other moon of Saturn about which much was known before the arrival of *Voyager* was Iapetus. Cassini, who discovered Iapetus, noted that it has the odd property of looking much brighter when on the west than when on the east side of Saturn. "He becomes visible in his greatest Occidental digression, and invisible in his greatest Oriental digression," Cassini wrote in 1672, with his characteristic combination of reliable content and florid style.[14] Today, thanks to spacecraft imaging, we know that Iapetus, 1,460 kilometers in diameter, is gravitationally locked—so that it always presents the same face to Saturn, as does the Moon to Earth—and is covered with material as black as soot on one hemisphere, a dichotomy that produces the dramatic differences in brightness that Cassini observed. The "soot," which lies on the bow side in terms of Iapetus's orbital motion, may be dust dislodged by meteors from the surface of Phoebe or another small outer satellite of Saturn. The rest of Iapetus reflects light brightly, leading the astronomer David Morrison and colleagues to conclude that "the zebra is revealed as a white animal with dark stripes, not the reverse."[15] Iapetus gets six times dimmer when east of Saturn, a phenomenon readily observable in amateur telescopes for those who care to follow it. Rhea, almost a twin of Iapetus, also displays some asymmetry in surface brightness, as does Dione.

The innermost bright satellites, Mimas and Enceladus, are small (their diameters are 394 and 502 kilometers respectively) and neighboring (their orbits are only 52,000 kilometers apart) but otherwise strikingly distinct from each other. Enceladus has a complex surface, part cratered terrain and part smooth, grooved plains—five distinct geologies in all. Yet Enceladus reflects light evenly from all these surfaces, suggesting that something may have redistributed ice across its globe, polishing it like a marble. Saturn's tenuous E ring—the outermost ring system—lies nearby, and shows an intensity peak near the satellite. Possibly the event that "polished" Enceladus also sprayed ice particles into neighboring space, creating the E ring.

Mimas, on the other hand, sports a fresh-looking impact crater fully 125 kilometers in diameter. Given that Mimas itself is less than 400 kilometers wide, the crater looks nearly as large in proportion as a pupil in a human eyeball, indicating that the impact that created it was almost robust enough to blow this little moon apart. If the rings are made of satellites that formed and subsequently were destroyed, Mimas is an example of one that barely survived.

Of interest to students of chaos is Saturn's satellite Hyperion. This potato-shaped moonlet, only 190 kilometers long by 114 kilometers wide, pursues an eccentric orbit between Titan and Iapetus. Probably a fragment of a larger, destroyed satellite, it tumbles wildly, its rate of spin and the direction of its polar axis changing with every few recurrences of its twenty-one-day orbit. These changes, which may have been set in motion by a collision and subsequently pumped by interactions with Titan, are inherently unpredictable: Forecasters on Hyperion would find it as difficult to predict how long the next day would be, or what star would next mark its north celestial pole, as terrestrial weather forecasters do in predicting the behavior of hurricanes.

Beyond the ice palace of Saturn with its moons and rings lie the still colder depths where Uranus, Neptune, and Pluto dwell.

Uranus is bright enough to be seen with the unaided eye—just barely so—

yet no pretelescopic observer seems to have noticed its regular wanderings among the stars. Then, on the clear night of March 13, 1781, William Herschel spotted it while charting the heavens from behind his house in Bath, with a six-inch telescope he had built by hand. Although Herschel was no planet hunter—his interest was in charting the heavens as a whole—he discerned that the object, unlike a star, increased in apparent diameter with increasing magnification.[16] He recorded in his journal that "in the quartile near [Zeta] Tauri the lowest of the two is a curious either Nebulous Star or perhaps a Comet."[17] When he returned to the object four nights later, Herschel confirmed that it had indeed moved, as one would expect of a comet, although it displayed no tail and moved very slowly—so slowly that Charles Messier, who knew his comets, was surprised that Herschel had even noticed that it changed position. Nevil Maskelyne, the Astronomer Royal, observed it three weeks later and considered that it might be a planet, since it was "very different from any comet" in his experience. Johann Bode checked historical observing records and found that Uranus had previously been charted by John Flamsteed, Johann T. Mayer, and James Bradley, among others, but that all mistook it for a star. Bode pronounced it to be "a hitherto unknown planetlike object circulating beyond the orbit of Saturn."[18] Herschel, his fame assured by his having made one of the great amateur discoveries in astronomical history, was soon propelled into a professional career.

Neptune, a near twin of Uranus but half again as distant from us and hence smaller and dimmer in the sky, was discovered half a century later and in a more modern way—by inferring its presence through the gravitational perturbations it causes in the orbit of Uranus.[19] The story of its discovery interweaves prodigal brilliance and plodding obduracy, sometimes within the same individual.

Its central figure is John Couch Adams, who while still an undergraduate at Cambridge University resolved to calculate where astronomers might find the hypothetical eighth planet from the Sun, the existence of which had been proposed in a general way by several researchers to explain observed oddities in the orbit of Uranus. Adams managed to do most of these calculations by the time of his graduation in the class of 1843, and in 1845 he called on the Astronomer Royal, George Airy, to present his prediction of the position of the theoretical "trans-Uranian" planet. Adams assumed that English observers, armed with his information, would look for the new planet, but Airy was methodical to a fault. (Unremittingly compulsive, Airy required observers to stay up through cloudy and even rainy nights, and would wander from dome to dome calling out, "You are there, aren't you?" The British astronomer Patrick Moore endorses "the tale that he once spent a day in the Greenwich cellars labeling empty boxes 'Empty.'")[20] The first time Adams turned up, Airy was out of town. The second time, Airy was out on an errand of some sort. Adams left a card and returned that afternoon, only to be told that Airy was having dinner and was never to be disturbed while dining. Adams left a letter, to which Airy replied, challenging a technical issue in Adams's calculations and generally pouring cold water on his claim that they pointed the way to an eighth planet.

Meanwhile, in France, similar mathematical calculations were being made by Urbain-Jean-Joseph Leverrier, a brilliant but legendarily irascible chemist turned astronomer. (In 1854 Leverrier became director of the Paris Observatory, only to be dismissed for "irritability." Moore notes that a contemporary "made the acid comment that although he may not have been the most detestable man in France, he was certainly the most detested.")[21] Leverrier published his calculations, which put the putative planet in almost precisely the same point of the sky as Adams's did, in 1845, while Adams was getting nowhere with Airy. Airy read the paper, and was persuaded by the work of the professional Leverrier although he'd dismissed the almost identical work of the amateur Adams. Without saying anything to Adams—even, it appears, when the two met by chance on July 2, walking across St. John's Bridge in Cambridge—Airy organized a search for the planet in the section of sky specified by both Adams and Leverrier. Unfortunately, he assigned the task to a plodder, James Challis, who set about methodically mapping dim stars at high magnification. "I get over the ground very slowly," Challis reported, gloomily but accurately, of his tortoiselike pace.[22]

Back in France, Leverrier on August 31, 1846, published another memoir, predicting that the putative planet should currently be located about five degrees east of the bright star Delta Capricorni. The potential discovery seemed so close that one could almost taste it: As the astronomer John Herschel wrote in a letter that was to be published that fall, "We see it [the new planet] as Columbus saw America from the shores of Spain. Its movements have been felt trembling along the far-reaching line of our analysis, with a certainty hardly inferior to that of ocular demonstration."[23] But if the "detested" Leverrier thought that French astronomers would pay him any heed and have a look east of Delta Capricorni, he was mistaken. Ignored by his countrymen, Leverrier handed the project over to the Germans, writing to Johann Gottfried Galle at the Berlin Observatory on September 18 and urging him to look for the planet. Unlike the plodding Challis, Galle actually had a star map—astoundingly, many major observatories at that time did not—and he promptly set to work, aided by an enthusiastic young student named Heinrich Louis d'Arrest. Galle, at the telescope, called out the positions of stars and d'Arrest checked them against the star map. Within minutes they came upon an 8th-magnitude object that was not on the map. The next night, September 24, 1846, they had another look and confirmed that it had moved. "The planet whose position you have pointed out actually exists," Galle wrote to Leverrier.[24] Today, in recognition of the prescience of their calculations, the young Adams and the regal Leverrier are generally credited with discovering Neptune, rather than Galle and d'Arrest, who were the first to see it.

A footnote to this story remains as a cautionary admonition to professional and amateur stargazers everywhere. The lackluster Challis, observing six days after Galle and d'Arrest had found Neptune and as yet unaware of their discovery, stumbled on the planet for himself and even had his assistant note in their log that "it seems to have a disk." But he was working at a magnification of only 116x and, incredibly, did not bother to insert a more powerful eyepiece to check whether the disk was real, as William Herschel had done on the night he discovered Uranus. Challis also neglected to follow up on it on subsequent nights, so his chance to detect its motion and make an independent discovery of Neptune was lost. Later he blamed his failure on his "impression that a long search was required to ensure success."[25]

Scholars looking over old observing logs have found that, as with Uranus, many earlier observers had spotted Neptune without recognizing that it was a planet. The Scottish astronomer John Lamont—known in Germany, where he worked, as Johann von Lamont—recorded the location of Neptune on three nights in 1845 and 1846 but never combed his observations for evidence of moving "stars." J.-J. de Lalande in France observed Neptune on two nights in May 1795, and saw that it seemed to move, but concluded that his first mapping of it must have been in error. Even Galileo saw Neptune. While observing Jupiter on December 27 and 28, 1612—just a week before Jupiter actually occulted Neptune, a spectacular event witnessed by nobody—Galileo recorded a "star" at Neptune's position. A month later, he again charted Neptune's approximate position, near the 7th-magnitude star today designated as SAO 119234, in Virgo, and on the following night noted that the two objects "seemed farther apart from one another."[26] But he did not follow up on the clue. Why? Perhaps because the nearly full Moon was close by, its glare making it difficult to observe an 8th-magnitude object through so small a telescope. Very often, when using a telescope, what we see—or don't see—is to a dangerous degree a product of our expectations.[27]

Uranus and Neptune turned out to be somewhat disappointing objects for telescopic observation. Although they are giant planets—with volumes, respectively, of sixty-six and fifty-eight Earths—they are so far away that most land-based observers perceive them as little more than featureless disks. (To make Uranus look the same size through a telescope that the Moon does to the unaided eye requires a magnification of 500x; for Neptune, 750x. Powers this high require steady seeing conditions, since they magnify turbulence in the air as well as the image passing through the air.) Both planets have dark ring systems, imaged by *Voyager* but detectable through terrestrial telescopes only when they occult stars.

Uranus has five satellites bright enough to be seen from Earth, of which two—Oberon and, closer in, Titania—were discovered by Herschel in 1787, six years after he found Uranus itself. As usual, Herschel was "sweeping" and charting the skies, this time with his 18.2-inch reflecting telescope. On January 11, he writes:

I selected a sweep which led to the Georgian planet [as Herschel originally called Uranus, after his patron, King George III]; and, while it passed the meridian, I perceived near its disk, and within a few of its diameters, some very faint stars whose places I noted down with great care. The next day, when the planet returned to the meridian, I looked with a most scrutinizing eye for my small stars, and perceived that two of them were missing. Had I been less acquainted with optical deceptions, I should immediately have announced the existence of one or more satellites to our new planet; but it was necessary that I should have no doubts. The least haziness, otherwise imperceptible, may often obscure small stars; and I judged, therefore, that nothing less than a series of observations ought to satisfy me, in a case of this importance. To this end I noticed all the small stars that were near the planet the 14th, 17th, and 24th of January, and the 4th and 5th of February; and though, at the end of this time, I had no longer any doubt of the existence of at least one satellite, I thought it right to defer this communication [that is, the report to the Royal Society from which this quotation is taken] till I could have an opportunity of seeing it actually in motion. Accordingly I began to pursue this satellite on Feb. the 7th, about six o'clock in the evening, and kept it in view till three in the morning of Feb. the 8th; at which time, on account of the situation of my house, which intercepts a view of part of the ecliptic, I was obliged to give over the chase:

and during those nine hours I saw this satellite faithfully attend its primary planet, and at the same time keep on, in its own course, by describing a considerable arch of its proper orbit.[28]

Herschel's account displays the hallmarks of a great astronomical observer
at work. First, he was acute enough to see the satellites in the first place—two
dim dots entangled in the glare of their planet, a sighting so difficult that no
other astronomer spotted either of them for fully ten years thereafter. Then he
was persistent, checking back on five other nights to make sure the satellites
were still there. Finally, although his own doubts had been laid to rest, he
devoted nine full hours, in the dead of an English winter, to tracking the
brighter satellite in real time, confirming that it both moved along with
Uranus against the background stars and described an orbital arc while doing
so. Any telescope user who has tried to concentrate on one object for a fraction
of this time can testify to just how exhausting such an effort can be. In the
process, Herschel also estimated the orbital periods of both satellites, obtaining 13.5 days for Oberon and 8.75 days for Titania—admirably close to the
correct values of 13.462 and 8.709 days.[29]

Although Uranus and Neptune are superficially twin planets, they are different enough to remind us—as do Venus and Earth—that we still have a lot
to learn about the mix of natural laws and historical accidents that formed the
planets and fashioned their destinies. The natural laws are complex, involving
the effects of gravitation and electrostatic fields on the cocktail of dust, ice,
and gas that composed the primordial solar nebula. The historical accidents
doubtless were complex, too. We don't yet know much about them, but it is
clear that an important role was played by impacts.

The polar axis of Uranus is tilted almost 98 degrees from perpendicular.[30]
Its satellites and its coal-black rings are tilted, too, suggesting that Uranus
was struck by a large planetoid while it was still forming and was knocked over
onto its side, like a galleon hove to on a beach. Had the satellites and rings
formed before the impact occurred, it is possible that a tilting Uranus, by
virtue of its equatorial bulge, could have tugged them along, but highly
unlikely that they would have remained so precisely aligned with its equator.
Uranus generates less internal heat than do Neptune, Saturn, and Jupiter, so
perhaps the impact mixed up its internal materials, homogenizing them so
that they released a lot of heat in a hurry rather than simmering on for eons.
We don't really know.[31]

Particularly dramatic evidence of the violent past may be found at
Miranda, a small (diameter 500 kilometers) satellite of Uranus that has been

called the weirdest object yet observed in the solar system. *Voyager* images show that Miranda's surface is a strange jumble of old, cratered terrain and newer material laid down haphazardly, as in the aftermath of a demolition. Perhaps Miranda was struck by an asteroid-class intruder so violently that it was blown to bits, but just gently enough so that many of the pieces, having not quite attained escape velocity, were able to reassemble themselves into the jumble we see there today.

Neptune, by contrast, seems relatively serene. Its axial tilt of 28 degrees and its rotational period of sixteen hours are unexceptional for a Jovian planet, and its almost perfectly circular orbit (eccentricity only 0.0097) indicates that it has never suffered a close encounter with another planet or a major chaotic event such as the passage of a star through the solar system. Calmly patrolling the outposts of the planetary system over the course of each 164.8-year Neptunian orbit, it answers well to its depiction by a tranquil, ethereal choir in the closing theme of Gustav Holst's suite *The Planets* (which otherwise evokes nothing of the planets as physical worlds, since Holst was concerned with astrology and mythology, not astronomy).

Through a telescope, Neptune looks nearly as featureless as Uranus but even smaller: Half again as far from us as Uranus is, Neptune displays a disk only 2.35 arc seconds wide. Many observers assumed that it closely resembled its putative twin, but when *Voyager 2* sped through the Neptune system on August 25, 1989, the pictures and data it transmitted back to Earth demonstrated once again that however similar planets may be in their gross characteristics, when examined in detail each proves to be unique. Whereas Uranus is almost featureless, Neptune displays bands and zones like those of Jupiter and Saturn, albeit with the greater subtlety suitable to a colder world. When *Voyager* visited Neptune, it even had a Great Dark Spot—an anticyclone comparable to Jupiter's Great Red Spot, adrift at a similar latitude in Neptune's southern hemisphere. When the Hubble Space Telescope took the next high-resolution images of Neptune, in 1994, the Great Dark Spot was gone and a new spot had appeared in the northern hemisphere. The agency responsible for these features presumably is heat generated from the core—2.6 times the energy Neptune receives from the Sun—which as we have seen is characteristic of the Jovian planets generally but absent in the case of Uranus. Uranus is greenish in hue, Neptune more blue: The colors of both planets have to do in part with the absorption of red light by methane gas, but it is not clear why Neptune is bluer than Uranus.

Neptune's icy satellite Triton is a strange, vividly colored world, adorned with splotches of red, pink, and blue-gray. It was determined by *Voyager* scientists to possess nitrogen geysers that blast material nearly ten kilometers into the "air"—which, in the case of Triton, is a thin fog of nitrogen and hydrocarbon gas—producing debris clouds that fall to the surface far downwind. (Triton's atmosphere, observed by *Voyager*, has also been detected by earthbound observers, during occultations of stars by Triton.) The geysers are thought to erupt when liquid nitrogen trapped beneath the surface percolates upward, decreasing in density, and turns into an expanding gas. This sort of behavior is usually found not among the icy satellites of the Jovian planets but on comets and Kuiper belt objects. Considering that Triton's low density resembles that of comets rather than Jovian satellites, and that its orbit is both retrograde and rather steeply (21 degrees) inclined to Neptune's equator, it seems likely that Triton is not a native Neptunian moon at all, but a captured visitor from the outer solar system.

Which brings us to Pluto, the one remaining planet—if it is a planet— still unexplored by any spacecraft. Pluto was discovered in 1930 by a young Clyde Tombaugh, a former amateur astronomer just off a Kansas farm who, although unable to afford a college education, had managed to land a job at Lowell Observatory right out of high school. As a skilled photographer, Tombaugh was assigned to search for "Planet X," the existence of which had been inferred from studies of the motions of Uranus and Neptune. Those inferences later turned out to have been wrong, but fortunately Tombaugh relied more on persistence and common sense than on sophisticated mathematics: Rather than concern himself with where Planet X was predicted to be, he proceeded in a fashion similar to that of the drunk who looks for his missing keys under a lamppost because that's where there's light enough to see by. Since all the known planets orbited on or near the ecliptic, and since any planet is brightest when in opposition, he set to work making time-exposure photographs of the part of the sky on the ecliptic opposite the Sun.

Thirteen constellations lie along the ecliptic—the twelve "signs" of the "Zodiac," plus Ophiuchus, which astrologers ignore since they regard thirteen as an unlucky number. Tombaugh made some photos in Gemini while testing the telescope and camera assigned to him, but by the time he started searching in earnest at the next dark of the Moon, Gemini was far west of the meridian, so he began his survey in Cancer instead. Murphy's law was in effect: Pluto was in Gemini. Tombaugh spent nearly a year working his way around the ecliptic, comparing thirty plates every lunar month, each of which contained the images of tens of thousands to hundreds of thousands of stars, looking for the one slowly moving point of light that could be a planet. He admitted that it was "grim" work, but he gained a lot of experience in this literally roundabout way, photographing twenty-nine thousand galaxies and eighteen hundred variable stars and discovering two comets.

"I was a perfectionist," Tombaugh recalled. "When I planted the kafir corn and milo maize [staples at the family farm] the rows across the field had to be straight as an arrow or I was unhappy. Later, every planet-suspect, no matter how faint, had to be checked out [by inspecting] a third plate—either yes or no, not maybe."[32] When the day came—February 18, 1930—that Tombaugh did at last find Pluto, he quietly rechecked his work and then, concealing his excitement, sauntered into the director's office and announced, "I have found your Planet X."[33] The Lowell astronomers initially downplayed young Tombaugh's contribution (their telegram to Harvard didn't mention his name, and the circular they distributed to the press stressed "the search program set going in 1905 by Dr. Lowell in conjunction with his theoretical work on the dynamical evidence of a planet beyond Neptune" while devoting only one sentence to the role played by "C. W. Tombaugh, assistant on the staff"), but his hard work would eventually cement his fame: Had he photographed Pluto at the very outset of the project, he might thereafter have borne the stigma of someone whose greatest accomplishment could be written off to beginner's luck.[34]

The world was galvanized by the news, and Lowell Observatory was inundated by suggested names for the new planet—among them Apollo, Artemis, Atlas, Bacchus, Cronus, Erebus, Idana, Osiris, Perseus, Tantalus, Vulcan, and Zymal. (Lowell's widow, Constance Lowell, preferred Zeus, Lowell, Percival—or Constance.) The name Pluto was suggested by Venetia Burney, an eleven-year-old schoolgirl in Oxford, England.[35] It helped that the first two letters are Percival Lowell's initials.

Although the discovery of Pluto is still sometimes cited as a monument to the calculations of theorists, the mathematics that inspired the search for Planet X in the first place were based on incorrect values for both the masses of Uranus and Neptune and their locations—and even if they had been right, we now know that Pluto is too tiny a world to have produced the claimed effects on their orbits.[36] Pluto was discovered not by mathematicians' inspiration but by Clyde Tombaugh's perspiration.

The case against categorizing Pluto as a planet is based on its small size, its tilted, eccentric orbit, and its composition. Only about 2,300 kilometers in diameter, Pluto is smaller than any planet—smaller than Neptune's satellite Triton (diameter 2,700 kilometers), which it resembles in many ways—and not much larger than a big asteroid like Ceres. Pluto's orbit is tilted seventeen degrees off the ecliptic, more than twice the tilt of any planet, and is so eccentric that it carries Pluto inside the orbit of Neptune for two decades out of each 249-year Plutonian year; none of the eight unambiguous planets has an intersecting orbit. Pluto's density—which could be calculated once its satellite, Charon, was discovered in 1978—is more like that of an icy comet or Kuiper belt object than a planet. Charon, fully half Pluto's size, is gravitationally locked to Pluto, as is the Moon to Earth; but, uniquely, Pluto is also gravitationally locked to Charon, so the two always keep the same sides facing each other. (In other words, Pluto's rotation period, 6.387 days, is the same as Charon's orbital period.) This situation is more like that of a binary asteroid than a planet. Pluto has an atmosphere, but it boils up when Pluto is relatively close to the Sun, then freezes and falls to the surface during its long excursions beyond Neptune—rather like a comet's. Pluto is probably best regarded as the largest, most conspicuous example of a growing group of objects that remain difficult to classify until more has been learned about the solar system. The asteroid belt contains several objects that pursue classic asteroidal orbits and were classified as asteroids until they boiled up, Pluto-style, and sprouted cometlike comas or tails. Further observations of Kuiper belt objects doubtless will aid in determining whether the belt is, as suspected, Pluto's native territory.

The assertion by some astronomers that Pluto is not a planet but an icy emissary from the Kuiper belt occasioned a surprisingly loud public outcry. When the renovated Hayden Planetarium reopened in the year 2000 as part of New York's glittering new Rose Center for Earth and Space, visitors to its solar system display reportedly were "perplexed" to find that Pluto had been "demoted from planethood." "In a move that has rocked the heavens, the Hayden has deplanetized Pluto," declared CBS News commentator Charles Osgood, adding that "visitors complain of a sense of depletion." "Pluto didn't make the cut," complained *The Santa Fe New Mexican*. "To some visitors who recall school lessons about nine planets in the solar system, the omission has been confusing and even distressing," reported *The New York Times*, quoting one of them, Pamela Curtice of Atlanta—who recalled learning the names of the planets by memorizing the phrase "My Very Educated Mother Just Served Us Nine Pizzas," for Mercury, Venus, Earth, Mars, Jupiter, Saturn, Uranus, Neptune—as lamenting, "Now I know my mother just served us nine—nine nothings."

The director of the Hayden Planetarium, Neil de Grasse Tyson, defended its division of the solar system into five families of objects—terrestrial planets, the asteroid belt, gas giant planets, the Kuiper belt (including Pluto), and the Oort cloud—as more instructive than merely learning the names of the planets. "Don't count planets, count families," Tyson pleaded. "In this way, you get to learn about the structure of the solar system." But he was in for some heavy sledding. Directors of other science centers sniffed that the International Astronomical Union still designated Pluto a planet: "We're sticking with Pluto," said Laura Danly, curator of space sciences at the Denver Museum of Nature and Science. Comet hunter David Levy, who wrote a biography of Tombaugh, said Tyson must be "in a different universe."[37]

But those who oppose "demoting" Pluto because they don't want to tarnish Clyde Tombaugh's reputation might take comfort in the prospect that he will instead be remembered as a pioneer in the exploration of the Kuiper belt, where swarms of inconspicuous solar orbiters remain to be discovered. Observers of the future who chart the belt will have to push their telescopes to their limits, seeking ever dimmer objects with ever slower motions in the sky and checking their results with a perfectionist's care to expand human knowledge of the Sun's family. I think Clyde would have liked that. As he said in his later years, "If I had to live my life over, I don't know if I would elect to do it much differently. . . . All those hours, tedious as they were—I have never regretted them."[38]

In any event, it seems appropriate that our exploration of the solar system ends in a question mark. Let us pause here, in the zone between the realm of the planets and that of the comets, where Pluto roams like a traveler with dual nationality, and look back. Huddled near the shrunken Sun and visible only in a telescope is Earth. Light from the Sun reaches Earth after eight minutes, and strikes Mars four minutes later, but takes forty-three minutes to reach Jupiter, over an hour to get to Saturn, two and a half hours for Uranus, and four hours for Neptune. What an expansive congress of worlds to observe, investigate, and even perhaps one day to explore: Eight planets, four of them small and rocky, four gas giants, and more satellites than most stargazers can memorize—among them geologically dead moons like Callisto, Rhea, Iapetus, Mimas, Oberon, and Umbriel; moons that have had at least some periods of geological activity, like Dione, Tethys, Ariel, Titania; Europa and Ganymede, with their deep-buried briny seas; active moons like fiery Io and the cold yet hissing Triton—each with its own history, only a few fragments of which we have as yet adduced. If this were all we could know—if, say, we were buried in a dark nebula where our black night skies held no stars—surely it would still be enough to reward curiosity, invention, and exploration for eons to come.

But it is not all. Around us is a vast galaxy, arrayed on scales so gigantic that galactic structure becomes discernible only once the solar system has dwindled to a dot the size of the period at the end of this sentence.

FROM THE OBSERVATORY LOG:
CHIMES AT MIDNIGHT

MARS IS COMING INTO OPPOSITION, forty-two Earth years since I first gazed at it through my 1.6-inch refractor back in 1956. The thought makes me feel a bit old, but it also testifies to the admirable reliability of solar system cycles, a solid left hand playing the stride piano planetary blues. Blustery winds howl through the pines like violinists tuning up—celestial music, Basho would say—and the stars are glittering as excitedly as eager students with their hands up. The Gemini twins are stepping down in the west, hand in hand, the crescent Moon beside them. The clock on the wall reads 11:59. I click on the shortwave radio and am surprised to hear these timely, timeless lines from Shakespeare's *Henry IV*:

> FALSTAFF: We have heard the chimes at midnight, Master Shallow.
> SHALLOW: That we have, that we have, that we have, in faith, Sir John,
> we have . . .

Either it's a remarkable coincidence, or somewhere a BBC producer has cleverly back-timed the tape to put this scene in the radio play on the air precisely at midnight Pacific time.

Blood-red Mars is disentangling itself from the tallest branches of the trees to the east, so I deploy a wind screen, aim the telescope, and have a look. My first impression is of how rapidly one's view of Mars improves as opposition approaches. The air is boiling and winds rattle the telescope, which has not cooled down sufficiently to perform properly, yet already I can see some dark, high-contrast Martian features. I stop down the telescope to six inches aperture, which reduces the effect of the air turbulence, and by 1:00 A.M. Mars has stabilized somewhat, although most of the time it still looks like a tomato omelet shimmering in butter in a frying pan. But it's rising in the sky all the while, further improving the view, and soon I am seeing more with each pause in the assailing winds.

By 2:00 A.M. the humidity has risen to 96 percent, lending a biting cold edge to the winds, which are rising, too, further scrambling the Martian disk. By three, Mars is sinking into fog-laden air above the western hills, but I feel no disappointment. It's a wild, beautiful night, Mars will be back, and meanwhile there are other things to see. I turn to the stars and galaxies.

III
THE DEPTHS

15.

The Night Sky

Rapturous is the night.
—Goethe

Ideas are to objects as constellations are to stars.
—Walter Benjamin

THE STARS ARE ALWAYS up there, of course, lost in the Sun's dazzle but shining on with their customary imperturbability. To become aware of this and not quite be able to put it out of one's mind is one way of getting into stargazing. The amateur astronomer David J. Eicher recalls that it happened to him after he was first shown a globular star cluster through a telescope:

> I came away from that night reeling with excitement, as if I'd suddenly been let in on a closely guarded secret. I hadn't changed. The world around me hadn't changed. But I felt profoundly different about walking down the street or tying my shoes or buying a soda at the supermarket. I was keenly aware as never before that out beyond the blue sky lay vast numbers of stars and countless worlds unseen. I couldn't help feeling differently. I had the whole universe on my mind.[1]

I was about fourteen years old when I started fixating on the fact that the stars don't go away in the daytime. The realization prompted some odd habits. Sometimes I would try to spot Venus or the bright star Sirius in broad daylight, usually by standing at a particular point in the schoolyard at recess and sighting off a corner of the roof, or at home by peering through a telescope or the cardboard tube from a defunct paper-towel roll. These efforts were rewarded, sometimes, with a glimpse of a point of white light amid oceanic blue, a reminder of our situation as inhabitants of an illuminated oasis amid

217

the greater darkness, as bedazzled as a lighthouse keeper polishing the lens of his beacon. At unlikely moments in the glaring Florida sunlight—tensioning reels on the pitching deck of a fishing boat, reading under a coconut palm, or knocked flat on my back on a football field—I would look up into the cobalt sky and think: *They're out there, waiting to be seen.*

Then at sundown, the curtain would rise and the show begin anew.

THE BEAUTY OF THE STARS AT NIGHT, which impressed our ancestors even when the sky was thought to be a spangled two-dimensional dome or lid, becomes all the more arresting when we awaken to its vast depths and the wealth of things it holds. As the painter Robert Henri observes in his book *The Art Spirit*, "No *thing* is beautiful. But all things await the sensitive and imaginative mind that may be aroused to pleasurable emotion at the sight of them. This is beauty."[2] Stargazing can be as much an aesthetic as an intellectual pursuit, its aim an informed attuning of our sense of beauty to the wider reality that surrounds us. With the naked eye alone you can see the phases of the Moon, solar and lunar eclipses, the ethereal dance of auroras, the colors and motions of the planets, the waxing and waning of bright variable stars, the faint glow of a few nebulae and nearby galaxies, and the stars—about two thousand of them at a time. A pair of binoculars multiplies the possibilities considerably, depending on their light-gathering power, and with a telescope the sky is no longer a limit.

So let's go outside and have a look.[3] Our signposts are the stars themselves and the patterns we make of them—the asterisms and constellations. Asterisms are conspicuous groups of stars, like the Big Dipper and the Pleiades. Constellations, stick figures drawn by connecting stars, were invented by shepherds and other early stargazers to aid in finding their way around in the sky. Their boundaries used to differ according to which star chart one consulted. Today they have official boundaries, designated by the International Astronomical Union, that parcel the entire sky into eighty-eight constellations. There is too much out there to cover, even superficially, in one chapter, so we will touch on just a few favorites here—and then, in the next chapter, resort to the telescope and probe parts of the sky in greater depth. Exploring it *all* is more than the work of a lifetime.

Our guide is a little ditty:

> Follow the arc to Arcturus,
> and on to Spica go;

Then turn northwest to Regulus,
the foot of the lion, Leo.

It's just that far to Gemini,
Where Castor and Pollux glow,
Near Rigel, and Capella,
And Sirius, down below.

The "arc" is the arc of the Big Dipper's handle. Known in England as the Plough, to the Chinese as the Northern Measure, to the Egyptians as the Bull's Thigh, and in modern Europe and ancient Rome as the Wagon or Cart, the Big Dipper is an asterism in the constellation Ursa Major, the great bear. Many asterisms consist of accidental alignments of stars that actually lie at varying distances, but five of the seven stars that make up the Dipper really are associated: They belong to a large, loose star cluster, called the Ursa Major moving cluster.[4] It measures 30 light-years in diameter and lies only about 80 light-years from the Sun, making it the nearest of all the star clusters. Its proximity to us and its high velocity—it's doing 14 kilometers per second, headed south toward Sagittarius—conspire to alter the shape of the Dipper over time. A hundred thousand years ago the Dipper had a squarer bowl and a straight handle, like a primitive implement, while at present it resembles a metal dipper in use in recent times, and one hundred thousand years from now it will sport a suitably futuristic, flattened bowl with its handle jauntily cocked into a right angle at the end.

Other star clusters visible to the unaided eye include the Beehive, in Cancer, the crab; the Hyades, in Taurus, the bull; and the Pleiades, a spectacular, dipper-shaped clutch of stars in Taurus, of which six are readily visible and the seventh is regarded as a test of good vision. On clear, dark nights, a sharp-eyed observer can see about a dozen stars in the Pleiades: Before the invention of the telescope, astronomers charted fourteen stars there. All these clusters are wonderful sights in binoculars, which have a larger field of view than most telescopes do.

Dubhe and Merak, at the end of the Big Dipper's bowl, are known as the Pointers, since a line drawn between them and extended north passes near Polaris, the north star. Looking through binoculars at Polaris you can see what some call its "engagement ring," a circle of 7th- and 8th-magnitude stars slightly to the south. (Of course, almost every star is south of Polaris, which stands less than one degree from the north celestial pole.)

The second star from the end of the Big Dipper's handle, Mizar, has a seemingly close companion, Alcor. The two were known as "Horse and Rider" to the Arabs of old, who tested the eyesight of prospective soldiers by determining whether they could resolve—"split"—them visually with the unaided eye.[5] (Most people today can easily pass the test, but it would have been useful back when eyeglasses were rare.) The pair is not a true double star but an accidental line-of-sight coincidence: Mizar is 88 light-years from Earth, Alcor 81 light-years. Mizar itself, however, is a true double star in its own right—the first to be discovered. The astronomer Giambattista Riccioli split it in 1650, as can any observer today with a small telescope. The dimmer of the Mizar twins is also double: This pair is too close to be seen in telescopes but can be discerned spectrographically, from changes in the positions of spectral lines caused by the companion's orbital motion.

Most stars belong to double or multiple systems, but only a few of them can be resolved with the unaided eye. Epsilon Lyrae, in Lyra, the lyre, is a good test for adults, although most youngsters can split it easily. Any double star that lies close to the limits of an observer's optics—whether a telescope, binoculars, or the naked eye—can look blurred and hence be mistaken for a nebula. Even Johannes Hevelius, who could see so well with the unaided eye that he preferred not to use the telescopes available to him—was deceived in this fashion, charting as a nebula what is actually a double star, 1.5 degrees northeast of Delta Ursae Majoris. Binoculars can provide splendid views of wide doubles like Albireo, in Cygnus, whose two bright stars are a vivid yellow and blue.

At the latitudes of Europe and North America, Ursa Major is a north circumpolar constellation, meaning that it never sets—a fact that ancient Greek navigators memorized with the phrase "the Bear never bathes." (The astronomer Robert Burnham called Ursa Major "a *polar* bear.")[6] This brings us to the old science of "spherical astronomy," in which one envisions the sky as a sphere centered on Earth and mapped by means of various lines and points. The points in the sky toward which the Earth's polar axis points are designated as the North and South Celestial Poles, so the altitude of the celestial pole above your horizon is the same as your latitude—and your latitude determines which constellations are circumpolar. At the North and South Poles, all constellations are circumpolar—the stars you can see in polar skies never set, and the ones you can't see never rise—while on the equator, none are.

The celestial equivalent of latitude is *declination*—measured in degrees, from 90° at the north celestial pole to 0° at the equator—and the celestial equivalent of longitude is *right ascension*. Lines of right ascension stretch from pole to pole, and are demarked in terms of hours, with twenty-four hours completing a circuit round the sky. The wobble of Earth's axis, called *precession*, slowly shifts this coordinate system against the stars, so star atlases are updated, usually every fifty years, to keep them accurate in the current epoch. Any object on the celestial sphere can be pinpointed by using the epoch plus these two coordinates. The coordinates of the star Arcturus, for example, are R.A. 14h 15m 43.7s, Dec. +10° 10' 25", epoch 2000, meaning that Arcturus is located at right ascension 14 hours, 15 minutes, and 43.7 seconds, declination 10 degrees, 10 arc minutes, and 25 arc seconds—in Bootes, the herdsman. Increasingly, computers handle the task of crunching these numbers: You tell the computer what you want to look at, and it looks up the coordinates and slews the telescope there. But when my telescope's venerable mount was retrofitted with a computer control system, I retained its old-fashioned "setting circles," a pair of large brass disks inscribed with right ascension and declination hash marks, if only for sentimental reasons.

Earth's motion in its orbit makes the Sun seem to grind eastward along the ecliptic, at a rate of a little less than one degree per day. As a result, constellations endlessly make their way westward, rising a little earlier on each succeeding night until they are gobbled up by the Sun and subsequently reappear in predawn skies. Hence constellations are associated with seasons. On winter evenings in the Northern Hemisphere, Orion is prominent. In spring, Leo has taken over and Orion is almost lost in the glare of the Sun. Summer skies are dominated by the "Summer Triangle" of the bright stars Vega, Altair, and Deneb. Meanwhile, Orion has re-emerged from behind the Sun. When I was a boy, eager for new things to observe with my tiny telescope, I would often get up before dawn to enjoy a preview of constellations that wouldn't reach evening skies for months. It was like time travel, and to this day, the sight of a fresh predawn sky fills me with a sense of new beginnings.

The two points at which the ecliptic crosses the celestial equator mark the spring equinox—designated, by convention, as zero hours right ascension—and the autumnal equinox, at twelve hours R.A. When the Sun passes through these points—which happens on about March 21 and September 22, depending on leap years and so forth—Northern Hemisphere spring and autumn begin. Equidistant between these two points are the summer and winter solstices, when the Sun—and therefore the ecliptic—strays farthest from the celestial equator. And that is all the celestial clockwork we should need for now.

Let's start our tour.

Follow the arc to Arcturus . . .

An arc of some thirty degrees extended away from the Dipper's bowl along the curve described by its handle lands us at the bright orange star Arcturus. The combination of its inherent luminosity (Arcturus is twenty-three times the diameter of the Sun and a hundred times more luminous) and its proximity (only 37 light-years from Earth) makes Arcturus the fourth-brightest star in the night sky. It was the first star to be seen through a telescope in the day-time—by the French astronomer Jean-Baptiste Morin in 1635. Its prominence is temporary, though, because Arcturus is a high-velocity star. Two million years ago it was a faint (6.7 magnitude) object 800 light-years away in Cepheus. Two million years from now, having hurried on its way, it will have retreated into the constellation Vela and dimmed to naked-eye invisibility again.

and on to Spica go . . .

Another thirty degrees along the same arc takes us to Spica, a 1st-magnitude star located near the ecliptic in the constellation Virgo. Like Arcturus, Spica is a massive star and intrinsically bright—more than two thousand times the luminosity of the Sun—but Spica is blue-white in color, while Arcturus is orange. The color of a star is a function of its surface temperature, which in turn results from its size and mass. Massive stars burn hot, pushing their colors toward the high-energy, blue-white end of the spectrum. But when they begin to exhaust their nuclear fuel, their outer layers expand and cool somewhat, turning their color to orange or red, like a white-hot iron taken from a forge and set aside to cool. Astronomers call massive blue-white stars "young" and the red giants that they evolve into "old," but the terms can be misleading. Massive stars live fast and die young: They can become red giants when they are only tens of millions of years old, whereas a modest star like the Sun burns for 10 *billion* years before turning into a red giant.

Astronomers catalog stars by their spectral types. The original system, developed at Harvard, classed stars alphabetically by the "complexity" of their spectra, but when it became clear that a color-based system made better astrophysical sense the letters got scrambled. Hence the spectral sequence in use today, proceeding from hot blue-white stars to cooler red ones, reads O, B, A, F, G, K, M—which generations of scientists have learned to memorize by phrases like "O, Be A Fine Girl, Kiss Me!" and "Oh Boy, An F Grade Kills Me!" Spica is a B star; Altair and Vega are A stars; Polaris is an F star; the Sun is a

middle-of-the-road type G; Arcturus a K type; and Betelgeuse, which we will
meet shortly, is a type M red giant.[7]

> *Then turn northwest to Regulus,*
> *the foot of the lion, Leo.*

Regulus—magnitude 1.4, distance 77.5 light-years—is the brightest star
in Leo. Its name means "the little king," referring to the king of the beasts.
Spectral class B, it is 3.5 times the diameter of the Sun. Part of a triple star
system, Regulus has an orange companion that can be seen easily in small telescopes—it's more than eight arc minutes away—which itself is accompanied
by a faint (13th magnitude) dwarf star only 2.6 arc seconds away and difficult
to discern in the glare of its brighter companion. A fourth star seen in the field
was cataloged Regulus D on the assumption that it, too, belonged to the Regulus system, but this evidently is a line-of-sight coincidence. Leo is a "zodiacal" constellation, so Regulus is occasionally occulted by the Moon and even
sometimes by a planet. Venus occulted Regulus on July 7, 1959, a rare event
that I'd like to have observed from Key Biscayne, but Regulus set before the
occultation occurred.

Five degrees west of Regulus one can *sometimes* see with the naked eye the
variable star R Leonis, which pulsates over a period averaging 310 days. Most
of the time R Leonis is a telescopic object of 6th to 10th magnitude, but at
peak brightness it can exceed 5th magnitude, bright enough to be visible without a telescope. It is classed as a Mira-type variable—after Mira (Latin for
"the wonderful") in Cetus, the whale. Mira caught the attention of the
ancients, who noticed that it attains naked-eye brightness for about a hundred
days at a time, sometimes getting up to magnitude 2.5, then dims to invisibility for months. The Mira-type variables are pulsating stars of average mass in
a late, unstable phase of their evolution. There are many other types of variable stars, among them the dramatic "flare" stars and "cataclysmic" variables.
Variables always visible without a telescope include Betelgeuse, which over
periods of six years can become nearly as dim as Aldebaran (magnitude about
0.9, itself a variable) or as bright as Rigel (magnitude 0.1); Delta Cephei, after
which the Cepheid-class variables are named—it ranges from magnitude 3.5
to 4.4 in less than a week—and Zeta Geminorum, another Cepheid, which
goes from magnitude 3.6 to 4.2 in ten days.

Eclipsing binary stars mimic pulsating variables, but actually are quite different. These systems happen to be oriented so that the one star periodically
passes in front of another, altering its apparent brightness. The prototypical

eclipsing variable is Algol (the "demon star") in Perseus. Algol is a triple system, but its variability results from mutual partial eclipses involving two of its stars, one bright blue one and the other a red giant, both brighter than the Sun. Although the distance of Algol is only 93 light-years, these two stars are too close together to be resolved by even the largest telescopes. They probably form what is called a contact binary system, so intimate that one star strips material from another. The generally accepted model of Algol has been derived from its light curve and from spectroscopic information revealing the types and velocities of its component stars. It indicates that the plane of the stars' orbit is tilted a bit from edge-on, so that most of the disk of the dimmer star is obscured when it passes behind the brighter one, and much of the brighter star obscured when the dimmer one passes in front of it. This produces a mild dip in magnitude when the dimmer star is eclipsed, and a pronounced one, from visual magnitude 2.1 to 3.4, every 2.87 days, when the brighter star goes into eclipse.

All the constellations we've visited so far—Ursa Major, Bootes, Virgo, and Leo—are rich in galaxies. A few of these distant islands of stars, among them M65 and M66 in Leo, are bright enough to be seen with binoculars. The reason so many galaxies are visible in these parts of the sky is that we are looking up out of the plane of our own galaxy. When we instead look along the galactic plane, gigantic clouds of dust and gas in the disk of our galaxy block our view, in visible wavelengths, of most of the galaxies beyond. Our next step takes us close to the Milky Way.

> *It's just that far to Gemini,*
> *Where Castor and Pollux glow . . .*

Moving west 37 degrees along the ecliptic to Pollux and Castor, the twin stars that mark the heads of Gemini, the twins, brings us into starry thickets along the edge of the Milky Way. As a result there are now a lot more local sights to see, with the unaided eye and especially through binoculars. Sweeping through the star fields of Gemini, we can spot several jewel-box–like open star clusters—such as M35, near Castor's left foot. And there's more to come.

> *Near Rigel, and Capella,*
> *And Sirius, down below.*

This last, fifty-degree leap takes us to Orion the hunter and his dog Canis Major, in the rich star fields of the Milky Way proper. Orion is the favorite constellation of many stargazers, and understandably so. For one thing, he actually looks like a hunter: You can readily perceive his broad shoulders, his right arm wielding a club and the left holding forth a shield—he appears to be challenging Taurus, the bull, a formidable foe—his running feet, and his magnificent belt and dangling sword. The stars not having bothered to arrange themselves into stick figures for our amusement, many constellations bear only a distant resemblance to their namesakes. (Is Ophiuchus a snake handler or a Dumpster; Virgo a virgin or a scow?) But there are a few, among them Scorpius the scorpion, Cygnus the swan, and little Delphinus the dolphin, that really stand out clearly enough to elicit delighted cries of "I can see it!" from first-time stargazers, and Orion is perhaps the most splendid of them all.

Orion has long figured in popular lore both as a mythological figure and as a guide to the seasons for farmers and mariners, who were expected to recognize this stunning constellation even if they knew few others. To the ancient Egyptians, Orion was Osiris, the god of the underworld, and the Milky Way the underground river down which he steered his boat. In the King James translation of the Old Testament, God asks Job, rhetorically, "Canst thou bind the sweet influence of the Pleiades, or loose the bands of Orion?" Virgil, Pliny, and Horace warn that Orion's appearance in the night skies coincides with the onset of winter storms, a time that Hesiod cautions is "when the winds battle with thunderous sound . . . and the sullen sea lies hidden in sable cloud."[8] They knew what they were talking about: As I write this, on an afternoon in January, the month when Orion rides highest, thunderclaps are rattling the windowpanes and the slate-gray surface of San Francisco Bay is being whipped into whitecaps by winds and driven rain.

Orion's two brightest stars present a distinct color contrast. Rigel, the beacon at the hunter's foot, is a blue-white supergiant. Betelgeuse, the ruby-colored lamp at his shoulder and the constellation's second-brightest star, is a pulsating red supergiant, its atmosphere so extended that if it were centered on the Sun it would embrace the orbit of Jupiter. The word "Betelgeuse," corrupted Arabic of disputed origin, has three correct English pronunciations, one of which, to children's delight, is indeed "Beetle-juice."

The brightness of a star in the sky (its *apparent* magnitude) is not a reliable guide to its intrinsic brightness (or *absolute* magnitude). Although Rigel and Betelgeuse have similar apparent magnitudes, Rigel is twice as far from Earth as Betelgeuse is. Sirius, the brightest star in Earth's nighttime skies and the pride of Canis Major, the larger of Orion's two hunting dogs, owes its imposing apparent magnitude—the word "Sirius" evidently comes from the Greek for "scorching"—to the fact that it is both intrinsically bright (twenty-three times as luminous as the Sun) and nearby, at a distance of only 8.6 light-years from Earth.

Similarly, the stars that form a constellation are not necessarily close to one another in three-dimensional space. The ragged line of stars that defines the leading edges of the wings of Cygnus all lie within 200 light-years of Earth, but Albireo, the admirable double star at the swan's eye, is 386 light-years from Earth; Sadr, the bright star at her heart, is 1,500 light-years away; and regal Deneb, the brightest star in Cygnus, is twice as distant as Sadr. Many of the stars in Orion do, however, inhabit the same general neighborhood, because they reside in a nearby spiral arm of our galaxy. The Milky Way galaxy is a typical large spiral, consisting of a few hundred billion stars and vast clouds of gas and dust.[9] Most of the stars and nearly all the gas and dust clouds lie along the galactic disk. The disk is wide—more than 80,000 light-years wide—but only a few hundred light-years thick. At its center is a protruding "bulge," as it is rather unimaginatively called, composed primarily of older red and yellow stars. (Take a yo-yo apart and stick its axis through the center hole in an old vinyl phonograph record: The yo-yo is the central bulge, and the vinyl record is the galactic disk.) The spiral arms contain vast star-forming regions, lit up by the multitudes of young, blue-white giant and supergiant stars that formed there recently in the course of cosmic history and have not yet run through their profligate careers and exploded.

When observing other spiral galaxies we can readily see the spiral arms, but those of our own galaxy are more difficult to map, since we are located in the disk that contains them. It's rather as if we were in a forest of decorated Christmas trees, some of which, delineating the spiral structure, had especially bright lights. The work of telling the forest from the trees has been aided by contributions from observations made in radio and infrared wavelengths, which can penetrate the intervening clouds. The resulting maps indicate that the solar system lies about two-thirds of the way out from the center of the galaxy to the edge of the glowing disk, situated near a spur between two spiral arms. The inner arm, which passes nearest to us where we look toward the galactic center, is known as the Sagittarius arm; we will visit it later in this tour. Right now we are looking in the opposite direction, away from the galactic center, toward the Orion arm.

The inner flanks of the Orion arm lie 1,500 light-years from Earth and are peppered with bright zones lit up by brilliant new stars. A time-exposure photograph of only ten minutes or so—I made a few myself, as a boy, using a motor-driven mount fashioned from plumbing pipes—shows that much of the constellation is entangled in these luminous clouds. The unaided eye can make out the brightest knot among them, a glowing hive at the middle of the scabbard dangling from the hunter's belt—the famous Orion nebula. It's a kind of blister—a region where gas has been ionized by the light of bright, newly minted stars within it—in a large cloud of gas and dust. With binoculars, one can see newborn stars entangled in the nebula, lighting it up like fireflies in a jar. One of these, Theta Orionis, is an easily resolved multiple star.

If we could speed up the passage of time so that millions of years passed in moments, we would find that bright blisters like the Orion nebula flare up briefly, when they make new batches of stars, and then subside, while at any given time most of the dust and gas in the spiral arm remains dark. These vast, dark clouds, many of them arced like the ribs of a beached and denuded whale, are not immediately evident to beginning observers, who may tend to assume that they represent empty space in gaps between the stars, but they become strikingly apparent once you know what you're looking for.[10] Long, dark tendrils wind through Orion itself, notably from alongside the scabbard up to his right shoulder, from which they extend up past the Rosette nebula, itself a spectacular star-forming "blister," located east of Betelgeuse in Monoceros the unicorn.

Dark nebulae were the special study of Edward Emerson Barnard, whom we have previously encountered making levelheaded Mars observations and discovering Jupiter's satellite Amalthea. As self-made an astronomer as this world has seen, E. E. Barnard was born into poverty in Nashville, Tennessee, in 1857, a few months after the death of his father. His widowed mother struggled to support her two children by making wax flowers, and the family often went hungry. During the Civil War, when a steamer loaded with provisions was sunk before it could reach the besieged city, Barnard dove into the Cumberland River and managed to fish out a box of hardtack to put dinner on the table. His boyhood was, he said, "so sad and bitter that even now I cannot look back on it without a shudder."[11] To help "soften the sadness," as he put it, Barnard lay out on warm nights in a weathered wagon bed, flat on his back, studying the stars. Few have beheld the night sky in a more perfect state of ignorance. Too poor to buy books, Barnard in all his youth had only two months of schooling and knew absolutely nothing of astronomy. He befriended a bright star that stood overhead on summer evenings but had no way of telling that it was called Vega. He memorized the nameless stars and "soon saw that a few of them changed their places among the others, but I did not know that they were planets."[12]

Shortly before his ninth birthday, having survived a cholera epidemic that swept through the city, Barnard was put to work at Van Stavoren's Photographic Gallery in Nashville. A photograph taken of him at that time shows a winsome, handsome boy with grimly set lips but large, hopeful eyes. In another photo, taken a year later, the mouth remains the same but much of the hope has drained from his gaze. His job was to crank a set of wheels to keep a giant, roof-mounted camera—the "Jupiter"—aimed constantly at the Sun, the light from which was employed by the photographer to make prints from negatives. Other boys had balked at this tedious task, or fallen asleep and let the Sun escape. But Barnard became fascinated by the fact that the Sun sometimes reached its apex in the sky at noon—as marked by the bell in St. Mary's church nearby—and at other times did so before or after noon, "the difference sometimes amounting to a considerable fraction of an hour. This set me to thinking and wondering."[13] Not until years later did he learn that what he had noticed is called the "equation of time," the difference between local and solar time that oscillates with the passing seasons. Making the long walk home at night, the boy befriended a yellow "moving star." There was no one to tell him that it was Saturn.

Barnard remained at the studio for seventeen years. There he gained a working knowledge of optics, and eventually managed to put together a telescope from found objects. Its tube came from an old ship's spyglass, its eyepiece from a broken microscope, and its tripod from a surveyor's instrument stand. He trained it on the night sky for hours at a time, and particularly liked to study Jupiter. His ignorance of astronomy ended on the day that an indigent friend borrowed two dollars, leaving a book as security for a loan that Barnard knew well would never be repaid. "I felt very angry, because the money was a large amount to me then, and it was some time before I would open the book," he recalled.[14] When he eventually did so, he was stunned to find that its subject was astronomy. Delighted, he pored over this find—Thomas Dick's *The Sidereal Heavens*. It contained a set of star charts, the first Barnard had seen, and he immediately went to work comparing them "with the sky from my open window, and in less than an hour had learned the names of my old friends; for there was Vega and the stars in the Cross of Cygnus and Altair and others that I had known from childhood. This was my first intelligent glimpse into astronomy."[15]

Barnard kept observing, discovered several comets, read extensively, engaged a mathematics tutor, graduated from Vanderbilt University with a degree in mathematics, taught there, and was hired by Lick Observatory after he showed up while the domes were still under construction to announce that he had quit his teaching post, sold his house, and was willing to work for nothing if only he could observe. At Lick, and later at Yerkes Observatory, his keen vision, superlative work ethic—colleagues said they never knew him to sleep—and knowledge of photography served him well. By the end of his career, he had published over nine hundred papers, discerned features on Jupiter's moon Io that were not to be confirmed until *Voyager* visited the giant planet nearly a century later, noted the uniquely rapid "proper" motion across the sky of "Barnard's star," discovered "Barnard's galaxy," and charted 349 dark nebulae, photographs of which he published—as original prints, pasted in by hand—in his exquisite 1927 *Photographic Atlas of Selected Regions of the Milky Way*.

Barnard thought that these black patches and lanes were holes in star fields, through which we see into the blackness of space beyond. This had also been the opinion of William Herschel, who called them "great cavities or vacancies" formed when stars were drawn away toward regions of higher density (and who seems, in turn, to have acquired this theory from James Ferguson, the onetime shepherd whose book *Astronomy Explained* first introduced Herschel to the stars). This wasn't really a plausible hypothesis—the "holes" would have to be long, thin tunnels or cracks of great depth, all of which just happened to be aimed at Earth—but whatever his weaknesses as a theorist, Barnard, a big, jovial man with a white walrus mustache, who despite his many sorrows literally whistled while he worked, was an excellent observer, some would say the best since Herschel himself.

Barnard's name today is a byword among all who take pleasure in tracing black rivers through the stars and pondering that, although frozen on human timescales, they actually are writhing, tumbling clouds, as dynamic as mountain storms. Discerning dark nebulae can be as disconcerting as fumbling for the light switch in a dark guest room. The eye has to learn to capture them, in something like the way one learns to spot the subtle glint of chanterelle mushrooms under carpets of leaves on the forest floor. But the rewards are splendid. Large dark nebulae can be observed with the naked eye—some observers shroud their heads with black hoods and peer through special filters that block out non-nebular light, a getup that makes them look like grim reapers with monocles—and smaller ones can be traced with binoculars and telescopes. Since more of the Milky Way is dark than bright, studying dark clouds affords an insight into galactic structure.

The largest of the dark nebulae, an inky archipelago known as the Great Rift, splits the glowing Milky Way like a spine, running from Cygnus and Aquila south to Sagittarius, the archer. In Sagittarius we are looking directly toward the center of the galaxy, and the Rift, having swollen like a river approaching the sea, obscures our view of most of the central bulge. The looming, spherical bulge protrudes to either side—it lies in the background, 28,000 light-years from Earth—while in the foreground Barnard's black chimneys and festoons arch above and below the disk, like waves of a crashing surf. The Great Rift is well placed for summer viewing from the Northern Hemisphere, while the central bulge is best seen from Southern Hemisphere latitudes where it reaches the zenith.

The intricate beauty of the Milky Way inspired a number of stargazers to map and draw its naked-eye appearance in the days before photography became equal to the task. In Aachen, Germany, Eduard Heis made painstaking pencil drawings of the Milky Way by peering through a blackened tube, a foot long and a foot wide, to improve his view of the dimmest objects. The Dutch astronomer Antonie Pannekoek drew long reaches of the Milky Way in exquisite detail, using an elaborate system that involved dictating the positions and brightness levels of nebulae by night, then constructing drawings the next morning based on his notes and recollections. His drawing of the Scorpius-Sagittarius region, made in 1926, stands up well in terms of accuracy when compared with modern photographs, and depicts glowing star clouds and dark dust lanes on scales too large to be appreciated in a telescope's narrow field of view.

Through binoculars, Sagittarius blossoms into an almost overwhelming wealth of star fields and nebulae. Its shoals of stars are particularly bright just a few degrees west and north of the 3d-magnitude star Gamma Sagittarii, where we are looking directly toward the heart of the galaxy, but they stretch all through the constellation and beyond. Prominent among the glowing clouds are the graceful Swan nebula; the Lagoon and Trifid nebulae, which appear in a single binocular field of view; and the ganged open star clusters M23, 24, and 25, along with M16, just across the border in the neighboring constellation Serpens the serpent. As Barnard put it in 1913, these clouds "are full of splendid details; one necessarily fails in an attempt to describe this wonderful region of star masses. They are like the billowy clouds of a summer afternoon."[16]

In life one sometimes has the feeling that something deeper is hidden behind the external, visible reality. Whether such is the case in spiritual terms is open to discussion, but it is literally true of the night sky. So let's get back to the telescope and go deeper.

Digital Universe:
A Virtual Visit to a Robotic Telescope

IT HAD BEEN CLOUDY for seventeen straight nights, I'd missed the dark of the Moon, and the observing list on my desk at the observatory was gathering dust, so I searched the Net, looking for time on a telescope where conditions were more favorable. This was still a shaky game in the year 2001. Many of the Web sites set up to offer remote observing on robotic telescopes—among them Case Western Reserve University's Nassau Station telescope, the Iowa Robotic Telescope Facility, and the Highland Road Park Observatory in Baton Rouge, Louisiana—were still under construction, some replete with nightmarish accounts of software glitches or hardware breakdowns that had brought work to a halt. Others, like the Global Telescope Network and the Automatic Telescope Network, were in planning stages. Some were listed merely as "concepts," and several had disappeared from the Web altogether. But I found two in working order, one in England and the other in California, and sent each an email requesting a ten-minute unfiltered CCD exposure of a pair of interacting galaxies, NGC 7805 and Arp 112, that I had been studying. I had never been to either observatory, and had no idea when the images would be taken. A computer would decide that, based on weather conditions and the competing requests of other observers.

Ever since Galileo, astronomers observed by using telescopes for themselves. When large telescopes were built on remote mountaintops, this often meant making pilgrimages of thousands of miles. Once up there, gasping in the thin air and fretting about the threat of bad weather and technical glitches, observers gave instructions to a "night assistant," an on-site technician familiar with the telescope's strengths and weaknesses, who served both to help the observer get the best results and to protect the instrument from damage. But even if the visiting astronomer wasn't actually running the telescope, he or she was on-site and directly involved.

Things began to change with the coming of computers. Once CCD imaging devices equipped with autoguiding mechanisms replaced cameras and film, observers no longer had to ride in cages high up in the telescope tube. Instead, they could repair to the warmth and relative comfort of the control room, where the telescope was aimed and the exposures checked on computer monitors. At extremely high-altitude observatories, like those atop Mauna Kea on the island of Hawaii, these facilities were located thousands of feet down the mountain, away from the debilitating and potentially dangerous effects of oxygen deprivation. The observers could see everything they needed on the computer screens anyway, and only a few night assistants had to remain up top. Meanwhile, *Hubble* and other space telescopes in orbit operated in a fully robotic mode, with no hands-on contact save for occasional maintenance and upgrade missions by space shuttle astronauts. As the advanced hardware and software systems that evolved for these uses percolated into other applications, a few fully robotic telescopes were installed at mountaintop observatories. Designed to perform for weeks or months with little or no direct human contact, they were given instructions via the Internet, and they sent back the resulting data the same way. Increasing numbers of observers, freed from the expense and travail of astronomical pilgrimages, began to rely on telescopes they have never even seen.

In the summer of 2000, I visited a small manufactory of robotic telescopes in Birkenhead, just across the Mersey River from Liverpool, and was given a tour by its managing director, Michael Daly. The facility, spanking new and fastidiously uncluttered, sat amid open fields swept by sea breezes, where coasting gulls traced lazy eights in the sky. Daly explained that this nonprofit enterprise, set up under the auspices of a local university, was funded by the British government and the European Union through a program aimed at developing high-tech jobs in economically depressed communities while offering European scientists a means of obtaining large telescopes without having to order them from Japan or the States.

We looked in on a carpeted room where technicians seated at computer-aided-design terminals were putting the finishing touches on digital blueprints for the company's initial run of four two-meter-class robotic telescopes. Then we entered the factory floor, where the telescopes loomed in various stages of construction, each a massive steel skeleton two stories high, resting on a concrete plinth. One was to be installed in the Canary Islands, another in India. The other two, bound for Hawaii and Australia, would be devoted in part to a British program enabling science students to conduct their own observations. "Kids start out fascinated by science," Paul Murdin, the astronomer who instigated the student program, had told me in Manchester a few days before, "but they drop away from it in their midteens. Something happens to them—puberty, presumably. We'll try to keep them interested by exposing them to real science."

Raising his voice in the echoing shed, Daly patted the flanks of one of the big telescopes and said, "When this one is on-site and in operation, each evening it will run its own self-test, query the weather status to determine if it's safe to open the dome, then look at its workload and elect what to do that night, based on weather conditions and efficiency, so that it's not moving all over the sky unnecessarily. We want the telescope to make the decisions, not the astronomers. Then, once you've taken the astronomers away, you ask, 'What are all these *engineers* doing here, halfway up a mountain in the cold and dark, a few hundred meters from the telescope?' To take *them* away, the design has got to be inherently more reliable, but as you get people out of there the facility starts getting smaller and simpler, and the cost of running it starts coming down."

A slew motor came to life with the sleek sound of oily gears interacting, and one of the big steel skeletons moved a few degrees. "Each of these telescopes weighs twenty-four tons and costs two to two and a half million pounds," Daly shouted over the din. "They move on hydrostatic oil bearings, using brushless DC motors. Fourteen computers operate each telescope. The goal is to try to get the computers positioned near where they're needed, so we can avoid massive cables, which are always a source of trouble. We use closed-cycle cooling systems and highly reliable instruments. The telescopes can run their own star tests to determine seeing conditions, and they can autocollimate their optics."

I looked up at the roof. It was shuttered, like a outsize version of an amateur's rolloff-roof observatory, so that the telescopes, when completed, could be tested on the stars.

"First we'll test each telescope as a machine, without the mirror," Daly said. "Then, once it passes those tests, we'll put in the mirror, roll the shutters back, and demonstrate that it will operate as a telescope. My hope is that one night we can invite the public in and run all four of them at once. The professional astronomers may turn up their noses at such a demonstration, but they'll remember it."

For professionals, robotic observing has distinct advantages. Night assistants like it because it gets the astronomers out of their hair. Instead of telling wicked stories about famed astronomers running telescopes into ladders, fracturing equipment they were trying to adjust, or dropping eyepieces (and, in one case, a peanut-butter-and-jelly sandwich) onto the primary mirror, the night assistants can relax and watch the telescope run itself—at least until they, too, are sent packing. The astronomers, for their part, get to stay home, write papers, and sleep at night while their observations are made for them. Administrators like the financial savings, and everyone benefits from the increased efficiency. In the old days, an astronomer taking high-dispersion spectra of intergalactic clouds in deep space was finished for the night once the Moon rose and the sky grew too bright to continue. A stellar spectroscopist could have worked during the moonlit hours, but if no stellar spectroscopist was around, the telescope tended to go idle. This was the dirty little secret of observational astronomy—that time on many good telescopes, the medium-size ones in particular, often was wasted simply because the funds weren't available to consistently put the right observers on them at all the right times. Robotic telescopes don't have that problem, since the observer need not be present for data to be collected.

Robotics gave amateurs access to more telescopes, and larger ones, than they had at home—including some of the underused professional ones. Now an amateur desperate to image a flare star or a new spot on Saturn didn't have to find an accommodating amateur or professional willing to help out. Instead, an email to a robotic observatory—or, soon, a clearinghouse managing time at several observatories—could handle the task. And amateurs could return the favor by putting time on their own automated telescopes at the disposal of professional projects of the sort that require lots of smaller telescopes rather than a few big ones. A new and more efficient—if less romantic—way of doing astronomy was emerging.

As robotic telescopes have come online, observing time has begun to turn into an Internet commodity—bringing the deep sky within reach of Tokyo office workers whose previous alternatives involved buying a telescope and lugging it to a distant, dark-sky site, or students in rural India who had no previous alternatives at all. Untapped brainpower is humanity's greatest asset, and future discoveries may frequently be made by observers who have never seen a major telescope and whose astronomical equipment consists of little more than a computer, a modem, and an active, educated mind.

A few days after I submitted my request, an image of the galaxy showed up in my email from one of the two telescopes, an off-the-shelf 14-inch Schmidt Cassegrain operated by the University of California, Santa Barbara. I went to work cleaning it up and looking for what I was after, which in this case was the glow of young stars along an intergalactic bridge. It was an effortless, pleasant way to observe—if uninspiring, compared to getting out under the stars and doing it for yourself.

16.

The Milky Way

A broad and ample road, whose dust is gold
And pavement stars, as stars to thee appear,
Seen in the galaxy, that milky way,
Which nightly, as a circling zone, thou seest
Powdered with stars.
    —Milton

Why should I feel lonely: Is not our planet in
the Milky Way?
    —Thoreau

MIDNIGHT, HIGH IN THE Chilean Andes, 1980. This lofty ridge-line, bathed in clean, stable air wafting in from the Pacific, has some of the world's best seeing. Three big observatories are arrayed along it, each just within sight of the next, like the torchlit signal towers of the ancient Romans. I was using one of the smallest telescopes in the northernmost of the three, sweeping at low power along the Milky Way, looking at star clusters and nebulae. Visual observers normally proceed in methodical fashion, aiming the telescope at specific targets and then studying them, but sometimes we just sweep, aimless as Huck Finn adrift on a river raft. And the skies of Earth hold no mightier river than the southern Milky Way, which on this night arched overhead near the zenith, splendid in a deep black sky.

I started in Aquila and rambled south, watching blazing stars troop by like jewels on velvet. Some were gathered in loose little aggregations, modest open clusters like NGC 6755 and NGC 6756. These made a passing impression, like looking out a night-train window at the paltry lamps of slumbering rural towns. A few glowing emission nebulae paraded past, their numbers slowly increasing. A vast, dark dust cloud intervened like a pregnant silence, then gave way to more stars, and the tight, warm glow of a globular star cluster, followed by a planetary nebula—ethereal, translucent, and subtly complex as a luminescent jellyfish.

The pace picked up as I passed through Scutum and crossed over the border into Serpens. The Star Queen nebula hove into view, a chandelier of light on a long, dark tendril of dust and gas, gaudy as a riverboat on a nighttime cruise down the Mississippi. In Sagittarius, scores of bright nebulae appeared, exotic as tropical blossoms. I lingered over the Swan nebula, which resembles smoke rising from a campfire on a still morning, though this plume stands ten light-years high. The Trifid nebula, a Chinese-lantern bubble, looked flat when the wind stirred but snapped into a three-dimensional glowing globe when the air steadied. Its companion, the murky Lagoon, faded into the surrounding space in successive folds of glowing gas, like bolts of drapery unrolled for display. There was much more to see—more coppery globulars, wispy nebulae, and glistening open star clusters—but it was getting to be almost too much to take in. I stepped away from the telescope to stretch and clear my head. As one is so often reminded when gazing into the depths of the night sky, nothing is more fantastic than the real.

ANCIENT OBSERVERS COULD HAVE GUESSED that the stars they saw in the night sky belong to the flattened disk of a galaxy. The great pretelescopic astronomer Tycho Brahe had in his observatory a giant brass celestial globe, five feet in diameter and adorned with ornate constellation figures, on which he engraved the positions of a thousand stars. Had he considered their distribution, he might have noticed that bright stars tend to be located near the Milky Way's glowing river in the sky, an obvious indication that there is some sort of association between the stars and the Milky Way.[1] But evidently neither Tycho nor any of his contemporaries made the connection. We see nothing naïvely—all perception is a dialogue between the seer and the seen, and depends as much on what you are looking *for* as what you are looking *at*—and Tycho, having no idea that the Sun belongs to a spiral galaxy, wasn't looking for evidence of this proposition. It remained for Galileo with his telescope to confirm that the Milky Way is made mostly of stars, and for twentieth-century astronomers to establish that it represents our view of a spiral galaxy from our position inside the disk.

As we noted earlier, massive stars burn much more furiously than the smaller ones do. The burn rate goes up by the fourth power of the mass, so a star with ten times the mass of the Sun is ten thousand times more luminous than the Sun. This cannot last: If you have ten times more money than I do but you spend it ten thousand times faster, you will soon go broke, and that's what happens to giant stars. Ordinary stars like the Sun shine steadily for some ten billion years, but supergiants of fifty solar masses start sputtering out when they are less than a million years old. Twentieth-century astrophysicists discovered that when stars are plotted on a graph, with one axis representing their colors and the other their luminosity, most lie on a clearly defined path—the "main sequence," where stars spend most of their lives. When they run low on fuel they become red giants, departing from the main sequence and turning redder as their outer atmospheres expand and cool. Eventually they slough off their outer shells of gas altogether, and the denuded core survives as a white dwarf. Although the night sky presents us with only one frame in a movie, we see around us stars of various masses and ages in various stages of their careers, and by studying them we can piece together the story of the galaxy's evolution. Part of the pleasure of using a telescope lies in appreciating the part played by each particular object in the long-running drama of the galaxy as a whole.

Planetary nebulae provide striking examples of such stellar snapshots in time. The term "planetary" was coined by their discoverer, William Herschel, in 1785. Herschel meant it metaphorically, his point being that many of these nebulae, since they have disklike shapes, might at first glance be mistaken for planets. He theorized that planetary nebulae consisted of some sort of "shining fluid," but he had no way of determining what this fluid might be. This temporary limitation—astronomers knowing where things were in the sky and what they looked like but not what they were made of—was seized upon by the philosopher Auguste Comte in 1835. Groping for an example of knowledge permanently beyond human ken—always a dangerous presumption—Comte declared that while humans might eventually learn the shapes, distances, sizes, and motions of celestial bodies, "never, by any means, will we be able to study their chemical composition."[2]

Comte's assertion was refuted just a few years after his death, when spectroscopes were trained on the Sun and stars by the physicists Joseph Fraunhofer, Gustav Kirchhoff, and Robert Bunsen, revealing their composition and ushering in the new science of astrophysics.[3] The English amateur astronomer William Huggins then took up the cause. The only child of a well-off London silk merchant, Huggins inherited his father's business at the age of eighteen but sold it twelve years later to devote himself to astronomy. He built a private observatory outside London, and at first contributed to the humdrum toil of logging star positions. But when he heard that Kirchhoff and Bunsen had identified spectral evidence of sodium in the Sun, Huggins obtained a

Clark refractor and began taking spectra of nebulae. He found that some of them, such as the planetary nebula NGC 6543, in Draco, consisted of a luminous gas, while others—spiral nebulae like those studied by Lord Rosse with his 72-inch reflector, known today to be galaxies—displayed stellar spectra. The Royal Society rewarded Huggins with the gift of a new telescope built for spectroscopy. His fortunes continued to improve when, while consulting with the Dublin telescope maker Howard Grubb, he met a young amateur astronomer and polymath named Margaret Lindsay Murray. They soon married, and for the remainder of Huggins's life they worked together on astronomical spectroscopy.

As a planetary nebula balloons out into space, ultraviolet light from the star it left behind pumps energy into the cloud, "exciting" its atoms—which is to say that photons absorbed by the atoms kick their electrons into higher orbits. When the electrons drop back into lower orbits they emit photons of their own, at wavelengths characteristic of each atom. In planetary nebulae, much of this energy is reradiated in the wavelength of singly and doubly ionized oxygen, so many stargazers use "OIII" filters, which repress light at other wavelengths, to improve their telescopic views of them.

Our galaxy contains a thousand visible planetary nebulae, plus an estimated ten thousand more, hidden behind the dust clouds of the galactic disk, that haven't yet been detected. Since most normal stars of up to ten or twelve solar masses are destined to go through a planetary nebula phase late in their lifetimes, and since there are billions of such stars in our galaxy, one might wonder why planetary nebulae are not more numerous. The answer is that they don't last long: It takes only about fifty thousand years for the shells to dissipate until they stop glowing and fade from view. On astronomical timescales, a planetary nebula is as brief and dramatic as the detonation of a firecracker.

Some planetaries, such as the Ring nebula, resemble disks or doughnuts. Others, like the Dumbbell nebula (a large, ragged-looking planetary in the small constellation Vulpecula, the little fox) or the Ghost of Jupiter (a spangled ring crossed by delicate wisps of gas, in Hydra, the water snake) display intricate patterns best seen with larger telescopes. The Blue Snowball, in Andromeda, is as spattered as a dropped drink. The Eskimo nebula, in Gemini, resembles a human face in a fur parka's hood. The Cat's Eye, in Draco, has intricate ringlets arrayed in an almost spiral pattern, while the dim planetary NGC 7139 shows just a few wisps, having all but disappeared. The Helix, in Aquarius, is the closest planetary to Earth, only 450 light-years distant, but is often overlooked because it looms so large in the sky: Fully thirteen arc minutes in diameter, it is best seen through binoculars or a wide-field, low-power telescope.

Stargazing is largely a matter of personal taste, and while most are content to admire the biggest and brightest planetary nebulae, a few seek out the many less conspicuous ones. Jay McNeil of Houston caught the bug while still in his twenties and was soon training his 16-inch reflecting telescope on exotic planetaries so small that abnormally high magnifications were required just to identify them. Stephen James O'Meara recalls that at a Texas Star Party McNeil "began expounding on the incredible beauty of these objects" and asked if O'Meara had ever seen Jonckheere 320, Peimbert-Batiz 4, and Manchado-Garcia-Pottasch 2. "I thought he was speaking a foreign language until he later handed me a typed list of his favorite planetaries—all 450 of them," O'Meara reported. "'Uh, no, I've never seen them,' I replied. 'In fact, I've never even heard of them.'"[4]

While planetary nebulae represent stellar endgames, the bright nebulae and glittering open clusters found in the star-forming regions of the Milky Way's spiral arms display the early stages of stars' birth and death. These regions typically contain star clusters that are still entangled in the nebula from which they condensed. (Such clouds, known generally as diffuse nebulae, are called emission nebulae when their atoms emit light, absorption nebulae when they absorb light, and reflection nebulae when they reflect it.) The Orion nebula is an excellent place to start investigating nebulae generally, since it lies nearby and displays a rich variety of emission, absorption, and reflection features.

The blossomlike emission part of the Orion nebula, its arms reaching out like petals spreading in the morning sun, is an endlessly fascinating sight in a telescope. "No amount of intensive gazing ever encompasses all its vivid splendor," ventured the veteran observer and astronomy writer Walter Scott Houston.[5] In small telescopes it looks gray, which is what the eye registers when light levels are too low to trigger its color sensors. Larger apertures turn on the color, showing traces of dusky ruby red in the tendrils and pockets of pale green toward the center. The red comes from the fundamental transition of hydrogen atoms, while the green comes from ionized oxygen, like that found in planetary nebulae. Whereas most atoms discharge a photon a fraction of a second after being "excited" by an incoming photon, oxygen-three (OIII) in space takes hours to "de-excite." The spectral lines produced in this leisurely fashion are called "forbidden lines," because they cannot normally be seen in laboratory gases, where excited atoms collide with other atoms before they de-excite. Nebulae can display forbidden lines because, although big and massive—the Orion nebula is 30 light-years wide and contains enough material to make ten thousand Suns—they are generally less dense than a laboratory vacuum.[6]

The energy pumped into the Orion nebula, ionizing its gas and exciting it to glow, comes from hot young stars that congealed from the cloud recently in cosmic history and remain embedded in it. The most spectacular of these is a quadruple star at the pistil of this celestial flower. Known as the Trapezium (after the Greek geometrical term for any four-sided figure; its stars make a kind of box), it is scarcely more than a hundred thousand years old and belongs to a young cluster containing perhaps a thousand stars. The Trapezium is readily visible at low magnifications, but high powers reveal some of the intricate structure of the surrounding nebula, which John Herschel likened to "the breaking up of a mackerel sky when the clouds of which it consists begin to assume a cirrus appearance."[7] All four stars of the Trapezium are hot and massive, and their powerful stellar winds have blown a bubble in the surrounding gas. It is because the bubble has burst through the near edge of the cloud that we have a clear view of the nebula's inner regions.

Similar sweepings-by-starlight have made visible many little stellar nurseries known as EGGs, for evaporating gaseous globules, some weighing in at only about one solar mass. These dark clouds consist of infant stars still wrapped in swaddling clothes of gas and dust. When blasted by particularly strong torrents of starlight and stellar winds, they glow on one side and trail off on the other, making them look like ice-cream cones. Discovered by astronomers using the Hubble Space Telescope, EGGs are too small to have been imaged with amateur telescopes, but amateurs can observe their larger counterparts, the Bok globules. Named after Bart J. Bok, a Dutch-American astronomer who studied the galaxy extensively and styled himself a "night watchman of the Milky Way," they range from a third to ten or more light-years in diameter, weigh as much as a thousand solar masses, and are thought to produce litters of ordinary, sunlike stars.[8] Many of the blackest clots in the dark, riverlike nebulae that E. E. Barnard and other stargazers have traced are Bok globules. Where they are silhouetted against bright emission nebulae, such as IC 2944, near Lambda Centauri, and the Eagle nebula, the globules stand out so sharply that they are easily mistaken in photographs for dirt on the lens.

Large dark clouds are woven through the Orion nebula and in places reflect the light of nearby stars. These reflection nebulae are often blue in color, for the same reason the daytime sky is blue—because blue light scatters in gas and dust more readily than red light. Totally dark clouds can be seen only in silhouette, against background stars or bright parts of the nebula. Looking toward the stem of the flower through a telescope, you can see dark material interceding between the Orion nebula proper, M42, and the adjacent nebula M43.[9] Another, wider, dust lane separates that complex from the nearby Ghost nebula, so named because it contains a patch of partially obscuring gas that resembles a portly, Casper-the-Ghost–like figure.

The nearby Horsehead nebula is a spectacular dark cloud shaped like the profile of a knight on a chessboard. It stands a few tens of light-years tall, etched against background sheets of glowing red gas. It is, however, rather dim, and even very capable observers have searched for it in vain when skies are less than pitch black. (A hydrogen-beta filter helps, but is so specialized that wags call it "the Horsehead filter" since it has few other applications.) The Horsehead is worth the trouble, though. There is something deeply moving about this brooding eddy, perched toward one end of the Orion star-making machine. It elicits a sense of the slow toilings of astronomical processes that draws one into further study, in something like the way that Edward Gibbon's wanderings in the ruins of ancient Rome inspired his inquiries, so he said, into the decline and fall of the Roman Empire.[10]

Like picnickers seeing animal shapes in cumulus clouds, stargazers have named other dark and bright nebulae after their evocative appearance. The Witch Head nebula, west-northwest of Rigel, looks sooty, smoky, almost industrial: Work is being done here. The North American and Pelican nebulae, in Cygnus, display a subtle and intricate geography drawn in shades of ashen gray. Getting a sense of the murky reaches of the California nebula, too rangy to take in through a single telescope view but too dim to see well in most binoculars, can be almost as challenging as trying to grasp all California by looking out of an airplane window. To best observe large nebulae like these, start with low-power eyepieces and examine them when they are close to the meridian, where the intervening air is thinnest.

Open star clusters adorn the Milky Way like spangles on a dove-gray scarf. Orion has several pleasing ones—notably NGC 1981, which runs right through the Orion nebula itself—but the greatest concentrations are in Sagittarius, Scorpius, and Scutum, where we are looking toward the heart of the galaxy. By graphing the colors and magnitudes of the stars in a cluster—to determine, for instance, how many have evolved into red giants—astrophysicists can estimate their ages. Such studies show that most open clusters are young: Lacking sufficient mutual gravity to hold together for long, they evaporate in time, ejecting stars through crack-the-whip interactions with one another and to the gravitational pull of passing stars and dense nebulae. Old open clusters tend to be sparse: Having already lost most of their stars, they are approaching total evaporation.

One of the youngest open clusters in our skies is the Double Cluster in Perseus, seven thousand light-years from Earth in the northern Milky Way. A shining new product of the Perseus arm star factory, these twin clusters stand out nicely against the tangled black jungle of the arm's dark nebulae. Much older is the Beehive cluster, in Cancer. It has been around for fully four hundred million years, and its long evolution has produced an enticing variety of blue, yellow, and red stars. The secret of the Beehive's longevity is that it ranks among the most massive open clusters, and so has been able to hold itself together gravitationally for longer than lesser clusters can. It is visible to the naked eye as a nebular patch—Hipparchos and Ptolemy mentioned it—but optics are required to resolve it into stars, an easy feat given its modest distance of only 580 light-years. Nearly as old is M37, in Auriga, home to about five hundred stars. Seen through a telescope, it calls to mind the lights of hillside houses in a port city, surrounded by dark seas, with one bright blue-white star standing out at the center like a lighthouse at the harbor bar.

Globular star clusters are the elder statesmen of the galaxy, their stars at least ten to twelve billion years old. The term "globular" was coined by William Herschel, who discovered thirty-seven of them, and as it suggests, these clusters are spherical or elliptical in shape. Harboring hundreds of thousands to millions of stars, they are much more massive than open clusters, and are distributed differently in space. Open clusters belong to the galactic disk, but globulars occupy a spherical volume of space, centered on the galactic bulge but ranging far above and below it and stretching out for tens of thousands of light-years. This suggests that the globular clusters formed when the young galaxy was collapsing from a spherical configuration to the flat plane it describes today.

Viewed through a telescope, globular and open star clusters look every bit as different from each other as their differing origins and histories would lead us to expect. Whereas open clusters are unpredictable and individual in appearance, globulars present us with a burnished, austere conformity.

Three globulars are bright enough to be seen with the unaided eye—M13, in Hercules, 47 Tucanae, and Omega Centauri. Each is a magnificent sight in a telescope or even binoculars—a grandiose city of chalk-white to yellow stars, with a scattering of red and golden giants emblematic of the cluster's antiquity. M13 is a favorite of Northern Hemisphere observers, with over one hundred thousand stars packed into a sphere only about 150 light-years in diameter. Southern Hemisphere observers can enjoy the splendors of 47 Tucanae, which stands in the foreground, a little under twenty thousand light-years from us, near one edge of the Small Magellanic Cloud, a satellite galaxy of the Milky Way. Omega Centauri, about the same distance away, is the largest known globular cluster in our galaxy and the most spectacular of them all, but its declination of 48 degrees South means that northern observers must venture to the subtropics to get a look at it. It was a favorite target of the little telescopes used by us members of the Key Biscayne Astronomical Association. With such instruments we could see the individual outer stars of major globulars, while those closer to the center melded into sheets of light. Larger telescopes resolve the centers as well, displaying ragged lines and loops of stars. Unlike the planar orbits of planets around the Sun and of stars in the disk of the Milky Way, globular cluster stars orbit in all planes: We see them frozen at various points in these orbits, like baseballs hit in batting practice, some in flat line drives and others in high pop flies.

The Milky Way galaxy is home to an estimated 200 globular clusters, of which 147 have been identified. (As with planetary nebulae, the rest presumably are hidden behind the nebulae and star clouds of the galactic disk.) Neophytes may assume that they all look alike, but to their devotees, globulars are as distinct as European cities. They differ in mass, from a few tens of thousands to millions of stars, and in density. Most are so chock full of stars that stargazers on planets near their centers would never experience a genuinely dark night, but would instead abide amid the multiple shadows cast on the ground by scores of blazing red and golden stars. Others are more diffuse; their night skies would be dark, although studded with many more bright stars than we see from Earth. Astronomers are still trying to figure out the nature-versus-nurture issue of which of these differences are due to disparate conditions when the globulars formed and which resulted from their subsequent experiences. On the "nature" side, tidal interactions with the galaxy limit the sizes of clusters when they form: If a cluster is born too big, the galaxy soon strips away its outer stars. On the "nurture" side, passages through the galactic disk can shake up a cluster and strip away its gas and dust, especially if it happens to encounter a giant molecular cloud like the one in which the Orion nebula is embedded. Students of such insults argue that some were sufficient to have dismembered globular clusters altogether; if so, the globulars we see today are the endangered survivors of a much larger primordial population.

These perturbations aside, globular clusters are relatively pristine laboratories of stellar evolution. A globular starts out with a single, original generation of stars that go through their careers along lines predestined by their masses, with the big ones soon exploding and leaving behind black holes or neutron stars, the average ones eventually turning into red giants and then white dwarfs, and the smaller ones shining steadily for eons. Few if any new stars form in globular clusters, since there is little interstellar material from which to make them: The debris from supernovae would suffice, except that it is generally ejected at speeds that exceed the cluster's escape velocity, and thus departs from the cluster before much use can be made of it. The gas shed by red giants and other aging stars is available for a while, but gets stripped out of the cluster each time its orbit takes it through the galactic plane. Hence globular clusters are thought to provide astrophysicists with something akin to the "untouched, unspoiled" tribes relished by anthropologists—populations that have remained relatively free from contaminating outside influences.[11]

Globular cluster aficionados with large telescopes may graduate from admiring big, bright clusters to hunting down small, dim ones. Some globulars look dim because they lie behind the dust clouds of the galactic plane, which diminish and redden their light. Using a 36-inch telescope, Barbara Wilson managed to see a particularly dim one called UKS 1: The faintest globular in the Milky Way, it had previously been discerned only on a photographic plate made with Cambridge University's 48-inch United Kingdom Schmidt camera. Others look dim because they are so distant. The Czech amateur astronomer Leos Ondra is fond of NGC 2419 in Lynx, a 10th-magnitude smudge of light that he describes as a "forlorn outpost of the galaxy." An outrider, it is located almost three hundred thousand light-years from the Sun, in an orbit so large that it takes an estimated three billion years to make one trip around the galaxy. Barbara Wilson's favorite "extreme halo globulars," as she calls them, include AM 1 and Palomar 4, both nearly four hundred thousand light-years away.

Observers who want to go even farther can seek out the globular clusters of other galaxies. The brightest extragalactic globular is G1, in the Andromeda galaxy. It is intrinsically bright—twice as luminous as Omega Centauri—and its orbit currently has carried it to a relatively conspicuous position well away from its parent galaxy. Reversing his perspective in something like the way that Kepler imagined himself to be on Mars and Leonardo on the Moon, Leos Ondra worked out which of the Milky Way's globular clusters would rank as the best and brightest "for Andromedan deep-sky observers."[12] The answer, surprisingly, was the outrider NGC 2419, a cluster uncelebrated on Earth but evidently prominent in Andromedan skies.

We stand now at the brink of intergalactic space, but before taking that plunge, let us pause to look back at the Sun's neighborhood, touching on a few of the many things necessarily overlooked in this brief overview of our galaxy as a backyard stargazer might explore it.

There are, first of all, our neighboring stars. Even a cursory examination of the solar neighborhood paints a quite different picture than has been suggested by our prior emphasis on the blazing supergiants and lurid nebulae of the star-forming spiral arms. Fewer than one percent of the stars in the solar neighborhood are giants: Ninety percent of them are ordinary main-sequence stars, and the rest are dim bulbs. Of the twenty stars closest to the Sun, only five are bright enough to be seen without a telescope. The others are mostly inconspicuous red dwarfs like Lalande 21185, Ross 154, Lacaille 9352, Ross 128, Groombridge 34, and Luyten's Star. Charles Darwin, asked what his research had taught him about the nature of God, replied that the deity seemed to have an inordinate fondness for beetles. Much the same might be said of the majority of our galaxy's several hundred billion stars, which go about their business without making a show of it. There may also be many failed stars, the so-called brown dwarfs, insufficiently massive to have fired up a thermonuclear furnace and glowing instead by the meager energy of gravitational contraction. It is difficult to take a census of the galactic population of brown dwarfs, since they are so dim, but two candidates have been located near the Sun—LP 944-20 and Gliese 876, each about sixteen light-years from us. Gliese 876 shows evidence of having a planet, about which little can yet be said except that it probably lacks good tanning beaches. Seventy percent of the nearest stars belong to double or multiple systems, and the percentage of such systems drops rapidly with distance, indicating that astronomers have underestimated the frequency of multiple systems simply because existing telescopes cannot yet detect their duplicity.

Extending our survey out to a distance of fifty light-years from the Sun, we find that the interstellar medium—the thin broth of gas and dust adrift in local space—is punctured by bubbles. The bubbles are thought to have been blown by stars that exploded as supernovae. The space inside them is ten times more vacuous than normal—about 0.05 atoms per cubic centimeter, versus 0.5 generally. Many are large enough to have blown all the way through the galactic disk, affording a clearer than average view of the wider universe to observers who are fortunate enough to live inside one. We are among those lucky observers: The Sun currently happens to be passing through a point almost at the center of what is called, predictably enough, the Local Bubble. It measures about twenty light-years wide and is marred by some internal debris (the "Local Fluff"), but is generally quite clear. Were we not so situated, we would still be able to see remote galaxies but their light would be reddened by interstellar dust, and at the short wavelengths of ultraviolet light our view would be rather foggy.

Double the radius of our survey and we find other, larger bubbles nearby, in Cygnus, Perseus, Orion, and Centaurus. Pull back farther and all this froth turns out to form part of an array of feathery clouds, angling off from the nearby spiral arms. Back off to a couple of thousand light-years and the grand design of the major spiral arms comes into view, with the Sagittarius arm inside the solar system's orbit and the Orion-Perseus arm outside it. An astronomer in a neighboring galaxy, viewing this part of the Milky Way through a superb telescope with a field of view stretching from the Sagittarius to the Orion arms, could see many of the prominent objects we have been describing in this chapter—among them the Orion nebula, the star clouds and bright and dark nebulae of Sagittarius, and perhaps the Pleiades and Hyades star clusters. But he—or she, or it—would be hard pressed to discern so humble a star as the Sun, much less make out even its largest planets, much less Earth.

Yet here we are, with our eyes and our minds and our curiosity, six billion passengers aboard a tiny blue boat, bobbing and wheeling our way around one vast Catherine wheel among many. Time now to leave the Milky Way, and explore the realm of the galaxies.

Blues Line:
A Visit with John Henry's Ghost

John Henry told the captain
A man ain't nothing but a man
But before I'll let your steam drill beat me down
I'll die with the hammer in my hand.
    —"John Henry" (traditional)

My hammer's on fire, boys,
My hammer's on fire.
    —Big Louisiana, "Let Your Hammer Ring"

I'M NO JOHN HENRY, I thought, lying on a bench at the observatory with my arms crossed behind my head, gazing idly into the Great Square of Pegasus while the telescope did all the work. He fought the machine, died trying; I'd given in to it. I used to hunt for supernovae visually—still do, sometimes—but now I was letting the telescope do the hunting. Humming, chirping, and burbling to itself, its gleaming black tube with the blinking red lamp of its CCD camera reared heroically against the night sky, it automatically took an image of a galaxy every few minutes, stored it in a computer, then slewed with a purposeful whine to the next galaxy on its list and repeated the process. The CCD could detect much fainter stars than my eye could, and I didn't even have to scrutinize all the images it produced. The computer would do that, checking each new one against others made of the same galaxy in the past and alerting me when it came across a new dot of light that might be an exploding star. It all made perfect sense.

So why was I uncomfortable about it?

Perhaps because it distanced me from the stars. The CCD camera plugs into the telescope where the eyepiece normally would be, so when it's looking through the telescope, I'm not. Thanks to modern technology, I could observe without being here at all—just as I could talk to my wife on the telephone without looking into her eyes. But was this an improvement?

Late one afternoon years ago, I was meditating on a boulder overlooking Tsankawi mesa in New Mexico while waiting for a group of picnicking students to arrive to hear a talk. The Anasazi, ancestors of the modern Pueblo, lived there centuries ago, in what amounted to a large apartment complex on the mesa's high grounds. They called the peaks at its outposts the four corners of their universe, and I could see what they meant. For all the complexities of the village ruins in the foreground and the bone-white and cactus-green ridges beyond, the mesa had a kind of isolated wholeness to it: Living there must have been a bit like colonizing a small comet or asteroid. Thinking about them, I inquired of their departed souls, "How may I talk about the universe, here in your former home, without needlessly offending your memory?"

The answer came instantly, in a voice of thundering silence.

"Your presence is not required," it said.

I was grateful for this marvelous injunction, combining as it did a wry comment on trespass with sound advice about how one actually ought to speak about cosmology. When the students arrived I tried to put it into practice in my talk, and I'd thought about it many times since. But tonight in the observatory, it seemed bleak and all too literal. The telescope was doing the observing. My presence was not required.

As for John Henry, he evidently was a real person—a former slave, it is said, physically spectacular, a gifted singer and banjo player and a leader of men, who dropped dead after drilling fifteen feet to the steam drill's nine.[1] His myth, however, is constantly being reinterpreted. Indeed, if you consider that John Henry may, like Socrates, have been complicit in his own death, it stands as his greatest composition, a memorable work of art too ambiguous to be distilled into any one message. John Henry bested the machine, but to what end? Steam drills were eventually adopted anyway, and they saved the lives of many railway workers—not on the Chesapeake & Ohio's Big Bend tunnel, where he reputedly met his death, but certainly on subsequent tunneling jobs—so why not lay down the hammer and let steam drills do the work? That was the practical moral that Mississippi John Hurt drew from the story, in a song about a steel driver who finds discretion the better part of valor:

> This is the hammer that killed John Henry
> But it won't kill me.[2]

In the end I took the way of John Hurt, not John Henry. Computers and CCDs were making almost all of the supernova discoveries now. Trying to beat them visually was hopeless. I'd give them a try.

The telescope needed renovation anyway. It had started out as a semi-portable affair that I lugged out of a storage shed on a wheelbarrowlike contraption fitted with a pair of rubber-rimmed wheels. Rolling it across bumpy, rocky fields, I'd tested various sites until I found a good one for the observatory. Once it was installed on the observatory's concrete pier, the telescope worked capably for years. Eventually, though, the design features that made it portable—its light skeleton tube and its aluminum rotating nose assembly, which put the eyepiece in a convenient position no matter where it was pointed in the sky—conspired to degrade its performance. It wouldn't hold collimation—using it was like playing a cello that won't stay in tune—nor could its mount be retrofitted with the motors required to automatically land galaxies in the camera's field of view. It was time for a change.

I decided to keep the primary mirror, fashioning a new, automated telescope around it. I consulted a number of engineers, equipment makers, and amateur astronomers, and a plan emerged. We would rebuild a massive old mount that had been crafted in 1972 by the master gear-maker Edward R. Byers, of Barstow, California. The mount had a checkered past—among other adventures it once spent three months underwater, at the bottom of a Hollywood swimming pool, after a party guest knocked it over while its owner was vacationing in Europe—but I could afford it, whereas a new one would have cost more than a Porsche. We'd fit the mount with a modern control system, marry it to a carbon-fiber tube—rigid, disinclined to shrink in the cold, and good at absorbing stray light—install the old mirror, and fire it up.

The project took a year longer than scheduled, during which time I learned more about the financial, psychological, and marital problems of some of the team members than I would have anticipated, but they proved to be as honest and hardworking as they were technically adept, and now the telescope was in operation—a black lacquer, chrome, and gold automaton pretty as a custom hot rod and capable of driving itself. Plugged into a computer that I'd built from shelf parts, in the back of a cramped electronics shop in a half-abandoned shopping center, it could be programmed to take images all night without human intervention.

Its power was seductive, but the more time I spent at the computer terminal the less I spent looking at the sky. My presence was not required.

All this was, of course, just one small footnote in the greater story of computers taking over the world. During the early years of the personal computer boom, when fears were voiced about whether we were unwittingly fashioning a species of artificial intelligence destined to dominate our own, people used to say, "Oh, well, you can always just unplug the computer." You didn't hear that now. Computers had become indispensable, and there were more of them every day. By the year 2001, a personal computer was being sold every *second* worldwide, and their users were eagerly linking them together into something resembling a planetary brain. Recently I'd talked with the physicist Paul Davies about the notion that we would eventually produce computers smarter than we are.

"Isn't that rather an appalling prospect?" I asked him.

"No," Davies replied, "because if it happened, we human beings would be so comfortable that we wouldn't care."[3]

This did not exactly set my mind at ease. Out under the stars, watching the computers run the telescope, I constructed paranoid fantasies about the shape of future history. I recalled that Alan Turing, who along with John von Neumann discovered the essentials of digital computation, had committed suicide by taking a bite from an apple that he'd injected with cyanide—thus re-enacting a scene from "Snow White," his favorite fairy tale. (Turing, whose work in cracking the Nazi Enigma code should have made him a war hero, had instead been sentenced to undergo "chemical castration" to prevent his engaging in homosexual conduct, then a crime in England.) The lethal apple was found on the bedside table next to Turing's body, a single bite taken out of it. And what was the corporate symbol of the company that had launched the home computer revolution? An apple with a bite out of it! Could the handwriting on the wall be any clearer?

In more rational moments, I consoled myself with the thought that what Turing and von Neumann had discovered was not just a technology but a fact of nature, as fundamental as nuclear fusion. The efficacy of computation indicates that natural processes really can be reduced, not just to numbers, but to the simplest possible numbers—the zeros and ones of binary computing. Quantum physics is full of systems that can have one of two states (spin up or spin down, charge positive or negative) and nothing in between—states, in other words, that can be represented by zeros and ones. DNA makes evolution possible by quantizing genetic information; it happens to use a four-based code, but it might as well be zeros and ones. Human thought is produced by the actions of billions of synapses, electrochemical switches each of which at any given moment either fires or does not—zeros and ones.

Starlight striking the chip in a CCD camera either fires a given pixel or does not—zeros and ones.

Ludwig Wittgenstein claimed that "the world is the totality of facts, not of things."[4] Computers agree, and support their argument by demonstrating in practice that all sorts of processes can be predicted, recorded, controlled, and interpreted by means of binary digits. The notion that the universe "is" a computer, though perhaps repugnant at first blush, really means that nature may best be viewed as being based, not on matter and energy or space and time, but on information—since all that we can know about the cosmos must necessarily consist of information, which in turn is reducible to the binary digits that computers thrive on. So maybe the burbling, autocratic, black-and-gold telescope was on to something. Maybe when we use computers to explore the sky, we are getting in touch not just with computers or the stars but with a deep and as yet only dimly comprehended principle that underlies them both.

Still, I like to look. So I stayed out there through the night, staring uselessly at the stars.

17.

Galaxies

Ships that pass in the night, and speak each other in passing,
Only a signal shown and a distant voice in the darkness;
So on the ocean of life we pass and speak . . .
. . . then darkness again and a silence.
   —Longfellow

Suns haste to set, that so remoter lights
Beckon the wanderer to his vaster home.
   —Emerson

ON A CRISP, SPLENDID NIGHT in the mountains, I had a look at the Andromeda galaxy through a small telescope with a wide enough field of view to capture most of its enormous glowing disk, which stretches nearly five degrees across the sky. Its melded starlight had a smoky, luminous quality, smooth and yet alive with scintillations just below the threshold of vision, like the campfires of a Roman legion viewed from a distant mountaintop. It stirred a restless welter of thoughts that, like the Andromedan stars, never quite resolved themselves. It's hard to get a handle on galaxies.

Needless to say, they're really big. Were the Sun a grain of sand, Earth's orbit would be an inch in radius, the solar system the size of a beach ball, and the nearest star another sand grain four miles away. Yet even on that absurdly compressed scale, the Milky Way galaxy would be a hundred thousand miles wide. Galaxies are so big that once you get up to their scale, the universe starts to take on an almost country-cottage intimacy. The larger galaxies in clusters like the Local Group, to which Andromeda and the Milky Way belong, typically lie only a couple of dozen galactic diameters apart from one another— comparable to dinner plates at the ends of a twenty-foot-long dining table. Add in the galaxies' halos of stars, globular clusters, associated hydrogen clouds, and dark outer disks, and they almost impinge on each other. On the same scale, the Virgo supercluster, of which the Local Group is an outlying member, comprises ten thousand plates scattered across an area not much larger than a football stadium, and the entire observable universe has a radius of only about twenty miles. From a galaxy's point of view, the universe isn't all that large.

The trouble is that it's difficult—probably impossible—for a human to make the mental leap to galactic scale. The very concept of space is inadequate for dealing with galaxies; one must invoke time as well. The Andromeda galaxy is steeply inclined to our line of sight, only fifteen degrees from edge-on. Since the visible part of its disk is roughly one hundred thousand light-years in diameter, the starlight reaching our eyes from its more distant side is about one hundred thousand years older than the light we simultaneously see coming from the near side.[1] When the starlight from the far side of Andromeda started its journey, *Homo habilis*, the first true humans, did not yet exist. By the time the near-side light started out, they did. So within that single field of view lies a swath of time that brackets our ancestors' origins—and that, like the incomplete dates in a biographical sketch of a living person (1944–?), inevitably raises the question of our destiny as a species. When the light leaving Andromeda tonight reaches Earth, 2.25 million years from now, who will be here to observe it? We think of Einstein's spacetime as an abstraction, but to observe a galaxy is to sense its physical reality.

THE SHEER SPLENDOR of galaxies, which has elicited passionate outbursts from many sober scholars ("Majestically beautiful," blurts out one professional astronomer, in the first sentence of an otherwise poker-faced encyclopedia entry),[2] derives in part from their immediate appearance—the elegant dusky arms of the spirals, the bald sterile glow of the ellipticals. But it has also to do with their enormous dimensions and the wealth of worlds they contain. As objects of study, galaxies are bottomless. If we spent eons observing the Andromeda galaxy with ever better equipment, we would, presumably, learn a great deal—indeed, one hopes that this will happen—but there would *always* be more to learn, if only because so many things keep changing there. To pick a literally glaring example, it is estimated that more than fifty thousand stars have exploded in Andromeda in the past two million years: The light from all those supernovae is already hurtling through space toward our telescopes, part of Andromeda's past and our future. A galaxy is not so much a thing as it is a grand, gorgeous exemplification of the scope of cosmic space and time.

Galaxies are gravitationally bound aggregations of stars, considerably larger than globular star clusters and displaying a greater variety of forms. They range in size from dwarfs with just a few million stars to giants with several *trillion*. (Large galaxies are greatly outnumbered by the dwarfs, but contain so many stars that if a species evolves by chance on a given planet, the odds are that it will find itself in a big galaxy rather than a small one.) In terms of their general appearance, galaxies fall into three broad categories: spirals, ellipticals, and irregulars. About a third of all the bright galaxies in the sky are spirals, flattened systems with a central bulge and a set of dusty, gas-laden, star-forming spiral arms. Ellipticals are spherical or oval in shape, contain relatively little dust and gas, and hence mint few new stars. The ranks of the ellipticals comprise nearly two-thirds of all cataloged galaxies, including some of the smallest dwarfs and virtually all of the largest giants. Irregulars constitute only a few percent of the known galaxies, although there are probably many dim ones that have to date escaped our cosmic census takers. Although irregulars lack any immediately evident structure, some turn out, on closer inspection, to have been spirals or ellipticals that were disrupted by tidal encounters with other galaxies.

This tripartite classification scheme, developed by Edwin Hubble in the early days of galaxy studies, has since been much refined and elaborated on. There are, for instance, barred spirals, whose spiral arms originate at either end of a bright bar that extends out from the central bulge like a tightrope walker's balancing staff, and ring galaxies, where the arms join to make a circle. There are also armless (or "S0") spirals, some of them hard to distinguish from ellipticals, and ellipticals so flattened that they impersonate edge-on spirals, as well as many sorts of almost unclassifiable "peculiar" galaxies.

Closest at hand are the Milky Way's satellite galaxies. Eleven are known at this writing, and more may be hidden from view behind the Milky Way's dusty disk. The nearest one yet identified, the Sagittarius Dwarf Elliptical, lies near the galactic plane on the far side of the bulge. Only eighty thousand light-years from Earth, it covers a patch of sky fully five by ten degrees across, but the Sagittarius star fields in the foreground obscure it so much that it wasn't even identified as a galaxy until 1994. A stream of its stars has been tugged away by the Milky Way, eventual absorption into which is the likely fate of the Sagittarius Dwarf Elliptical as a whole. In contrast, the large and small Magellanic Clouds, a pair of Southern Hemisphere irregular galaxies located 160,000 and 180,000 light-years from Earth, respectively, appear likely to maintain their independence for billions of years to come, although they will probably merge with us eventually.

The Milky Way's other satellite galaxies, in order of increasing distance, are the Ursa Minor, Sculptor, Draco, Sextans, Carina, and Fornax galaxies, followed by Leo II and Leo I. The last of these takes us out to a distance of 830,000 light-years.[3] All are dim dwarf ellipticals, the largest only three thousand light-years in diameter, so they make for rather disappointing sights in a telescope if one notices them at all. The Sculptor galaxy looks like a somewhat flattened globular cluster, and is so dim that visual observers classify it as a "challenge object."

The Local Group is dominated by two big spirals, the Andromeda galaxy (M31) and the Milky Way, with most of the Group's forty-one other known galaxies clustered around them. Twice as massive as the Milky Way, Andromeda is one of five galaxies bright enough to be seen distinctly with the unaided eye. (The others are the Milky Way itself, the two Magellanic Clouds, and M33, a spiral near Andromeda.) Al-Sufi in Persia described Andromeda in A.D. 964 as a "little cloud," and the German astronomer Simon Marius, who observed it with a telescope in 1611 or 1612, memorably wrote that its glow resembled "the light of a candle shining through horn." (He was thinking, presumably, of the alabaster candles popular at the time.) Through a telescope, Andromeda presents us with the full spectacle of a major spiral galaxy viewed at close range.

Thousands of photographs of Andromeda have been published, but almost all of them overexpose the central bulge in order to bring out the spiral arms. Observing it visually through a telescope, one is surprised to see just how bright its almost starlike nucleus is, and how steep the falloff in luminosity as one looks out from the nucleus to the rest of the bulge. If the disk were not there, the bulge would resemble an elliptical galaxy with a bright, compact core. But of course the disk *is* there, a vast array of glowing star-forming regions and clusters of massive young stars lining the spiral arms, interspersed with archipelagos of dark clouds that twist out, in turn after turn, from the center to the almost boundless outer reaches. One dark arm stretches dramatically across the front of the bulge, inviting comparison with our view of the Sagittarius arm of our own galaxy. Bright star-forming regions can be seen along the arms, and experienced visual observers can pick out some of Andromeda's globular star clusters as well.

The disk of this magnificent galaxy is warped a bit, probably by the gravitation of its two largest satellite galaxies. One can readily see them in a small telescope—M32, peeping up over the south edge of the disk, and NGC 205, a bit more in the clear to the north. Both are dwarf ellipticals. Weighing in at three billion and ten billion solar masses, they are substantial enough to have bent Andromeda on their own, although other, smaller nearby galaxies may have contributed to the process.

Off toward the edge of the Local Group resides its only other spiral, M33, its disk spread out across a full degree of sky, lovely as a lotus blossom though rather dim. Its highlight is an enormous star-forming nebula, NGC 604, thirty times larger than the Orion nebula. M33 may be a distant satellite of M31. For that matter, the Milky Way itself is, in a sense, a satellite of Andromeda: The two big spirals are gravitationally bound together, and are currently approaching one another. Some astronomers theorize that eventually all the galaxies of the Local Group are destined to meld into one system, in which case nearly everything in the Milky Way will become part of a future mega-Andromeda. There are some remarkably large galaxies, here and there in the universe, that could have got that way by gobbling up their siblings.

The Local Group is one among many compact clusters of galaxies. The compact cluster nearest to it, at a distance of ten million light-years, is the Sculptor Group—also called the South Polar Group, since it lies near the part of the sky toward which points the south pole of the Milky Way. Its brightest member is the lovely "Silver Coin" galaxy, discovered by Caroline Herschel, William Herschel's sister and collaborator, on the night of September 23, 1783. Nearly half a degree wide and quite bright at 7th magnitude, it does display something of an old coin's hue. The group also contains the small but beautiful spiral NGC 300, and the strangely jumbled NGC 55, which seems to be a perturbed spiral. Slightly more distant, in the northern constellation of Ursa Major, resides the M81 group, named for its dominant galaxy, a beautiful 8th-magnitude spiral and a favorite among stargazers. M81 has a 9th-magnitude companion, M82, that has been racked by an episode of massive star formation. Evidently tidal interactions with M81 recently created density waves in M82 that triggered the fireworks.

When we look out to the NGC 5128 group, twelve million light-years away, we encounter something very different. Its dominant galaxy, NGC 5128, is a brightly glowing sphere of light bisected by a dark belt—"a most wonderful object," as John Herschel described it in 1847.[4] A powerful radio emitter, it is also called Centaurus A, meaning that it is the strongest source of radio noise in the Southern Hemisphere constellation Centaurus. Closer examination shows that the sphere is a large elliptical galaxy and that the belt is the dusty disk of a spiral galaxy in the process of being swallowed up by it. The dust lane obscures the center of Centaurus A from view in optical wavelengths, but it can be probed with radio telescopes and spaceborne X-ray telescopes. For students of galaxy evolution, Centaurus A is as valuable as a prosecutor's being presented with a snapshot of a holdup in progress: Giant ellipticals are thought to have grown fat by eating spirals, but this is the most dramatic case of such galactic cannibalism that we can see at close range.

Another nearby instance of galaxy interaction is found in the case of M51, the Whirlpool, a spectacular face-on spiral located 37 million light-years from Earth. A smaller galaxy, NGC 5195, recently swung past M51, distorting it somewhat and firing up its spiral arms. The encounter jumbled NGC 5195 so badly that astronomers cannot tell what its shape used to be.

Near the M51 group lies the M101 group, dominated by its photogenic namesake, a loosely sprung spiral nearly half a degree wide, its glowing star-forming regions standing out like insects trapped in a spider's web. Some of these bright nebulae are located so far out on the arms that they are easily mistaken for independent nebulae. The M101 group contains at least three other spirals, plus assorted irregulars and scores of dwarfs.

Farther out, we continue to find galaxy groups dominated by spirals (although dim ellipticals could be lurking in some of them, unseen at these distances). NGC 2841, a large spiral with exquisitely subtle, detailed arms, hosts a group that includes four or five other spirals and perhaps a dozen dwarfs. NGC 1023, ruler of the next roost, is an armless—or S0—galaxy with five spiral companions. Twenty-five million light-years from Earth we encounter a pretty triplet of interacting spirals, M66, M65, and NGC 3628, all visible within the one-degree field of a typical low-power eyepiece.

Although most of the galaxies in our neighborhood belong to groups, there are a few isolated (or "field") galaxies that have no companions at all, or at least none bright enough to have yet been detected. Astronomers used to think that most galaxies were isolated, but as telescopes and detectors improved and dimmer companions were discerned, it became evident that true field systems are rare. NGC 404, an S0 galaxy located midway in the sky between M32 and M33 but farther from us, seems to live alone, but may instead be an unrecognized fringe member of the Local Group. The barred spirals NGC 2903 in Leo, NGC 6946 in Cygnus, and 6744 in the southern constellation Pavo are said to be solitary, as is NGC 1313 in the far southern constellation Reticulum—but the shattered appearance and violent activity of NGC 1313 leads one to suspect that it has been interacting with *something*, perhaps another galaxy currently hidden behind it.

A great many other engrossing nearby galaxies are to be found within thirty million light-years of Earth, but this is a good point at which to step back and consider the wider picture, because we're now halfway to the Virgo cluster.

If we limited our survey of galaxies to those we've encountered so far, we might conclude that the galactic population of the universe consists of spirals, dwarf ellipticals and irregulars, and a few large ellipticals, organized into small clusters that are scattered pretty much at random in all directions. That impression changes once we train a telescope on Virgo. Here one is treated to the astounding sight of scores of galaxies densely strewn across some fifty degrees of the sky. The Virgo cluster is home to some two thousand galaxies, of which a couple of hundred are bright enough to be conspicuous in amateur telescopes. Its galactic population is distinctly different from what we've seen closer to home: An initial glimpse can make you feel like a country bumpkin on his first trip to the big city. The cluster core is dominated by two big ellipticals, M84 and M86.[5] At least three other galaxies appear in the same low-power field of view, where one can see, in the words of the amateur astronomer Alan Goldstein, "the combined light of at least a trillion stars."[6] A line drawn eastward through these two giants encounters a string of galaxies—"Markarian's chain," named for the Armenian astronomer Benik Markarian—along which lie four spirals, two ellipticals, and an S0.[7]

Nearby, just southeast of the cluster core, glows the giant elliptical M87, home to an estimated 2.5 trillion stars. Extremely long-exposure ("deep") images show that its halo, rich in globular star clusters, extends across an area of the sky wider than the full Moon. The plasma jet that protrudes from the nucleus of M87—we encountered Barbara Wilson observing it, at the Texas Star Party—is one of a pair thought to have been ejected from the swirling accretion disk surrounding a supermassive black hole. The far-side jet has never been detected visually—it's on Wilson's list of "impossible" challenge objects—but has been imaged by radio telescopes.

The abundance of large elliptical galaxies in central Virgo, and their scarcity out where we reside, suggest that the present-day appearance and behavior of galaxies in big clusters is determined by the ecology of the cluster in which they find themselves. We will see more evidence of this in the next chapter, when we examine what goes on at the centers of densely populated galaxy clusters.

The Virgo cluster lies near the north galactic pole, as far as one can get from the disk of the Milky Way galaxy, so our view of it is besmirched by relatively few foreground stars. This can make it difficult for observers to find their bearings at first, but it also provides an unusually close approximation to what galaxies look like in their natural habitat, as glowing spheres and disks set against the unabridged blackness of intergalactic space. At these distances, galaxies do not look as large and bright as do those in nearby clusters, but their abundance and variety is memorable. In the words of Leland S. Copeland, an assiduous amateur deep-sky observer prominent in the first half of the twentieth century, they are intriguing "not for what they seem to be, but for what they are. Each is a distant Milky Way, seen by light millions of years older than prehistoric man. They can help us gain true perspective—we and our world are the minutiae and curiosa—galaxies are the grand realities."[8]

To find your way around in Virgo, use a good chart, start with lower-power eyepieces to identify landmark galaxies, then switch to higher magnifications to examine the dimmer ones. Its many splendors include M90 and M88, both big, brooding spirals with only a scattering of foreground stars, and the Siamese Twins, NGC 4567 and 4568, symmetrical spirals that appear close together but lack the signs of obvious disruption that close interactions normally produce: Possibly the twins are a binary pair, akin to the Milky Way and the Andromeda galaxy, that happen to be aligned almost along our line of sight. Another intriguing Virgo pair is NGC 4435 and 4438, called by Copeland "the Eyes," two narrowed ovals that really do seem to be peering at us, and rather eerily at that. NGC 4565, a prominent edge-on spiral, lies to the near side of the cluster and so spans more sky than do its compatriots, measuring a generous sixteen minutes of arc across. Other pleasing near-side Virgo galaxies are M104, the Sombrero galaxy, its portly bulge and bisecting dust lane so prominent that I have seen them with a small telescope low over the lights of Hollywood Boulevard, and M64, known as the Black Eye or Sleeping Beauty galaxy. M64 shows evidence of having undergone two recent spasms of star formation, and it contains a substantial amount of material that is rotating in the opposite direction from its main disk, anomalies that suggest that it has recently cannibalized another galaxy. While in Virgo you may also wish to see M99, bright enough at magnitude 9.8 to be readily visible even in small telescopes, and M100, whose spiral arms are discernible through telescopes of about eight inches or larger aperture.

Virgo is a typical example of an irregular cluster of galaxies in which ellipticals occupy the center while spirals predominate in the outer regions. It forms, in turn, the central mass concentration of what is called the Virgo (or Local) supercluster, an aggregation of tens of thousands of galaxies scattered across a volume of space 150 million light-years wide. Other than the Virgo cluster itself, the supercluster consists mainly of small families of galaxies like the Local Group and our neighbors the Pavo-Indus, Fornax, and Dorado groups.

All of which presents us with another of astronomy's infamous lessons in humility. The Earth is not at the center of the universe, as the pre-Copernicans assumed, nor is the Sun, as Copernicus thought. Instead, the Sun lies out in the suburbs of an average (well, larger than average) spiral galaxy, in an ordinary group, which in turn resides out toward the fringes of a supercluster. If the supercluster were scaled down to cover the surface of Earth, our galaxy would be, say, Boston; the Andromeda galaxy would be about the size and distance of New York City; and the brightest city lights in this part of the universe, those of downtown Virgo, would be in Los Angeles.

But even from out here in the boondocks there's plenty to see. Within 100 million light-years of Earth—well within the reach of visual observers with medium-sized telescopes, and of small telescopes with CCD cameras—reside 160 groups comprising 2,500 large galaxies and perhaps 25,000 dwarfs, containing something on the order of 500 trillion stars. Beyond that things get dim and distant, but not yet invisible.

Big Science:
A Visit with Edgar O. Smith

IN DECEMBER 2000, I spent a night using the 1.2-meter Calypso telescope at Kitt Peak National Observatory outside Tucson to image bridges of stars and luminous gas linking interacting galaxies. Calypso is the only private telescope on the peak; all the others are operated by universities and the government. It was designed and built by Edgar O. Smith, an entrepreneur turned astronomer who invited other observers to use it as well, noting that a fraction of its time suffices for his own research, since "data come off it like water out of a fire hose."

I've known many amateurs whose passion for astronomy led them up what the ancient Romans called "rugged ways to the stars"—sleepless nights of observing, long days spent tweaking gear and processing data, telescope-building projects that entangled them in second mortgages and marriage counseling, elaborate expeditions to the remote sites of occultations and eclipses—but none who took a steeper path or hiked it more vigorously than Edgar O. Smith. When we'd first met, nine years earlier, Edgar was a bachelor millionaire with an enviable wine cellar, an elegant art collection, and a restless intellect, who was observing with an off-the-shelf 14-inch Schmidt-Cassegrain telescope in the back yard of his country place in Connecticut. Like many amateurs, he wanted to get a bigger telescope and learn more about astronomy so he could use it properly. What made Edgar unique was the loftiness of his aspirations. His new telescope, he decided, would be a one-of-a-kind meter-class instrument equipped with a sophisticated adaptive optics system, located on a high mountaintop and capable of the best resolution attainable this side of low-Earth orbit. Learning enough astronomy to make proper use of it meant, for Edgar, enrolling as a graduate student at Columbia University and earning a doctorate in astrophysics—notwithstanding the fact that he was well into his fifties and had a business to run.

I kept in touch with Edgar over the years as he pursued these goals. We visited the plant where his telescope was being built and discussed his progress through the graduate program in astronomy and astrophysics at

Columbia University, where the other students made him feel welcome, although he was twice their age, and he was hazed by few of the professors. Not until his doctorate was in hand, his thesis published (as three papers in the *Astronomical Journal*, concerning stellar evolution in the Fornax galaxy and certain globular star clusters), and his telescope in operation at Kitt Peak did he agree to sit down for an interview. We met at his corner office at the headquarters of E. O. Smith & Co. in the Seagram Building on Park Avenue. More reticent than he looks—he's a big, fit man with a rugged, open face and bright eyes—Edgar spoke softly, as if musing aloud.

He told me that he had been born in Washington, D.C., but grew up on a dairy farm in Doylestown, Pennsylvania, where his father, formerly a field engineer for the Civil Aviation Authority, had bought some land. "My father was a restless, aggressive, rough guy, hard-drinking and violent, a very difficult man," Edgar said. "He knocked me around a lot. A defining moment in my life came when I was about thirteen years old. We had a guy working on the farm who was kind of a club boxer, and he and I used to put out hay bales to create a ring and box. One night my father came by and said, 'What are *you* doing?' and insisted on putting the gloves on. He was going to show me, but I had a lot of resentment, and I knocked him cold. I hit him so hard he had a headache for two days.

"That was the end of the violence, but I was still rebelling against it, heading for juvenile delinquency. They kicked me out of high school three times, and I was later told that the deportment of the entire school improved each time I left it. Fortunately, they had very good high school football programs in Pennsylvania, and one day the line coach said, 'Gosh, Edgar, I see you hanging out on the corner with those bums, smoking cigarettes; I guess you're afraid to come out for the football team.' He was a smart man. I took his challenge and joined the football team, and it was a metamorphosis. I went from being a troubled, troublemaking, smoking, fistfighting, girl-chasing kid—and worse—to being an ideal student and crazy about sports. I played tackle, threw the discus, and wrestled. It takes a lot of energy to be a bad boy, and I found that the energy was better channeled into sports."

Edgar's father refused to help with his college applications—"He tried to trip me up"—and Edgar graduated high school with no prospect of further education, although he'd been offered football scholarships at several universities. Taking matters into his own hands, he went to see the dean of admissions at the University of Pennsylvania that summer and talked his way in. "The first semester I was in the top third of my class," Edgar told me. "I majored in economics, took a lot of literature courses, and was named to the two honor societies for students who contributed the most to the campus. I lettered in football, wrestling, and track at Penn.

"At Harvard Business School, I similarly never applied. Instead I went up there and talked to the assistant dean of admissions, James L. 'Leslie' Rollins. He was a bit of a maverick himself, and one of the things he did was look for unusual students. I told him that I would do this and I would do that and they ought to take me, and finally he said, 'Stop! One thing you're going to have to learn is how to keep your mouth shut, because you were in five minutes ago.' I loved it there: They pushed me really hard."

After earning his Harvard MBA and learning the ropes in a New York financial firm, Edgar went into business for himself, acquiring small, family-owned industrial manufacturing companies. He did well at first, but "then the economy changed and we couldn't get anything done. My friends were moving ahead with their careers, and there I was, stuck, twenty-eight years old and barely able to keep the wolf from the door. I got frustrated and depressed, but I met a great lady, a Ph.D. student in art history at Columbia—very, very wise—who said, 'Edgar, don't quit.' I took her advice and sure enough, I soon did a large deal—God knows how—that put me on the map. With the money I bought my first Navajo textile. I'd had some German expressionist woodcuts before, but when I ran out of money I had to sell them. I'd had to sell everything."

When I asked Edgar how he first got involved with astronomy, he told me about two incidents that both seemed like dead ends at the time. The first was a visit with his mother to the Naval Observatory in Washington, D.C., when he was perhaps five years old. An astronomer there showed them the big refracting telescope and invited them to come back by night to have a look through it. But they never returned. "When I consider the conditions under which my mother lived, it was remarkable that she ever took me there at all," Edgar remarked. Then, when the family had moved to the farm, Edgar's father came home with an aluminum tube and a copy of Albert G. Ingalls's classic *Amateur Telescope Making.* Edgar was beside himself with excitement—"My eyes popped out and I thought, 'Oh boy!'"—but nothing came of that, either. Instead, "we bought more and more cows, and pretty soon all we were doing was working hard on the farm. When my father died, I went down and cleaned out that barn, and among the things I found was that tube, and the Ingalls book on the shelf."

Astronomy remained on the shelf for decades, until Edgar installed the telescope at his country house and conceived of becoming an astronomer. "I'd

done OK in life but I wanted something new," he said. "It gnawed at me a little that when I picked up, say, *Scientific American*, I couldn't read it right
through. I was afraid that if I tried to get a Ph.D. in astronomy I might not be
able to do the work, but a friend told me, 'If a challenge doesn't have some risk
to it, it's not worth doing,' and I bit on that. I worked very, very hard at
Columbia, getting up early in the mornings to study, working every weekend,
staying up there late at night—sometimes all through the night. Astrophysics
can be a monastic, lonely way of life, one that requires tremendous dedication,
but I was thrilled that I had the opportunity. I pushed myself really hard, and
got through pretty fast."

Building the new telescope cost Edgar a substantial portion of his net
worth ("Nearly all such telescopes wind up costing three times as much money
and twice as much time as you expect, and this one was no exception") and
involved so many frustrations that he dubbed his crew of engineers the Don
Quixote team. He named the telescope Calypso because "Calypso snared
Odysseus for seven years—which is about the time it took to build it—and
was considered to have unusually good sight." Once it was in operation, on a
site at Kitt Peak that Edgar negotiated for after considering several other
observatories, he embarked on his new career as an astrophysicist, using
Calypso's sharp vision to separate stars crowded together at the centers of
globular clusters and obtain data on their brightness and color. "There's not a
lot left of me at this point," Edgar said, laughing. "I feel like I've walked across
a desert and haven't had a drink of water in a long time. I'm starting to feel a
sense of excitement, but it's been exhausting from a time aspect—and finan-
cially, too."

Around midnight on my run at Calypso, the telescope froze. Accompanied
by the observatory manager, a French astronomer named Adeline Caulet, I
climbed three flights of steel-mesh outdoor stairs to the lofty perch where the
telescope stood in the open—to maximize its thermal acclimation to its surroundings, it is housed in a shed that rolls completely away at night—and we
pushed it to its parked position so that the computers could be rebooted. Adeline went back down to the control room but I remained behind for a time,
feeling at home on this open perch with no companionship but the telescope
and the starry sky.

Every good observatory embodies a coherent central philosophy, and this
one was dedicated to the behavior of the air. The platform under my feet was
made of steel mesh, to allow free circulation of air. It stood 34 feet high, to get
it above the local boundary layer—the zone above which air flows freely and

below which it is churned by friction with the ground—and thus avoid "ground effects" like those Clyde Tombaugh had warned me about. The pier supporting the telescope was tapered toward the top, to slice through the prevailing winds rather than rolling them down to where they could stir up ground-level air. A massive evacuation system drew air away from the telescope through big white ducts and expelled it far below and downwind, to minimize local puddling and the effects of heat rising from telescope components. Not that the telescope generated much heat: Its CCD cameras were chilled by liquid nitrogen, the mount was made of thermally inert materials selected for minimum heat production, and the rolloff shed, when closed, was pneumatically sealed and kept air-conditioned by day so that the telescope would start work each night at the ambient air temperature. There was even an "experimental airfoil"—deployed, as Edgar wrote in his thesis, "to accelerate incoming wind downward."

The telescope itself—with its black skeleton tube, its multitude of flat-black light baffles (selectively perforated with ventilation holes to discourage them from clawing at the air), and an exotic adaptive optics system capable of adjusting to changing seeing conditions a thousand times per second—looked like a space probe sent here by Martians. Its performance lived up to its futuristic appearance. Professional telescopes are generally considered to be operating well if they point to within ten arc seconds of the desired target; this one had a pointing accuracy of one arc second. An excellent primary mirror might be figured to an overall accuracy of one-tenth of a wave of sodium light, a superb one to one-twentieth-wave; this primary mirror was figured to one-fiftieth wave. A combination of telescope and environment capable of reaching a resolution of one arc second is regarded as first-rate; this one was already performing four times better than that, resolving to a quarter of an arc second on good nights. ("Nature is so humbling, though," Edgar had said with a sigh. "If you open the telescope up and it's a bad night, that's that.") A technophile in my line of work gets to study some pretty elegant machines, from spaceships to race cars and stealth fighter planes, but this was as good as any of them.

Responding to Adeline's remote commands, the telescope stirred itself and glided toward the coordinates of the next galaxy on our list. Watching it move in the soft red wash of the night-vision lamps, I realized that it amounted to a work not just of science but of art—a modern equivalent of the first-phase Navajo rug that hangs in Edgar's bone-white Manhattan apartment. During our interview, I asked Edgar what attracted him to astronomy rather than some other science. "I'm not a conventionally religious person," he replied. "But I think that, for me, astronomy held a wonderment, and was perhaps a spiritual foray, a substitute for religion. You know, I have not done things in a very conventional way—but hell, my whole life is a statement that I don't care what other people think."

18.

The Dark Ages

I recognize the signals of the ancient flame.
  —Dante

This is *our* Universe, our museum of wonder and beauty,
our cathedral.
  —John Archibald Wheeler

L ONG PAST MIDNIGHT at Rocky Hill, I took a few minutes to gaze
through the telescope at a little point of light I'd found with the aid of
a hand-drawn chart. It looked like a star but it was a quasar, the glowing, high-energy nucleus of a distant galaxy. This one, 3C273, is fairly
bright—usually about magnitude 12.8, sputtering up to 11.7 and down to
13.2—yet its light has been traveling through space for more than two billion
years, since the time when the first bacteria were emerging on Earth. If you
climbed aboard an H. G. Wells–style time machine, one that hurled you backward at the dizzying rate of one century per second, you'd be riding for eight
months before you got back two billion years. Yet here in the calm of the
predawn night, all it took was to point the telescope at the right point of light
in the sky. I stared for some time at the antique light of 3C273, wondering
whether it is by chance or necessity that one keeps such an appointment with
the prehuman past.

FEW STARGAZERS TODAY look at many things more distant than the galaxies of the Virgo supercluster—just as, a century ago, few looked at much
beyond the Milky Way. But this is changing, and among today's visual
observers and CCD virtuosos there are some who find satisfaction in the contemplation of galaxies long ago and far away. These intrepid investigators

principally explore our four neighboring superclusters, in the constellations Centaurus, Perseus, Coma Berenices, and Hercules.

Superclusters are made of clusters of galaxies—one or more big clusters plus lots of compact clusters or groups. The Virgo supercluster with its tens of thousands of galaxies is regarded as a modest specimen, since the Virgo cluster is its sole major constituent. It is dwarfed by the Centaurus supercluster, which lies 250 million light-years from Earth, rambles along from Centaurus to the constellations Hydra and Antlia, and harbors three big clusters— Hydra I, Centaurus, and IC 4329. Each of these clusters has its own dusky charms. Hydra I makes for pretty good viewing, considering its distance: A medium-size telescope under dark skies will reveal the twin ellipticals at its core, a couple of perturbed-looking spirals nearby, and perhaps a half-dozen other galaxies in the immediate vicinity. The Centaurus cluster provides an arguably better view, presumably because it lies toward the near side of the supercluster: Its core is a dim but imposing array of giant ellipticals interposed with strings of spirals and S0 galaxies. In the IC 4329 cluster, about a dozen galaxies can be seen within a one-degree field of view, among them curiosities like the Seashell, an interacting pair that resembles an opened oyster shell, and the cluster's namesake, IC 4329A, an energetic galaxy with a nucleus so bright that it could be mistaken for a star. There are also many lesser clusters worth exploring in the Centaurus supercluster. All are in southern latitudes, so the farther south your location, the better the viewing of this, our next-door-neighbor supercluster.

Looking to the northern constellation of Perseus—and a bit farther out, about 300 million light-years—we encounter a long, fat chain of galaxy clusters belonging to the Perseus supercluster, which covers 90 degrees of sky and 200 million light-years of space. Its centerpiece, the Perseus cluster, is a compact gaggle of ellipticals and S0s. The amateur astronomer Steve Gottlieb was able to discern fifty-eight of them with a 17.5-inch telescope. Flocked like birds in a gyre, they convey a vivid sense of the pervasive gravitational fields that bind galaxies together in a cluster core. Indeed, it was by studying the dynamics of the Perseus cluster that astronomers learned that large amounts of material that generate gravity but not light—"dark matter"—are concentrated near the cores of galaxy clusters. Four other rich clusters lie along the backbone of the Perseus supercluster. One of them, NGC 383, displays a striking chain of six galaxies in a row.

Our next stepping-stone, the Coma supercluster, contains only a few substantial clusters but is popular with observers because one of them is Coma, the core of which places more visible galaxies in one field of view than any other cluster in the sky. A low-power (62x) eyepiece with a 0.75-degree field of view on my 18-inch reflector embraces something like three dozen galaxies in the center of the Coma cluster, though higher powers are required to see the dimmer ones. Gottlieb, using a telescope of almost the same aperture at magnifications of 220x and 280x over several nights, was able to detect eighty-eight Coma galaxies in all.

The Coma cluster is a prime specimen of what the American astronomer George Abell, a pioneering investigator of large-scale cosmic structures, called a "rich," "spherical," or "regular" cluster. Abell identified hundreds of clusters on all-sky survey plates and divided them into two main classes—regular ones like Coma and irregular ones like Virgo. Irregular clusters are scattered-looking, contain mostly spirals, and leave their galaxies plenty of elbow room. Regular clusters are roughly spherical in shape, are populated almost entirely by elliptical and S0 galaxies, and are much more dense. The Coma cluster (Abell 1656) is so rich that it crams thousands of galaxies within a radius only a few times that of the Local Group.

Astronomers today place Abell's classes at either end of a continuum, and stress the general rule that the higher a cluster's density, the greater the proportion of its galaxies will be ellipticals and S0s. In high-density clusters, ellipticals and S0s outnumber spirals by better than three to one, while in low-density clusters their abundances are about equal. This finding suggests that the differing environments of richer and poorer clusters determine how their galaxies evolve. Spirals in rich, spherical clusters, orbiting at the higher velocities dictated by the stronger gravitational fields generated by all that mass, have passed near the cluster core several times in the course of cosmic history, and so are more likely to have collided with one another or to have encountered clouds of intergalactic material, stripping them of gas and dust and transforming them into ellipticals and S0s. Galaxies in poor, irregular clusters, however, have been downtown only infrequently—the outlying ones may not yet have made it there even once—and the central precincts of irregular clusters are relatively sparse anyway, so many of these galaxies have been able to retain their original spiral morphology. Downtown Coma, as one would expect of a compact regular cluster, consists mainly of featureless round galaxies, while spirals are found in the cluster halo and in the foreground. There's foreground space aplenty—four hundred million light-years of it between Coma and us.

Even dimmer, near the limits for most visual stargazers, is the Hercules supercluster, 500 million light-years out and home to ten prominent clusters with over four hundred member galaxies. Its centerpiece, the irregular Hercules cluster (Abell 2151), is a magnet for observers who like to pick out faint objects, with a pleasing variety of elliptical and spiral galaxies, many of them interacting. Gottlieb lists them as "faint," "very faint," and "extremely faint," but nevertheless has been able to see dozens of them.

At these distances, otherwise reliable star maps and catalogs begin to break down into briar patches of error. Help in resolving the discrepancies has come from the NGC/IC Project, a collaboration of amateur and professional astronomers dedicated to observing all the objects listed in the New General Catalog (NGC) and the supplementary Index Catalog (IC), to verify their many accurate entries and repair the erroneous ones. The project's elder statesman, Harold G. Corwin Jr., estimates that "there are at least a thousand known or potential identification problems in the NGC itself, and certainly that many again (perhaps more) in the IC's." He writes that the group's motives include "straightening out the confusion, and the simple fun of doing so."[1] Using the best and most up-to-date catalogs, amateur observers like Ken Hewitt-White, who totes his 17.5-inch reflector to remote, dark skies to "hit the 'Hercules Highway'" and "wander the side streets" of the supercluster's bright clusters, have been able to find many of the obscure galaxies listed in the Morphological Catalog of Galaxies (MCG) and the Uppsala General Catalog (UGC). "Success at tracking them down," Hewitt-White notes, "hinges on aperture, sky conditions, and the tenacity of the observer."[2]

One can go still deeper. The Corona Borealis cluster, jewel of the "CorBor" supercluster, lies more than a billion light-years from Earth, yet some visual observers have risen to its challenge. Using high-power eyepieces just "to put some space between the individual members," Steve Gottlieb managed to make out six of its galaxies from a mountaintop seven thousand feet high in the Sierra Nevada. "Linger several minutes on each faint glow," he advises. "Savor the satisfaction of detecting photons that started their journey over one billion years ago."[3]

A billion light-years is a considerable distance—over five percent of the radius of the observable universe—so this may be a good point at which to consider how superclusters fit into even larger frames of reference. Within a billion light-years of Earth reside about eighty superclusters, composed of some 160,000 galaxy clusters containing 3 million large galaxies and an estimated 30 million dwarfs. To discern patterns on these huge scales requires gathering and interpreting information about tens of thousands of galaxies. This feat has become possible in recent decades, thanks to survey telescopes with fiber-optics systems capable of simultaneously analyzing the light from many galaxies at a time. Such studies indicate that the universe ultimately is organized into bubblelike formations roughly 250 million light-years in diameter, and that massive superclusters consist of dense regions where the walls of two or more bubbles intersect. The Coma supercluster is part of one such region, called the Great Wall. A half-billion light-years long and skewed to our line of sight, like a ruler viewed with one end held close to the eye, it ranges in distance from about 350 million light-years in Leo to 500 million light-years in Hercules, where it encounters the Hercules supercluster. The Perseus-Pisces supercluster may be another bubble intersection, more readily discerned since it presents its flank to us.

Given all these hierarchies of structure—galaxies in groups and clusters, which in turn belong to superclusters, which lie on the walls of bubbles—one naturally wonders whether the bubbles belong to even larger objects. The answer, it seems, is that they do not—that bubbles are the biggest things. If you average across many bubbles, the universe looks smoothly homogeneous. Consider a forest, sampled first at close range and then from far away. Running through the woods in a straight line, you experience a clumpy environment: You either pass between trees or bump into them, and sometimes you cross open meadows only to plunge into woods again on the other side. But shoot a picture of the forest from orbit, lay a grid across the photograph, count the trees in a sufficiently large sample of grid boxes, and you may find that the overall density of trees is homogeneous. In this model the trees are galaxies, the meadows are the voids inside cosmic bubbles, and the forest is the universe.

This finding was a relief to cosmologists, who had long assumed that if you sampled large volumes of space, the universe would prove to be homogeneous (the galaxies evenly distributed) and isotropic (so that all observers see homogeneity in all directions). The inhomogeneities that make life possible— the local concentration of matter into planets, stars, galaxies, clusters, superclusters, and bubbles—ultimately resolve into a smooth, even distribution on the cosmological scale.

But if the cosmos overall is as smooth as tapioca, where did the lumps come from? This is one of the more intriguing questions in cosmology, and it leads us to consider the big bang and the expansion of the universe.

When Edwin Hubble was photographing galaxies from Mt. Wilson—aided by his observing partner, the former observatory janitor Milton Humason— he found that galaxies generally are moving apart from one another, at rates directly correlated with their distances. Hubble and Humason were uncertain what to make of this "velocity-distance relationship," known today as the Hubble law, but they learned that it had been predicted by Einstein's general theory of relativity, which implied that the universe must be either expanding or contracting. What relativity and the Hubble law tell us is not that galaxies move through space—although they certainly do some of that, for instance in their orbits around the centers of galaxy clusters—but that cosmic space is itself expanding, carrying the galaxies with it. Hence the Hubble law: The farther a remote galaxy, the faster expanding space is carrying it away. Suppose you take a rubber band and mark a series of evenly spaced ink marks on it, one inch apart from one another, like this:

A   B   C   D

Each point represents a galaxy. Now slowly stretch the rubber band, to mimic the cosmic expansion of space, until, one minute later, the adjacent points are two inches apart:

A   B   C   D

Since the distance between galaxies A and B increased from one inch to two inches in one minute, they have a mutual recession velocity of one inch per minute. But note that the distance from A to C has doubled, too, in the same interval—it went from two to four inches—so the mutual recession velocity of galaxies A and C is two inches per minute, twice that of the A-B pair. Similarly, the velocity of A relative to D is three inches per minute. So we have a linear velocity-distance relationship: For every inch farther apart two points are when we make our measurement, their relative velocity is one inch per minute faster. Every other observer—those on B, C, and D—observes the same effect, for every other galaxy. Substitute three-dimensional space for the one-dimensional lineup on the rubber band, and you have the expanding universe.

If that were all there was to it, determining the cosmic expansion rate—the "Hubble constant"—would be a straightforward matter of measuring the velocities and distances of a few nearby galaxies. Instead, the clumpishness of matter gravely complicates the process. Clusters of galaxies, even poor ones like the Local Group, are gravitationally bound: The space inside them doesn't expand, so examining our compatriot galaxies in the Local Group tells us nothing about the cosmic expansion rate. The Virgo supercluster expands, but at a rate that is retarded by its own gravitational web. Caught in this web, the Local Group and the Virgo cluster are moving apart from one another more slowly than would otherwise be the case. To observe unfettered cosmic expansion—"pure Hubble flow"—one has to look farther out, to other superclusters, but when looking that far away it becomes difficult to accurately measure the distances of galaxies. For these and other reasons, astronomers still aren't sure about the precise value of the Hubble constant.

Which is unfortunate, since the expansion rate indicates the age of the universe—the amount of time that has elapsed since all these distant places, from the Centaurus supercluster to the Great Wall and beyond, were the same place. The faster you stretch the rubber band, the less time it takes to reach any given length, so high values for the Hubble constant (meaning a more rapid cosmic expansion rate) imply a younger universe than do low values. The estimates of most astronomers studying cosmic expansion yield an age for the universe of more than ten billion but less than twenty billion years. This fits reasonably well with the ages of the oldest known stars, those in globular clusters, which are around twelve billion years old.

The beginning of cosmic expansion—the so-called big bang—is well understood in some ways but mysterious in others.

The well-understood parts come mainly from high-energy physics. Theorists start the clock running a fraction of a second after the beginning of expansion and calculate forward from there. Unsurprisingly, considering that all the energy that today is arrayed across the universe was at that time compacted in a volume smaller than a golf ball, one is dealing with a very hot primordial soup, in which matter could not yet form any stable structures above the level of subatomic particles. High-energy physicists, adept at calculating how such a simple system evolved over time, have been able to test some of their calculations against observations made in particle accelerators and out among the galaxies.

One of their successes has to do with the relative abundance of various elements. The calculations of the physicists indicate that as the fireball expanded and cooled, about a quarter of the hydrogen in the universe should have been converted, through nuclear fusion, into helium. And, sure enough, the universe is about one-quarter helium. Stars fuse hydrogen into helium, too, but there has not been anywhere near enough time as yet for stars to have made that much helium—which in any event is also abundant in intergalactic clouds, where few stars could have contributed it—so evidently the big bang did the job.

Another success concerns the cosmic microwave background radiation—light set free three hundred thousand years after Time Zero, suffusing the universe with a faded glow that can still be observed today, although it has been stretched by the subsequent expansion of space to the longer wavelengths of microwave radio. This background radiation has been observed, from space and by detectors in high-flying balloons, and it has the luminosity and spectral characteristics predicted by the theory. Moreover, maps of the background radiation show inhomogeneities on the expected scales, the seeds of galaxies and cosmic bubbles to come. The big bang theory is taken seriously precisely because its predictions have been borne out—in the compatibility of the expansion rate to star ages, the helium abundance, and the presence and spectrum of the cosmic microwave background—and fit together to make a coherent picture.

The big bang did not occur in a pre-existing space, but was itself an expansion of space from an infinitesimal pinpoint to the grand array of galaxies we see scattered in all directions today. Today as in the beginning, no place is any closer or farther from the location of the big bang: It happened here, there, and everywhere. Hence nobody occupies the center of the universe, any more than anybody occupies the center of the surface of the Earth. Every observer everywhere has pretty much the same view—of receding galaxies that, with increasing distance, are seen ever farther in the past, with the glow of the background radiation hanging behind it all.

Yet many mysteries remain. We don't yet know what triggered the big bang (a bubble nucleating off the space of some other universe, perhaps) or what planted the seeds of cosmic structure (quantum flux events, presumably, the random emergence of subatomic particles out of the vacuum), or even what most of the universe is made of. (The rotation rates of galaxies and their orbital velocities in galaxy clusters indicates that 90 to 99 percent of the matter they contain emits no light; nobody yet knows what this "dark matter"— or dark energy—may be.) There's a lot we don't know—and anyway, what have all these theoretical dream-weavings to do with the activities of stargazers at their telescopes?

Not much, say many amateur astronomers. They note that most of the great advances in astronomy have resulted less from theorizing than from observers getting their hands on better telescopes and detectors. They tend to be impatient with talk of inflation (the theory that the universe originally ballooned at a much faster rate than it does today), quintessence (an exotic energy field that may repel subatomic particles from one another), string theory (the idea that matter is made of strings fashioned out of hyperdimensional space), and brane theory (in which the strings are more like noodles, and in one version of which, as the physicist Neil Turok explains, "our current universe is [a] four-dimensional membrane embedded in a five-dimensional 'bulk' space").[4]

The amateurs have a point: Cosmology and every other science ultimately comes back to, and indeed often starts from, observation. But observations are never entirely innocent of theory: One must usually have at least a notion of what one is looking for in order to see much of anything at all. That's why beginners are often disappointed with the views they get on first peering through a telescope, at least of anything less dramatic than the rings of Saturn or the mountains of the Moon: They don't see much because they don't know what to try to see. And once you start looking *for* something, you have necessarily invoked a theory—or at least an assumption, which is what a theory looks like to someone who doesn't recognize it as a theory.

Amateur astronomers can make cosmologically useful observations. One popular way to do this is by hunting for exploding stars in other galaxies. When a star in a distant galaxy goes supernova, its light curve—a rapid rise followed by a long decline—and its spectrum can in some cases be used to measure the distance of the galaxy in which it occurred, which aids in determining the size and expansion rate of the universe.

There are several types of supernovae, none of them perfectly understood. The two of principal concern to amateur astronomers are called Type II and Type Ia. Type II supernovae occur when giant stars that have squandered their fuel collapse, triggering an explosion. (The falling outer layers of the star hit its iron-dense core and rebound, like a rubber ball hurled against the pavement.) They are found in the star-forming regions of spiral galaxies, occasionally in irregular galaxies, and seldom if ever in ellipticals. Type Ia supernovae are found in all types of galaxies and show no preference for spiral arms. They are thought to belong to binary star systems that started out consisting of a massive star—big, but not big enough to go supernova—orbited by a less massive companion. The massive star evolves rapidly to its red-giant stage, sheds its outer envelope, and subsides into retirement as a white dwarf. The companion star, being less massive, evolves more slowly, but eventually it, too, becomes a red giant. If the two stars are orbiting close enough together, the white dwarf will then begin capturing gas from the outer atmosphere of the blowsy red giant. This thievery persists until the white dwarf grows to 1.4 solar masses—the Chandrasekhar limit, equivalent in astrophysics to the "critical mass" of a thermonuclear weapon—whereupon the star detonates.

The beauty of this process to cosmologists is that while Type II supernovae involve stars of various masses, Type Ia supernovae evidently all occur when stars reach the same critical—Chandrasekhar—mass. So all Type Ia supernovae should explode with about the same force and attain approximately the same brightness. The process isn't quite that simple—these things never are—but there does seem to be an underlying cookie-cutter similarity to Type Ia supernovae. If so, astronomers who had an accurate distance for just one such supernova could in principle chart the distances of all the others. This makes Type Ia supernovae valuable as "standard candles" for measuring the distances of their host galaxies and estimating the cosmic expansion rate: If you know how bright a given supernova actually is, you need only compare that value to its apparent brightness in the sky to learn its distance.

Searching for supernovae means observing galaxies—visually, photographically, or with a CCD detector—and looking for stars that have not been seen there before. If you find one that really is a supernova—and not, say, an asteroid passing through the field of view—and nobody has noticed it before, you will have contributed to science and discovered something of cosmological significance. (You won't get your name on it, though; supernovae are cataloged by the year in which they occurred plus a letter designating their order of discovery. Supernova 2001a, discovered by a pro-am team consisting of Berkeley astronomers and Michael Schwartz, a crack amateur supernova hunter in Cottage grove, Oregon, was the first bagged that year; 2001b, detected at Beijing Observatory, was the second.)

Few supernovae were detected before the advent of the telescope, since few get bright enough to be seen with the naked eye, but the ones that did made a lasting impression.

Ancient observers in China, Japan, Korea, Europe, and the Arab dominions left records of a supernova in the year 1006 that was bright enough to be visible during the day, as did Chinese observers half a century later, in 1054. (Much brighter than Venus, the supernova of 1054 was visible in broad daylight for twenty-three days, and remained a nighttime naked-eye object for a total of 653 days.) Because early observers took the trouble to accurately describe the positions of these apparitions in the sky, modern-day astronomers were able to determine that the supernova of 1006 produced a radio-bright remnant, today cataloged as PKS 1459-41, and that the one in 1054 produced the Crab nebula. Readily visible in small telescopes, the Crab nebula is currently some ten light-years in diameter and is continuing to expand so rapidly that its diameter grows visibly in the course of a human lifetime. (Carl Otto Lampland, of Lowell Observatory, first detected its growth in 1921. High-resolution images taken through the Hubble Space Telescope show structural changes in the nebula over a matter of days.) At its center sits a pulsar—the rapidly spinning collapsed core of the detonated star, no larger than a city but as massive as the Sun, emitting pulses of light and radio thirty-three times per second. Amateur CCD images show the pulsar as a 16th-magnitude point of light, but since it pulses so rapidly—faster than the thirty-frames-per-second refresh rate of a TV picture—it looks as steady as a normal star.

Two bright supernovae were instrumental in awakening Renaissance astronomers to the fact that, Aristotle to the contrary, the starry heavens are not changeless. Tycho was stopped in his tracks during an after-dinner stroll in 1572 by the sight of a "new star" in Cassiopeia. "Since I had, from boyhood, known all the stars of the heavens perfectly, it was quite evident to me that there had never been any star in that place of the sky, even the smallest, to say nothing of a star so conspicuous and bright as this," he recalled. "I was so astonished at this sight that I was not ashamed to doubt the trustworthiness of my own eyes."[5] Ever the empiricist, Tycho accepted the reality of the "new star" only after friends to whom he pointed it out verified that they saw it, too. Johannes Kepler studied the supernova of 1604, in Ophiuchus. Similarly skeptical, he allowed that it might, as many assumed, be a genuinely new star, "but before we come to [that] I think we should try everything else."[6] The supernovae produced remnants today cataloged as the radio sources 3C 10 and 3C 358 but known colloquially by Tycho's and Kepler's names.

Detection of the vast majority of supernovae requires telescopes. During most of the twentieth century their discovery was almost entirely the work of professionals—two of whom, Walter Baade and Fritz Zwicky at Mt. Wilson, coined the term "super-nova," in 1933, to describe what looked like a brighter version of a nova, a star that flares up intermittently. Amateurs who wanted to get into the game were frustrated by a lack of widely available photographs or charts against which they could compare the sky to determine if any but the brightest stars in view were "new." Such charts were generally so expensive that few amateurs could afford them, and making repeated photographs of galaxies for themselves was costly and time-consuming. Not until 1968 was the first amateur supernova discovery of the twentieth century made, when John Caister Bennett, a government employee in South Africa, spotted a 9th-magnitude exploding star in the galaxy M83 while sweeping for comets.

The Reverend Robert Evans, of Coonabarabran, New South Wales, Australia, addressed the problem by collecting charts and photographs from every source he could find. Gifted with the ability to memorize the patterns of stars seen in the foreground of hundreds of galaxies, he hopped quickly from one to the next, often covering more than a hundred galaxies an hour. So efficient was Evans that he picked out seventeen supernovae from his front yard from 1986 to mid-1991, during which period an automated search by Berkeley astronomers, using a CCD-equipped telescope with an aperture nearly twice that of Evans's, discovered only twenty. This prompted Evans to speculate that "if a visual search had been conducted with a telescope like the one used at Berkeley, and assuming that the same amount of telescope time was available which the Berkeley team enjoyed, perhaps similar results might have been obtained."[7] To test his idea, Evans obtained observing time on a large professional telescope, the 40-inch at Siding Spring. Despite inaccuracies in its control system, which made finding galaxies slower than he normally could by pushing the tube of his own 16-inch telescope around by hand, Evans was able to observe nearly sixty galaxies an hour, and to detect existing supernovae at a rate that equaled or exceeded that attained by an automated search then being conducted at Perth. "The value of visual searching has again been demonstrated," Evans concluded, in a scientific paper evaluating his results.[8]

Visually searching for supernovae is in some ways its own reward—after all, you're looking at galaxies—but the odds against success are steep. An average large spiral galaxy produces a supernova every thirty to fifty years, so if you were to examine thirty to fifty suitable galaxies every night for a year you would have a reasonable chance of seeing a supernova. But to see one first means besting the competition, and there's more competition out there all the time. It takes nothing away from the estimable reputation earned by the Reverend Evans to note that when he set his records there were probably too few amateur supernova hunters in the world to get up a soccer game. By the year 2001 there were enough to form an entire soccer league, and most of them had CCD cameras. A CCD surpasses the sensitivity of the human eye on a given telescope with exposures of just a few seconds, and with longer exposures exceeds the capacities of almost any visual observer with almost any telescope. CCDs work well in bright moonlight, too, when visual observers are hobbled, and when attached to a fully computer-controlled telescope they can search scores of galaxies while the "observer" gets a night's sleep.

Dedicated amateurs using CCDs have mounted supernova search projects that rival professional efforts and in some ways surpass them. Michael Schwartz operates dedicated search telescopes in Oregon and Arizona, the latter a futuristic-looking 32-inch Ritchey-Chretien with more light-gathering power than the equipment employed by his professional collaborators at Lick Observatory. He estimates that he discovers a supernova in about one out of every 1,400 CCD images he takes. Tim Puckett, a former earthmoving-equipment salesman, spent ten thousand hours in the workshop at his home in Mountaintown, Georgia, building a 24-inch computer-controlled Ritchey-Chretien. "I didn't really have the money to go out and buy an expensive telescope," he recalled, "so I raised enough to buy the optics, and then for years I would go to salvage yards every month. Back then they didn't recycle everything, so you could find large steel plates and stuff like that. I kept revising my design to accommodate the parts I was able to find. It was a massive project."[9] Using this telescope, with friends helping him examine the CCD images that kept pouring in, Puckett discovered thirty-one supernovae in twenty-four months. He then built an even larger telescope, planning to use its greater light-gathering power to shorten exposure times and image more galaxies per night, rather than to go deeper into the sky. "Each magnitude dimmer that you go means 2.5 times more stars and galaxies," he noted, "so at some point you have to say, 'Hey, enough is enough.' The important thing for science is that the professionals take spectra of these things, and generally they can't get good spectra below about magnitude 19.5 anyway. I can get dimmer than that with an exposure of only one minute.

"I can't really tell you why I do it," he added. "I like to be able to sit down and look at several hundred galaxies; it means that every day is different. The actual hunt I don't consider to be pure science—the real science means taking spectra, and for that you need a one- to three-meter-class telescope and a spectrograph—but it's nice to make some contribution to science. And, of course, ego has something to do with it. Anybody who hunts these things knows that there's some prestige associated with discovering one."[10]

Supernovae hunters lacking access to telescopes the size of Puckett's may seek to improve their chances by adopting clever search strategies, although every strategy involves tradeoffs. Supernovae in nearby galaxies are the easiest to see, but lots of other people are looking at those same galaxies, too, if only for fun, and that reduces the likelihood that you will be the first to detect a supernova there. Ellipticals and edge-on spirals are a relatively poor bet— ellipticals because they don't produce Type II supernovae, and edge-on spirals because dust in the disk will conceal supernovae that don't happen to occur on the near edge of the galaxy.[11] But your competitors, knowing that, may tend to neglect ellipticals and edge-on spirals, so perhaps you could gain an advantage by concentrating on them. And so it goes.

Chance favors the prepared mind, as Pasteur used to say, but supernovae have been discovered by observers who were hardly prepared at all. In 1994, after amateur astronomers discovered a supernova in the Whirlpool galaxy, an even earlier image of it was found on a CCD image that two students, Heather Tartara and Melody Spence of Oil City High School in Pennsylvania, had requested be taken by a 30-inch robotic telescope at Leuschner Observatory as part of "Hands On Universe," a science outreach project administered by the University of California, Berkeley. (Carl Pennypacker, the astrophysicist who runs the program, told me that "all sorts of students get into supernova searching—from valedictorians to gang members and dyslexic kids with fourth-grade reading levels. I'm not sure how much they're learning, but they all learn something.")[12] The image, which Ms. Tartara said they'd ordered because the Whirlpool "just seemed interesting," turned out to be one of the earliest ever made of a supernova caught in its period of rapidly increasing luminosity.

On finding a supernova, the observer sends word to the authorities—normally the Central Bureau for Astronomical Telegrams—and waits to see whether it had already been detected elsewhere. As one would expect, the process is fraught with potential disappointment.

In 1993, A. William Neely, an amateur astronomer in Silver City, New Mexico, took a CCD image of the prominent galaxy M81 but failed to notice that a bright supernova had appeared in one of its spiral arms. Less than fifteen hours later it was discovered visually by another amateur, Francisco García Diaz, in Madrid, using a ten-inch telescope.[13] Neely could at least console himself that his CCD image provided a valuable data point on the magnitude of the exploding star, caught while it was still rising in brightness.

On February 23, 1987, light reached Earth from a star that had exploded at the edge of the Tarantula nebula, in the Large Magellanic Cloud, 168,000 years ago. Bright enough to be seen without a telescope, it was a historic event, the first naked-eye supernova since Kepler's in 1604. Oscar Duhalde, a night assistant on the 40-inch telescope at Las Campanas Observatory in the Chilean Andes who had gone outside to check the clarity of the night sky, noticed that the Tarantula region seemed brighter than usual. He returned to the dome, meaning to mention this fact to the astronomers inside, but they were busy, his help was urgently required, and in the press of events he neglected to say anything. Meanwhile Ian Shelton, a former amateur astronomer who was doing graduate work and living on the mountaintop, took a photograph of the Tarantula region with a ten-inch "astrograph," a telescope devoted to photography. Rising winds blew the shed's roof shut, so Shelton closed up shop and developed his plates. He immediately spotted the "new" star, went outside and saw it in the sky, and reported it to the astronomers in the dome. Shelton went into the history books as the discoverer of Supernova 1987A. Duhalde did not. Nor did Robert McNaught, who had photographed it from Australia but hadn't yet developed his film, or Albert Jones, who independently spotted it from New Zealand a few hours later.

Jones, a veteran amateur astronomer with more than half a million variable-star observations to his credit, had looked at Tarantula a little earlier that evening and seen nothing unusual there. This seemingly routine observation wound up playing a vital role in testing an exotic "neutrino cooling" theory of supernovae. The explosion that causes a supernova occurs deep inside the doomed star, where the collapsing atmosphere hits the dense core. Astrophysicists theorized that most of its energy is released in the form of neutrinos, which are low-mass subatomic particles. Neutrinos interact so weakly with matter that they can fly through planets as if they weren't there—half of the trillions of neutrinos that are passing through your body as you read this sentence are coming up from below, having sailed right through the Earth—so they should exit the exploding star and fly into space in moments, while the light from the blast is still fighting its way to the star's surface. According to the theorists' calculations it takes the light about two hours to reach the stellar interior and be released into space. If so, neutrinos coming from a supernova will reach Earth two hours before its light appears in the sky. That was the theory, but nobody had yet made any such observation.

When Supernova 1987A occurred, neutrinos coming from it were indeed picked up, by underground detectors in Cleveland, Ohio, and Kamioka-cho, Gifu, Japan. This event, the first observation of neutrinos from deep space, occurred on February 23, at 7:35:41 Universal Time (UT, the same as Greenwich Mean Time). When the physicists learned of Shelton's supernova discovery and realized that their neutrinos might have come from the exploding star, the question they needed to answer was just when its light first appeared. If the theory was correct, the light should have first arrived at 9:35 UT, two hours after the neutrinos did. Robert McNaught's photograph, the first to show the supernova, was made at 10:30 UT, a little less than three hours after the neutrinos were detected. That confirmed the theory at one end—at least two hours had elapsed—but what about the other end: Had the light arrived sooner than expected, before 9:30 UT? Luckily, it was right at about 9:30 UT— early evening in New Zealand—that Jones made his visual observation of the Tarantula nebula and saw nothing unusual there. Like Sherlock Holmes's dog that did not bark, his failing to see the supernova meant that its light must indeed have arrived no earlier than two hours after the neutrinos did. So a key theory of exploding stars had been confirmed, thanks to the unwitting collaboration of physicists working at underground neutrino detectors in Cleveland and Kamioka-cho, an amateur astronomer checking variable stars in New Zealand, another amateur photographing the sky from Australia, and an amateur turned pro making photographs of his own in Chile. If one were to choose a date on which astronomy shifted from the old days of solitary professionals at their telescopes to a worldwide web of professionals and amateurs using a polyglot mix of instruments adding up to more than the sum of their parts, a good candidate would be the night of February 23–24, 1987.

How far can an amateur astronomer see just by looking through a tele-scope? Billions of light-years, if the object being observed is a quasar.

Some galaxies have—or had—nuclei bright enough to outshine all their stars combined. These bright nuclei are known as quasars. The term stands for "quasi-stellar objects." It came into use because the first quasars to be identi-fied looked like odd, isolated stars. Quasars were too distant for existing tele-scopes to reveal anything of their surroundings, and for many years there was dispute about whether they belonged to galaxies at all—a point not fully resolved until *Hubble* flew, and made quasar images that showed each to be located near the center of a galaxy. A quasar evidently is powered by matter that heats up and glows as it spirals in toward a massive black hole of the sort now known to occupy the nucleus of most galaxies.

The spectral lines of quasars are shifted far toward the red end of the spec-trum, indicating that they are very far away. (That's the Hubble relation: In an expanding universe, the more distant an object, the faster its recession veloc-ity and, hence, the more its spectrum is transposed toward the low frequen-cies of red light, much as a car horn drops in pitch when the car is speeding away.) One doesn't have to take the astronomers' word for it: The French ama-teur astronomer Christian Buil is one of several stargazers who has taken his own spectrum of 3C273 and seen for himself that it has a recession velocity of 45,000 kilometers per second, indicative of a distance of some 2.7 billion light-years. The farther one looks, the more abundant the quasar population becomes—up to a point.

Peering far into space means looking deep into time gone by. This phenom-enon, known as "lookback time," makes historians of stargazers. "Astronomers have an advantage over other historians," writes the American astronomer Alan Dressler. "They can observe history *directly*—if not their own, at least someone else's."[14] The fact that quasars become more abundant at high look-back times means they are denizens of the earlier universe. Evidently young galaxies had more interstellar matter near their cores to feed their quasars, which after billions of years exhausted the available fuel and quieted down. People sometimes ask stargazers whether they can see stars that aren't there anymore. This is unlikely to be the case with the stars in the Milky Way, since the time it takes their light to reach us—a few hundred to a few thousand years—is only a fraction of the stars' lifetimes. But it's true of quasars: The quasars we see almost certainly have stopped shining during the billions of years that have elapsed since their light left them.

Quasars are so luminous that if there were one in action in a Local Group galaxy, its brilliance would surpass that of the full Moon. Even at their great distances, more than a dozen can be seen through small telescopes. They don't look like much, just points of light, but there is a certain allure in their sheer remoteness.

A few quasars happen to lie behind a galaxy, or cluster of galaxies, whose gravitation has bent their diverging light beams back toward us, creating multiple images of the same quasar in the sky. Since most quasars vary in brightness—due, presumably, to the inconstancy of material spiraling into the black hole that powers them—astronomers can check the differing arrival times of light variations in these "gravitationally lensed" quasars to determine the difference in the length of the various paths the light took in reaching us. Such observations are difficult to make, but in theory can yield direct, geometrical measurements of the distances of both the quasar and the foreground galaxy. Amateur astronomers with larger telescopes have visually observed gravitationally lensed quasars in Ursa Major, Leo, and Canes Venatici. The Cloverleaf quasar, in Bootes, presents four images of the same quasar, but they are separated by so little sky—only 1.36 arc seconds—that they are extremely difficult to resolve. Einstein's Cross, in Pegasus, likewise presents four images of the same quasar; two of them are magnitude 17.4 and the other two even dimmer at magnitudes 18.4 and 18.7. A 14th-magnitude galaxy, designated CGCG 378-15, sits at the center of the cross. Barbara Wilson saw the lensing galaxy and the two brighter components with a 20-inch telescope that she had set up in the parking lot of McDonald Observatory, and glimpsed the other two through a 36-inch Dobsonian. By using "averted vision"—looking slightly away from the sought-after object, to put it on the most light-sensitive parts of the eye—she could spot one or two of the ghostly images at a time, but never all four at once. "I have never seen Einstein's 'Cross' itself," she reported, "because to actually see the 'Cross,' one would have to see all four images simultaneously."[15]

And beyond that? Studies conducted with big professional telescopes show that the quasar population becomes constantly denser with increasing lookback times until one gets back to about a billion years after the big bang—but thereafter their numbers abruptly drop off, even though they are bright enough to be seen much farther out. The cause of the dropoff is that one is looking back to the original dark ages, a time before which galaxies and their central quasars had not yet formed. That's why the sky is dark at night. If every line of sight eventually terminated at the surface of a star, the sky would be a blazing wall of light. Instead, some ten billion to twenty billion light-years out, darkness descends.

All observers in the universe see modern times in nearby space, ancient times farther away, and then the inky black curtain of the dark ages. Beyond that, discernible in radio wavelengths, resides the soft glow of the cosmic microwave background, the dying light of the big bang itself. And that is the universe we live in—a set of nested, spacetime eggs, rising to brilliance near the edge and then falling off to darkness and the background glow.

Years ago I was visiting with the physicist John Archibald Wheeler in Austin, Texas. He drew a little sketch of a point, representing the big bang, with a curving, U-shaped, expanding throat extending from it and terminating in an eye that looked back at the original point. It represented his continuing effort to make sense of our role in the universe, as observers and thinkers who have arisen from cosmic processes and yet are able to distance ourselves from them and analyze them—as if from the outside, even though we cannot really ever get outside the universe. In one of his books Wheeler compares our situation to a legendary dialog between Abraham and Jehovah: "Jehovah chides Abraham, 'You would not even exist if it were not for me!' 'Yes, Lord, that I know,' Abraham replies, 'but also You would not be known if it were not for me.'"[16]

"It's still not quite right," Wheeler said, tapping his index finger on the drawing, and the same could be said of every other attempt to understand the relationship between the human mind and the wider universe, although some are more helpful than others. That, I suppose, is what it means to be human. We observe, and try to understand, and formulate ideas that, if we're honest with ourselves, we will admit to be "not quite right." But we keep trying, knowing that we'll never figure it all out but trusting that if we persevere we shall keep doing better.

Life, like the universe, rounds off to darkness where it runs out of time, and contemplation of one's death is perhaps the mainspring of astronomy and other human strivings. It is to celebrate the audacity of life confronted with death that we cherish admirable dying words like those of Thomas Hobbes ("I am about to take my last voyage, a great leap in the dark"), Ludwig Wittgenstein ("Tell them I've had a wonderful life"), and Su Tung-P'o (who, having been advised that one must try to attain the afterlife, replied, "It's a mistake to try"). Perhaps the key to dying well—or living well—is to have laid in a stock of worthy memories. To that end, when darkness is falling for good, it is well to have in mind, in addition to memories of human love and loss and of the natural splendors of this world—of birdsong at dawn, the roaring spray of the surf, the sweet smell of the air in the eye of a hurricane, the workings of bees in the throats of wildflowers—a few memories of the other worlds as well. If you have seen plasma arches rising off the edge of the Sun, yellow dust

storms raging on Mars, angry red Io emerging from the shadow of Jupiter, the golden rings of Saturn, the green dot of Uranus and the blue dot of Neptune, the glittering star fields of Sagittarius and the delicate tendrils connecting interacting galaxies, have watched auroras and meteors writing silent signatures in the sky—if, in short, you have seen not only this world but something of the other worlds, too—well then, you have lived.

So, while life is in us, and we are in it, let's keep our eyes open.

FROM THE OBSERVATORY LOG:
MINERVA AT DAWN

A S THE FIRST FINGERS of sunlight touch the mountain peaks, I linger over the sight of the Whirlpool galaxy, a bone-white spiral floating on the brightening blue sky. Minerva, the local owl, sounds one last hoot and then falls silent. She has lived here since before the observatory was built, and I named her after Hegel's aphorism, "The owl of Minerva flies only at dusk," perhaps the only line of Hegel's that has meant much to me.[1] Taking her silence as a signal that the night is over, I heel the telescope to its stowed position, cap the main mirror, put the cold, weighty eyepieces away, close the roof, and walk down the hill. No supernovae were discovered here tonight, but light rays from millions of them are out there, rushing toward us as fast as anything can travel, and perhaps some night I will be the first to catch sight of one. A late-homing bat speeds across the whitewashed sky and dives into the dark heart of a stand of oaks, bringing to mind Blake's lines:

> The Bat that flits at close of Eve
> Has left the Brain that won't Believe.[2]

Were Blake to ask me, "What do *you* believe?" my unsatisfactory answer would be that I'm not really sure. I believe that the wheeling galaxies are really out there, that they are not figments of the imagination: My imagination is insufficient to have conjured them. I believe that we are involved with them, somehow, but as to how that might be—what, in other words, is the true relationship between mind and matter, once the silt of Cartesian dualism is wiped away—I remain uncertain. Perhaps it suffices to approach life in the spirit of the artist and the scientist, to be what Einstein called an "unscrupulous opportunist," seizing upon what one can use. Blake himself said as much: "I will not Reason & Compare: My business is to Create." But he also said, in the line immediately preceding that one, "I must Create a System, or be enslav'd by another Man's."[3] Have I created a system, or am I enslaved by the scientific system that Blake scoffed at as the merely knowing, unbelieving outlook of what he scornfully called the Natural Man? The natural man sees a "guinea Sun," he said.

The real Sun is rising now, in a noble cloak of Chinese red, gilding the bare tree branches, and it does not look like a guinea coin to me. Wittgenstein, fighting in the Great War on the Eastern Front, volunteered day after day for the most dangerous duty, perched in an observer's tower vulnerable to snipers, and there wrote in his notebook:

I know that this world exists.

That I am placed in it like my eye in its visual field.

That something about it is problematic, which we call its meaning.[4]

Years later he was still wondering: "I have a world picture. Is it true or false? Above all it is the substratum of all my enquiring and asserting."[5]

"This business of inquiry is damned hard!" he once exclaimed, to a little band of students gathered for class in his rooms at Cambridge. So it is, I reflect, kicking a stone down the hill: Were it not so hard we would not love it so. The business of inquiry stretches the primate mind from here to the quasars: If there are limits to human thought, why are they so difficult to find? *Mens aeterna est quaternus res sub specie aeternitatis*, wrote Spinoza: "The mind is eternal insofar as it conceives things from the standpoint of eternity."[6] But whose is it, this eternal mind?

Back home, a few coals still glow in the fireplace. I warm my hands over them, then slip into bed. The last thing I see against the darkness of my closed eyelids is the Whirlpool galaxy, afloat on a blue ocean, indelible.

APPENDIXES

Then be it ours with steady mind to clasp
The purport of the skies—the law behind
The wandering courses of the Sun and Moon
—Lucretius

Observing Techniques

As with many other outdoor activities, from bird watching and trout fishing to canoeing and mountain climbing, your involvement in stargazing can be as casual or as committed as you like, producing anything from a few happy memories to a lifelong passion. These tips are meant to help in getting started.

Naked-Eye Stargazing

A good way to begin is by learning the major constellations—or by teaching them to children or friends. All you need is a star chart, a red-lensed flashlight, and access to the night sky. The simple star charts in this book will do for starters. (To find better charts see Appendix F, Further Reading.) A wide variety of red-lensed flashlights are available commercially, ranging up to fancy LED models with rheostat brightness controls, but an ordinary flashlight can be converted for night-vision use by painting its lens with red nail polish or wrapping a piece of red plastic around the front end and securing it with a rubber band. (I've used the wrapper from a loaf of bread for this purpose, and am told that red taillight repair tape also works.) The resulting light should be just bright enough to read a star chart by, and not so bright that it interferes with your night vision. If you find that after using your light you cannot see faint stars as well as before, make the light dimmer by adding another layer of red plastic or a second coat of nail polish.

It takes most people at least twenty minutes to become completely dark adapted, meaning that they can see optimally at night, and even a momentary exposure to a bright light can ruin dark adaption. To preserve your night-vision investment, avoid any use of unfiltered flashlights (once dark adapted, you can usually find your way around by starlight alone) and plan ahead so you don't have to go into a lighted room or open a car door and trigger its interior lights while you are stargazing. (If you must for some reason be exposed to a bright light, keep one eye closed, so that it, at least, will remain dark adapted.)

Fatigue, uncomfortable viewing positions, and the use of tobacco and intoxicants can reduce your night-vision capacity. (Smoking reduces blood oxygen supply to the eyes; drinking slows pupil dilation times and reduces the maximum diameter to which the pupil dilates.) So does low blood sugar, so don't go hungry on observing nights.

Genuinely dark-sky sites are hard to come by these days, owing to light pollution—an international problem that wastes billions of dollars annually to illuminate, as one wag put it, the bellies of birds and low-flying aircraft. Leslie Peltier predicted as much decades ago. "The Moon and the stars no longer come to the farm," he wrote, in *Starlight Nights.* "The farmer has exchanged his birthright in them for the wattage of his all-night Sun. His children will never know the blessed dark of night." But perfectly dark skies are not required to see prominent constellations, provided that you can find shelter from the direct glare of the brightest nearby lights. If your neighbors are the culprits, have a friendly talk with them about the aesthetic and money-saving virtues of shielding their exterior lights to illuminate what they're intended for rather than the sky (which often means the same results can be obtained with a lower-wattage bulb) and plugging security lights into motion sensors (which improves security while cutting the electric bill). Dealing with brilliant streetlights calls for more innovative solutions. The astrophotographer Robert Gendler, who makes observatory-grade CCD images with a telescope rolled onto the driveway of his home in a Connecticut suburb, starts each observing run by climbing a ladder to drape a black cloth over his local streetlight. A few determined amateur astronomers have persuaded municipal authorities to shield a particularly intrusive lamp or even to put a switch on it. Ultimately, the best solution is for municipalities to replace wasteful street lighting with efficient, shielded lamps that save tax dollars and bring back the stars. One community that took that step years ago is Tucson, Arizona, where the Milky Way is often visible even from commercial downtown intersections; eighty percent of the residents say they prefer the new, shielded, low-pressure sodium streetlights to the glaring high-pressure ones they replaced. To learn more about light pollution, consult the International Dark-Sky Association (darksky.org).

While learning your way around the constellations, you may wish to familiarize yourself with the basic angular measurements used in astronomy: A fist held at arm's length subtends about ten degrees; the pointers of the Big Dipper are five degrees apart; and the full Moon is half a degree wide. This is also a good time to note the colors of the stars. Stars do have colors, but their perception is somewhat subjective and it is interesting how many different colors may be described by different observers of the same star—especially when it's near the horizon, where atmospheric refraction can introduce spurious colors.

Warm summer nights are ideal for casual stargazing, but cold winter skies are often the most spectacular. To remain comfortable while stargazing when temperatures drop, dress more heavily than you might think necessary: It's a lot easier to stay warm in the first place than to get warm once you've become chilled. My customary winter observing costume consists of long underwear, wool socks, wool pants, a padded flannel work shirt, a down vest, and a wool cap—to which, on the coldest nights, I add an arctic parka, a scarf, and finger-less gloves. The most important item is the cap, since the body vents heat upward, chimney-style, from the top of the head.

The brilliance of the full Moon can banish all but the brightest stars from the sky, so try to observe when the Moon is below the horizon or less than about half-illuminated (that is, before first quarter in the evenings or after third quarter in predawn skies). The eleventh-century Chinese poet Su Tung-P'o liked to spend full-Moon nights boating with friends, taking along a jug of wine and reciting poetry; I can attest that this is still an excellent way to spend bright moonlit nights.

Large artificial satellites such as the International Space Station can be seen even in indifferent skies, and in dark conditions many more are visible. The best times are within two hours after sunset or before sunrise, when the sky is dark but satellites in low orbits have not yet been engulfed by the Earth's shadow. To find when bright satellites will pass over your site, consult NASA's "Liftoff to Space Exploration" Web site, liftoff.msfc.nasa.gov.

Zodiacal light—sunlight reflected off dust particles along the ecliptic—can be discerned on some dark nights under good conditions, particularly when the ecliptic stands high in the sky: This is always the case in the tropics, while at higher northern latitudes the best times are on spring evenings and fall mornings. When twilight has faded, or before it begins in the morning, look for a softly glowing pyramid with its base near the horizon and its peak about twenty or thirty degrees (two or three fist diameters) up along the eclip-tic. The much fainter Gegenschein, light reflected from a point opposite the Sun, can sometimes be seen under excellent conditions. Look for a glowing ellipse about ten degrees (one fist) wide, near the zenith around midnight.

Auroras set the sky aglow when particles from solar flares encounter Earth's magnetic field. Most auroras occur near the magnetic poles, and are best seen from far northern and southern latitudes. Nearly two hundred auroras occur annually in Alaska and northern Canada, but the southern United States have auroras only about five or ten times per year, and only once or twice per *decade* are auroras visible from as far south as the Caribbean or as far north as Peru. Observers at suitable latitudes are advised to check solar activity Web sites, such as spaceweather.com, for news of the large solar flares that trigger auroras.

Meteors will more often enliven naked-eye stargazing sessions if you observe during a periodic meteor shower. See Appendix B for a list of the major ones.

Stargazing with Binoculars

Binoculars are rated by their magnifying power and the aperture of their objective lenses: A 7x35 binocular, for instance, magnifies seven times and has objective lenses 35 millimeters in diameter. For astronomical purposes, light-gathering power is generally more important than magnification. What counts is the area—πR^2—of the objective lens: Doubling the diameter of the lens yields four times the light-gathering power, so 50-mm binoculars gather twice as much light as 35-mm binoculars do. In my experience 7x50s usually offer the best combination of light-gathering power and ease of use. Optical quality counts for a lot, though: The wide-field 7x35-mm binoculars I purchased at age sixteen, by making monthly payments from my lawn-mowing money for two years, outperform all but the most exotic 7x50s I have since been able to find.

Binoculars with 75-mm and larger objective lenses gather lots of light but tend to be heavy, discouraging prolonged viewing unless they are mounted on a tripod of some kind. Very large binoculars, like the 25x150s preferred by serious comet hunters, require a mount but can provide breathtaking views under good conditions. A few amateur astronomers have even ganged pairs of large reflecting telescopes together to form gigantic binoculars that can produce staggering images if properly aligned.

Center-focus binoculars let you change the focus on both eyepieces at once, while one eyepiece has individual focus to accommodate differences in one's eyes. To focus such binoculars properly, cover the objective lens on the side that has individual focus, use the main wheel to focus, then cover the other lens and adjust the individually focusing eyepiece. Avoid closing one eye, as this strains eye muscles and can result in less than ideal focus. Binoculars that use individual focusing on each eyepiece are less expensive and will suffice for astronomy, since you are always focusing to infinity anyway: The money saved can be invested in better optical quality.

Tripod-mounted or image-stabilized binoculars can afford rewarding views of the Moon and planets, while hand-held binoculars are good for surveying prominent comets, big star clusters like the Pleiades, and the Milky Way. You can learn a great deal about Milky Way structure by studying it through tripod-mounted binoculars while referring to a competent star chart. Start with the bright star clusters and nebulae listed in the Messier catalog (Appendix D), then browse around to take in the large-scale star fields and nebulae.

Using a Telescope

Telescopes are tools, and the best telescope is the one that will give you the best views of the things you want to see from the site where it's most often deployed. To get a sense of what sort of telescope this might be, attend the stargazing sessions held by your local amateur astronomy club and have a look through a variety of different telescopes. Ask questions, and ask yourself what you want from the instrument. A big Dobsonian reflector gathers a lot more light than a small refractor, but it also may require frequent collimation to keep the optics tuned up, and dragging it outdoors repeatedly could become a chore. A small telescope that you use frequently is preferable to a big one that gathers dust in a closet.

Large-aperture telescopes can support higher-power eyepieces, but magnification is only one aspect of telescope performance: If the optics are of poor quality, high-powers will merely deliver larger images of fuzzy-looking blobs. Avoid purchasing telescopes that are advertised by their power ("Magnifies up to 1,000x!"), as this is usually a sign of inferior optics being foisted on untutored consumers. Any telescope can magnify an image 1,000x, but unless it has a large enough aperture and sufficiently good quality to support that much power, it won't be worth looking through.

Big telescopes are at least as vulnerable to bad seeing as small ones are, so if the air is typically unsteady at your observing site you may be better off with a modest instrument. Similarly, if your skies are bright, the ability of your telescope to capture dim objects like distant galaxies may be limited by the sky rather than by optics. Experienced observers test a site by stopping down a large telescope to various smaller apertures: If it performs as well at, say, six inches as at twelve inches, then a six-inch telescope may suffice at that site and you can put your money into quality rather than aperture.

If you travel frequently, a small portable telescope may be advisable. I tote around a diminutive Maksutov in a carry-on case that also holds a couple of eyepieces and a small tripod, and through it have seen more memorable sights in the night sky than with anything else this side of the big observatory instruments. Again, the most rewarding telescope is the one you use often.

A telescope, like a stereo system, is only as good as its weakest link, so a few high-quality eyepieces are preferable to a passel of mediocre ones. A Barlow lens inserted in front of an eyepiece multiplies its power, doubling the range of magnifications available with the eyepieces you have. For able long-term performance, take good care of your optics. Keep the eyepieces in a sealed case and cap the telescope when it's not in use, and you won't often have to clean the optics. I clean my eyepieces with ethanol and distilled water, using optical tissues and genuine (not synthetic) cotton wads. It's simpler than it sounds and it does the job without scratching the lens surfaces—as will, say, breathing on an eyepiece and wiping it with your shirttail. Objective lenses and primary mirrors, if cared for properly, can go for years without cleaning. When you do have to clean a primary mirror, wash it in the sink with detergent, rinse with distilled water, and dry it with a hair dryer; then reinstall and collimate it. The star-side surfaces of refractor objective lenses can if necessary be cleaned with the same technique used for eyepieces; the inside surface seldom calls for maintenance. Repeated exposure to dew can besmirch mirrors and lenses, so if dewing is a problem use a hair dryer or an electric heating collar to keep the optics dry.

An equatorial mounting equipped with a clock drive tracks astronomical objects by compensating for the Earth's rotation—which looks a hundred times faster when you're observing with a 100-power eyepiece—while simpler altazimuth mounts, though lighter and less costly, offer little or no tracking ability. For low-power viewing of galaxies and nebulae, an altazimuth may be all you require. Dobsonians are typically altazimuth mounted, but can be fitted with drive motors to provide a short-term tracking capability. For a first telescope, it may make sense to invest in good quality optics on a simple altazimuth mount; if necessary, you can always get an equatorial mount later on. The most important consideration is that the mount moves smoothly and is massive enough to hold the telescope steady in a light wind. Patrick Moore's advice is to "work out the absolute maximum mass you need for the mounting, then multiply by three."

To point your telescope, consult a star chart, use the small, low-power finder scope to get a bright star in the field of view, then "star hop" to dimmer objects by working from the chart. Computerized "go-to" telescopes make star hopping unnecessary, but it's still an excellent method for learning your way around. Get to know the field of view of the eyepieces you use most frequently. Each eyepiece has a stated focal length and an inherent field of view.

Its focal length, divided into that of the telescope, yields its power: Hence a 25-millimeter eyepiece magnifies 40x when used on a telescope with a focal length of 1,000 mm, and 120x on a telescope with a focal length of 3,000 mm. The eyepiece's true field of view is its inherent field divided by its magnifying power: An eyepiece with a 40° inherent field, operating at 80x, delivers a true field of view of 40 divided by 80, or 0.5°. The same eyepiece, put on a longer telescope where it magnifies 120x, has a field of view of 40 divided by 120 or 0.33°, a third of a degree. Modern computerized star programs will do these calculations for you, generating field-of-view overlays on the display that let you easily match the telescope view to the chart.

When looking through an eyepiece, cover the other eye or just ignore it, rather than closing it; this makes for more relaxed and perceptive viewing. Binocular viewers let you use both eyes at once, but good ones are expensive, require matched pairs of eyepieces, and unlike real binoculars do not actually increase the telescope's light-gathering power: Indeed they reduce it slightly, since the light beam is split and run through a set of prisms. Still, their psychological effect can be profound, especially on planets, globular star clusters, and the Moon. Observing lunar craters with a binocular eyepiece setup on a high-resolution telescope is akin to orbiting the Moon in a spacecraft.

Telescopes are like musical instruments, in that you get out of them what you put into them. The visual virtuoso William Herschel cautioned that seeing is "in some respects an art which must be learned" and noted that he achieved his own dexterity through "constant practice." Be patient: Prolonged study of a planet or galaxy will result in more rewarding and memorable views than just taking a quick look. When observing dim objects near the limit of visibility, try looking slightly away from the object: The eye's peripheral vision is more sensitive to light than is direct vision, which employs the less sensitive cells of the central fovea. For best results, put the object above or on the nose side of your central vision; avoid the ear side, as there is a blind spot there, in the data port where optic nerves go through the retina to the inner brain. The Blinking Planetary (NGC 6826, in Cygnus) provides a striking example of the advantage of using "averted vision," as this technique is called: Look directly at it and you see a small, fuzzy ball of light; look slightly away, and this planetary nebula suddenly becomes much larger.

Observing sessions benefit greatly from a little preparation. Consult a star chart and make up an observing list: That way you can spend your time outdoors actually looking at things rather than figuring out where next to point the telescope. Take the phase of the Moon into account: When the Moon is too bright to permit good viewing of nebulae and galaxies, you can concentrate on planets, double stars, and the Moon itself. A little study pays large rewards; the more you know about the objects you are viewing, the more meaningful the experience will be.

Many stargazers keep a journal or log of their observing sessions, recording the date and time, describing the objects observed, and evaluating the atmospheric conditions. The two main atmospheric considerations are seeing (the steadiness of the atmosphere) and transparency (its clarity). Some observers use a scale of 1–10, from worst to best, for both, while others find a 1–5 scale sufficient, for seeing at least: They note that it's pretty hard to decide whether seeing is, say, a 7 or 8, especially since the turbulence of the atmosphere can vary from hour to hour and in different parts of the sky at the same time. Experienced observers rate transparency by noting the magnitude of the dimmest stars they can see with the unaided eye in the region of Polaris—where the stars are always in the sky and are always at about the same elevation above the horizon.

Extended deep-space objects like galaxies and nebulae may show up to good advantage on nights of high transparency but poor seeing, while for planets, seeing is more important than transparency. Some of the best views I've had of Mars came on nights when a thin fog made for poor transparency but stabilized the air, bringing out details on the red planet's disk. Keep in mind that the distorting effects of Earth's atmosphere are minimized at the zenith, so the most favorable telescopic views of an object are likely to come around the time it crosses the meridian, since it is then at its nearest approach to the zenith.

And remember: Never point an unfiltered telescope at the Sun.

Taking Pictures

Astrophotography used to be relatively uncomplicated. You had but a few choices of film formats, emulsions, and a standard retinue of darkroom techniques, and that—plus the question of how good a telescope you could get your hands on—was about it. Today there are many more options. Traditional photography is still popular, the emulsions are better than ever, and many high-quality telescopes suitable for astrophotography are on the market. CCD imaging is burgeoning, with new opportunities opening up all the time, while digital still and video cameras are also producing promising results. There has never been a better time to get into astrophotography. Which also means, alas, that the subject is much too large to be dealt with properly here.

One point, however: If you are interested in taking astronomical pictures, give it a try with whatever gear you have around or can readily borrow. An ordinary camera mounted on a tripod can be used to make colorful time-exposure photos of star trails caused by Earth's rotation, and of meteor showers and auroras. The same camera, piggybacked on a clock-driven telescope, can take exposures of the sky that show much more than the naked eye sees— or, if simply pointed into the eyepiece of a telescope aimed at the Moon, may at least produce a decent snapshot. It's great to make astronomical images with a liquid-nitrogen–cooled, research-grade CCD on a big telescope, or with a broadcast-quality video camera on a rare-earth lens that costs more than college tuition, but it's not necessary. What matters most is the talent and dedication of the photographer and the conditions under which he or she is shooting. Don't assume that it's pointless to photograph a lunar eclipse or a conjunction of Venus and the Moon because other, more experienced photographers are doing the same thing with better equipment. As with visual observing, your result may prove to be unique in all the world—but you'll never know unless you try.

Finally, whatever your involvement in stargazing may be, have fun. There's a whole universe out there, and it's where you live, so don't let it intimidate you. You're coming home.

APPENDIX B

Notable Periodic Meteor Showers

Shower Name	Duration	Maximum	Comments
Quadrantids	Jan. 1–6	Jan. 4	Variable; can be rich
Lyrids	Apr. 19–25	Apr. 21	Modest; variable
Eta Aquarids	Apr. 24–May 20	May 5	Rich in Southern Hemisphere
Delta Aquarids	Jul. 15–Aug. 20	Jul. 28	Has two peaks
Perseids	Jul. 23–Aug. 20	Aug. 12	Rich, reliable summer shower
Northern Taurids	Oct. 12–Dec. 2	Nov. 4–7	Two overlapping showers;
Southern Taurids	Sept. 17–Nov. 27	Oct. 30–Nov. 7	sparse but long-lived
Orionids	Oct. 16–27	Oct. 22	Unpredictable maximum date
Leonids	Nov. 15–20	Nov. 17	Sometimes spectacular
Geminids	Dec. 7–16	Dec. 13	Often rich
Ursids	Dec. 17–25	Dec. 22	Usually modest

Representative Bright Stars

Star Name	Apparent Magnitude	Absolute Magnitude	Distance
Negative magnitudes			
Sun	−26.8	+4.8	8 light-minutes
Sirius	−1.4	+1.5	9 light-years
(The "Dog Star," in Canis Major, the big dog)			
Zero magnitude			
Vega	+0.0	+0.6	25 light-years
(Prominent blue-white star in Lyra, the lyre)			
First magnitude			
Antares	1.1	−5.8	604 light-years
(The red giant "rival of Mars")			
Second magnitude			
Polaris	1.9	−4.1 (variable)	431 light-years
(The North Star)			
Third magnitude			
Phekda	3.0	−0.1	147 light-years
(Gamma Ursa Minor; marks the southwestern corner of the Big Dipper)			
Fourth magnitude			
Delta Cephei	4.1	−4.4 (variable)	950 light-years
(The first Cepheid variable to be identified, and the brighter component of a beautiful double star)			

APPENDIX D

Messier Objects by Season

THE MESSIER CATALOG, which lists bright objects suitable for observing with binoculars and small telescopes, is here organized for handy viewing on evenings during each season. The columns contain the Messier number of each object; the constellation in which it is located; the type of object (see the key below); its (epoch 2000) coordinates, in hours (h) and minutes (m) of right ascension, and degrees (°) and minutes (') of arc for declination; its visual magnitude (mag.); and its angular size in minutes of arc (')—or, for planetary nebula, in seconds of arc ("). Entries marked "!!" are especially spectacular "showpiece" objects.

Key to object types:

OC = open star cluster
GC = globular star cluster
PN = planetary nebula
EN = emission nebula
RN = reflection nebula
E/RN = combination of emission and reflection nebula
SNR = supernova remnant
G = galaxy (E = elliptical, I = irregular, SA = normal spiral, SB = barred spiral, S0 = lenticular, pec = peculiar galaxy. Elliptical galaxies are classified by their roundness: those with low numbers like 0 and 1 are nearly circular; those with higher numbers are more elongated.)

THE WINTER SKY

M#	Constellation	Type	RA (2000) h m	Dec ° '	Mag.	Size '	Remarks
1	Taurus	SNR	5 34.5	+22 01	8.4	6 x 4	!! Crab neb. supernova remnant
45	Taurus	OC	3 47.0	+24 07	1.2	110	!! Pleiades
36	Auriga	OC	5 36.1	+34 08	6.0	12	Use low power
37	Auriga	OC	5 52.4	+32 33	5.6	20	!! Very rich
38	Auriga	OC	5 28.7	+35 50	6.4	21	Look for small cluster NGC 1907 ½° S
42	Orion	E/RN	5 35.4	−5 27	—	65×60	!! Orion nebula
43	Orion	E/RN	5 35.6	−5 16	—	20×15	Detached part of Orion nebula
78	Orion	RN	5 46.7	+0 03	—	8×6	Bright featureless reflection nebula
79	Lepus	GC	5 24.5	−24 33	7.8	8.7	200-mm telescope needed to resolve
35	Gemini	OC	6 08.9	+24 20	5.1	28	!! Look for cluster NGC 2158 ¼° S
41	Canis Major	OC	6 47.0	−20 44	4.5	38	4° south of Sirius; bright but coarse

50	Monoceros	OC	7 03.2	−8 20	5.9	16	Between Sirius & Procyon
46	Puppis	OC	7 41.8	−14 49	6.1	27	!! contains planetary nebula NGC 2438
47	Puppis	OC	7 36.6	−14 30	4.4	29	Coarse cluster 1.5° west of M46
93	Puppis	OC	7 44.6	−23 52	≈6.2	22	Compact, bright cluster; fairly rich
48	Hydra	OC	8 13.8	−5 48	5.8	54	Large, sparse cluster

THE SPRING SKY

M#	Constellation	Type	RA (2000) h m	Dec ° '	Mag.	Size '	Remarks
44	Cancer	OC	8 40.1	+19 59	3.1	95	!! Beehive or Praesepe; use low power
67	Cancer	OC	8 50.4	+11 49	6.9	29	One of the oldest star clusters known
40	Ursa Major	2 stars	12 22.4	+58 05	8.0	—	Double star Winnecke 4; separation 50"
81	Ursa Major	G-SA	9 55.6	+69 04	6.9	24×13	!! Bright spiral visible in binoculars
82	Ursa Major	G-I0	9 55.8	+69 41	8.4	12×6	!! The "exploding" galaxy; M81 ½° S
97	Ursa Major	PN	11 14.8	+55 01	9.9	194"	!! Owl nebula; distinct gray oval
101	Ursa Major	G-SAB	14 03.2	+54 21	7.9	26×26	!! Pinwheel galaxy; diffuse face-on spiral
108	Ursa Major	G-SB	11 11.5	+55 40	10.0	8.1×2.1	Nearly edge-on; paired with M97 ¾° SE
109	Ursa Major	G-SB	11 57.6	+53 23	9.8	7.6×4.3	Barred spiral
65	Leo	G-SAB	11 18.9	+13 05	9.3	8.7×2.2	!! Bright elongated spiral
66	Leo	G-SAB	11 20.2	+12 59	8.9	8.2×3.9	!! M65 and NGC 3628 in same field
95	Leo	G-SB	10 44.0	+11 42	9.7	7.8×4.6	Bright barred spiral
96	Leo	G-SAB	10 46.8	+11 49	9.2	6.9×4.6	M95 in same field
105	Leo	G-E1	10 47.8	+12 35	9.3	3.9×3.9	Bright elliptical near M95 and M96
53	Coma Berenices	GC	13 12.9	+18 10	7.5	12.6	150-mm telescope needed to resolve
64	Coma Berenices	G-SA	12 56.7	+21 41	8.5	9.2×4.6	!! Black Eye galaxy; needs big scope
85	Coma Berenices	G-SA	12 25.4	+18 11	9.1	7.5×5.7	Bright elliptical
88	Coma Berenices	G-SA	12 32.0	+14 25	9.6	6.1×2.8	Bright multiple-arm spiral
91	Coma Berenices	G-SBb	12 35.4	+14 30	10.2	5.0×4.1	
98	Coma Berenices	G-SAB	12 13.8	+14 54	10.1	9.1×2.1	Nearly edge-on spiral
99	Coma Berenices	G-SA	12 18.8	+14 25	9.9	4.6×4.3	Nearly face-on spiral near M98
100	Coma Berenices	G-SAB	12 22.9	+15 49	9.3	6.2×5.3	Face-on spiral with starlike nucleus
49	Virgo	G-E2	12 29.8	+8 00	8.4	8.1×7.1	Very bright elliptical
58	Virgo	G-SAB	12 37.7	+11 49	9.7	5.5×4.6	Bright barred spiral; M59 and M60 1° E
59	Virgo	G-E5	12 42.0	+11 39	9.6	4.6×3.6	Bright elliptical paired with M60
60	Virgo	G-E2	12 43.7	+11 33	8.8	7.1×6.1	Bright elliptical with M59 and NGC 4647
61	Virgo	G-SAB	12 21.9	+4 28	9.7	6.0×5.9	Face-on two-armed spiral
84	Virgo	G-E1	12 25.1	+12 53	9.1	5.1×4.1	!! w/ M86 in Markarian's Chain
86	Virgo	G-E3	12 26.2	+12 57	8.9	12×9	!! w/ many NGC galaxies in Chain
87	Virgo	G-E0-1	12 30.8	+12 24	8.6	7.1×7.1	The one with famous jet
89	Virgo	G-E	12 35.7	+12 33	9.8	3.4×3.4	Elliptical; resembles M87 but smaller
90	Virgo	G-SAB	12 36.8	+13 10	9.5	10×4	Bright barred spiral near M89
104	Virgo	G-SA	12 40.0	−11 37	8.0	7.1×4.4	!! Sombrero galaxy; look for dust lane
3	Canes Vn.	GC	13 42.2	+28 23	5.9	16.2	!! Contains many variable stars
51	Canes Vn.	G-SA	13 29.9	+47 12	8.4	8×7	!! Whirlpool galaxy; superb in big scope
63	Canes Vn.	G-SA	13 15.8	+42 02	8.6	14×8	!! Sunflower galaxy; bright, elongated
94	Canes Vn.	G-SA	12 50.9	+41 07	8.2	13×11	Very bright and cometlike
106	Canes Vn.	G-SAB	12 19.0	+47 18	8.4	20×8	!! Superb large, bright spiral
68	Hydra	GC	12 39.5	−26 45	7.7	12	150-mm telescope needed to resolve
83	Hydra	G-SAB	13 37.0	−29 52	7.6	16×13	Large and diffuse; superb from far south
102	Draco	G-SA0	15 06.5	+55 46	9.9	6.6×3.2	
5	Serpens	GC	15 18.6	+2 05	5.7	17.4	!! One of the sky's finest globulars

The Summer Sky

M#	Constellation	Type	RA (2000) h m	Dec ° ′	Mag.	Size ′	Remarks
13	Hercules	GC	16 41.7	+36 28	5.7	16.6	!! Hercules cluster; NGC 6207 ½° NE
92	Hercules	GC	17 17.1	+43 08	6.4	11.2	9° NE of M13; fine but often overlooked
9	Ophiuchus	GC	17 19.2	−18 31	7.6	9.3	Smallest of Ophiuchus globulars
10	Ophiuchus	GC	16 57.1	−4 06	6.6	15.1	Rich globular cluster; M12 is 3° NW
12	Ophiuchus	GC	16 47.2	−1 57	6.8	14.5	Loose globular cluster near M10
14	Ophiuchus	GC	17 37.6	−3 15	7.6	11.7	200-mm telescope needed to resolve
19	Ophiuchus	GC	17 02.6	−26 16	6.7	13.5	Oblate globular; M62 4° south
62	Ophiuchus	GC	17 01.2	−30 07	6.7	14.1	Asymmetrical; in rich field
107	Ophiuchus	GC	16 32.5	−13 03	8.1	10.0	Small, faint globular
4	Scorpius	GC	16 23.6	−26 32	5.8	26.3	Bright globular near Antares
6	Scorpius	OC	17 40.1	−32 13	4.2	33	!! Butterfly cluster; best at low power
7	Scorpius	OC	17 53.9	−34 49	3.3	80	!! Excellent in binoculars
80	Scorpius	GC	16 17.0	−22 59	7.3	8.9	Very compressed globular
16	Serpens	EN+OC	18 18.6	−13 58	—	35×28	Eagle neb. w/ open cluster
8	Sagittarius	EN	18 03.8	−24 23	—	45×30	!! Lagoon nebula w/open cluster
17	Sagittarius	EN	18 20.8	−16 11	—	20×15	!! Swan or Omega nebula
18	Sagittarius	OC	18 19.9	−17 08	6.9	10	Sparse cluster; 1° south of M17
20	Sagittarius	E/RN	18 02.3	−23 02	—	20×20	!! Trifid nebula; look for dark lanes
21	Sagittarius	OC	18 04.6	−22 30	5.9	13	0.7° NE of M20; sparse cluster
22	Sagittarius	GC	18 36.4	−23 54	5.1	24	Spectacular from southern latitude
23	Sagittarius	OC	17 56.8	−19 01	5.5	27	Bright, loose open cluster
24	Sagittarius	Star-cloud	18 16.5	−18 50	4.6	95×35	Rich star cloud; best in big binoculars
25	Sagittarius	OC	18 31.6	−19 15	4.6	32	Bright but sparse open cluster
28	Sagittarius	GC	18 24.5	−24 52	6.8	11.2	Compact globular near M22
54	Sagittarius	GC	18 55.1	−30 29	7.6	9.1	Not easily resolved
55	Sagittarius	GC	19 40.0	−30 58	6.4	19.0	Bright, loose globular cluster
69	Sagittarius	GC	18 31.4	−32 21	7.6	7.1	Small, poor globular cluster
70	Sagittarius	GC	18 43.2	−32 18	8.0	7.8	Small globular 2° east of M69
75	Sagittarius	GC	20 06.1	−21 55	8.5	6	Small and distant; 59,000 ly away
11	Scutum	OC	18 51.1	−6 16	5.8	13	!! Wild Duck; the best open cluster?
26	Scutum	OC	18 45.2	−9 24	8.0	14	Bright, coarse cluster
56	Lyra	GC	19 16.6	+30 11	8.3	7.1	Within a rich starfield
57	Lyra	PN	18 53.6	+33 02	8.8	>71″	!! Ring nebula; an amazing smoke ring
71	Sagitta	GC	19 53.8	+18 47	8.0	7.2	Loose globular; looks like an open cluster
27	Vulpecula	PN	19 59.6	+22 43	7.3	>348″	!! Dumbbell nebula; a superb object
29	Cygnus	OC	20 23.9	+38 32	6.6	6	Small, poor open cluster 2° S of Gamma Cygni
39	Cygnus	OC	21 32.2	+48 26	4.6	31	Very sparse cluster; use low power

The Autumn Sky

M#	Constellation	Type	RA (2000) h m	Dec ° ′	Mag.	Size ′	Remarks
2	Aquarius	GC	21 33.5	−0 49	6.4	12.9	200-mm telescope needed to resolve
72	Aquarius	GC	20 53.5	−12 32	9.3	5.9	Near the Saturn nebula, NGC 7009
73	Aquarius	OC	20 59.0	−12 38	8.9p	2.8	Group of four stars only; an "asterism"
15	Pegasus	GC	21 30.0	+12 10	6.0	12.3	Rich, compact globular
30	Capricornus	GC	21 40.4	−23 11	7.3	11	
52	Cassiopeia	OC	23 24.2	+61 35	6.9	12	Young, rich cluster
103	Cassiopeia	OC	1 33.2	+60 42	7.4	6	Three NGC open clusters nearby
31	Andromeda	G-SA	0 42.7	+41 16	3.4	185×75	!! Andromeda galaxy
32	Andromeda	G-E5 pec	0 42.7	+40 52	8.1	11×7	Close companion to M31

M#	Constellation	Type	RA (2000) h m	Dec ° '	Mag.	Size '	Remarks
110	Andromeda	G-E3 pec	0 40.4	+41 41	8.1	20×12	More distant companion to M31
33	Triangulum	G-SA	1 33.9	+30 39	5.7	67×42	Large, diffuse spiral; requires dark sky
74	Pisces	G-SA	1 36.7	+15 47	9.4	11×11	Faint, elusive spiral; tough in small scope
77	Cetus	G-SAB	2 42.7	−0 01	8.9	8.2×7.3	Seyfert galaxy with starlike nucleus
34	Perseus	OC	2 42.0	+42 47	5.2	35	Best at low power
76	Perseus	PN	1 42.4	+51 34	10.1	>65"	Little Dumbbell

APPENDIX E

Planets and Their Satellites

Physical Elements of the Planets

These tables list each planet's diameter; its oblateness (the degree to which it is flattened rather than spherical); its mass (Earth=1); its overall density in metric tons per cubic meter (that is, its density compared to that of water, which weighs in at one metric ton per cubic meter); the gravitational force at its surface, relative to Earth's; its escape velocity (the speed required to escape the planet's gravitational field, if launching from its equator, in kilometers per second; to convert to miles per hour, multiply by 2,160); its rotation period in days; and the inclination, in degrees, of its equator to its orbital plane. The Sun and Moon are included for purposes of comparison.

Object	Equatorial Diameter km	Oblate- ness	Mass Earth=1	Den- sity t/m³	Gravity Earth=1	Escape Speed km/s	Rotation Period days	Incl. °
Sun	1,392,000	0*	332,946	1.41	27.9	617.5	25–35†	—
Mercury	4,879	0	0.055	5.43	0.38	4.2	58.646	0.0
Venus	12,104	0	0.815	5.24	0.90	10.4	243.019	177.4
Earth	12,756	1/298	1.00	5.52	1.00	11.2	0.9973	23.4
Moon	3,475	0	0.012	3.34	0.17	2.4	27.3217	6.7
Mars	6,794	1/154	0.107	3.94	0.38	5.	1.0260	25.2
Jupiter	142,980‡	1/15.4	317.833	1.33	2.53	59.5	0.4101§	3.1
Saturn	120,540‡	1/10.2	95.159	0.70	1.06	35.5	0.4440	25.3
Uranus	51,120‡	1/43.6	14.500	1.30	0.90	21.3	0.7183	97.9
Neptune	49,530‡	1/58.5	17.204	1.76	1.14	23.5	0.6712	28.3
Pluto	2,300	0?	0.003	1.1	0.08	1.3	6.3872	123.

* The Sun is very slightly oblate—about 36 km flattened at the poles, or ~1 part in 40,000.
† Depending on latitude.
‡ At 1 bar atmospheric pressure.
§ For the most rapidly rotating part of Jupiter, the equatorial region.

Orbital Elements of the Planets

These tables list each planet's average distance from the Sun in astronomical units (1 A.U., the distance between Earth and Sun, equals 150 million kilometers or 93 million miles); its orbital eccentricity (the extent to which its orbit differs from a perfect circle, which has zero eccentricity); the inclination of its orbit (in degrees, relative to Earth's); and the time (in Earth years) it takes to complete one orbit.

Planet	Mean Distance AU	Eccentricity °	Inclination °	Period years
Mercury	0.387	0.206	7.00	0.241
Venus	0.723	0.007	3.39	0.615
Earth	1.000	0.017	0.00	1.000
Mars	1.524	0.093	1.85	1.881
Jupiter	5.203	0.049	1.30	11.864
Saturn	9.586	0.058	2.49	29.675
Uranus	19.155	0.048	0.77	83.835
Neptune	29.97	0.011	1.77	164.050
Pluto	39.28	0.246	17.17	246.177

Satellites of the Planets

This table lists the solar system's known natural satellites as of the time of publication. Columns show each satellite's diameter in kilometers; its magnitude, as seen from Earth at an average opposition; its albedo (or reflectivity); its average distance from its planet at opposition (stated as both actual distance, in kilometers, and angular distance on the sky in seconds of arc); its orbital period, in Earth days; and information concerning its discovery.

Name	Diameter km	Visual Magnitude	Albedo	Distance from Planet 10^3km	"	Orbital Period days	Discovery
Satellite of Earth							
Moon*	3,476.	−12.7	0.11	384.5	—	27.322	
Satellites of Mars							
I Phobos*	21.	11.6	0.07	9.4	25.	0.319	A. Hall, 1877
II Deimos*	12.	12.7	0.07	23.5	63.	1.263	" " "
Satellites of Jupiter							
XVI Metis	43.	17.5	≈0.05	128.	42.	0.294	S. Synnott, 1979
XV Adrastea	16.	18.7	≈0.05	129.	42.	0.297	D. Jewitt et al., 1979
V Amalthea*	167.	14.1	0.05	180.	59.	0.498	E. Barnard, 1892
XIV Thebe**	99.	16.	≈0.05	222.	73.	0.674	S. Synnott, 1979
I Io*	3,630.	5.	0.6	422.	138.	1.769	Galileo, 1610
II Europa*	3,140.	5.3	0.6	671.	220.	3.551	" " "
III Ganymede*	5,260.	4.6	0.4	1070.	351.	7.155	" " "
IV Callisto*	4,800.	5.6	0.2	1885.	618.	16.689	" " "
1975J1ᴬ	≈4.	20.	—	7507.	2462.	130.	C. Kowal, 1975†
XIII Leda	≈15.	20.	—	11,110.	3640.	240.	C. Kowal, 1974
VI Himalia***	185.	14.8	0.03	11,470.	3760.	251.	C. Perrine, 1904
X Lysithea***	≈35.	18.4	—	11,710.	3840.	260.	S. Nicholson, 1938
VII Elara	75.	16.8	0.03	11,740.	3850.	260.	C. Perrine, 1905
2000J11ᴬ	≈3.	20.5	—	12,557.	4119.	287.	S. Sheppard et al., 2000
2000J3ᴬ	≈4.	20.1	—	20,210.	6629.	585.	" " "
2000J7ᴬ	≈5.	19.8	—	20,929.	6865.	616.	" " "
2000J5ᴬ	≈3.	20.2	—	21,132.	6931.	624.	" " "
XII Ananke***	≈30.	18.9	—	21,200.	6954.	631.	S. Nicholson, 1951
XI Carme	≈40.	18.	—	22,350.	7330.	692.	S. Nicholson, 1938
2000J10ᴬ	≈3.	20.3	—	22,988.	7540.	716.	S. Sheppard et al., 2000
2000J6ᴬ	≈3.	20.5	—	23,074.	7568.	719.	" " "
2000J9ᴬ	≈4.	20.1	—	23,140.	7590.	723.	" " "
2000J4ᴬ	≈3.	20.4	—	23,169.	7599.	723.	" " "
VIII Pasiphae	≈50.	17.1	—	23,330.	7650.	735.	P. Melotte, 1908
IX Sinope***	≈35.	18.3	—	23,370.	7660.	758.	S. Nicholson, 1914
2000J2ᴬ	≈4.	20.1	—	23,746.	7789.	751.	S. Sheppard et al., 2000
2000J8ᴬ	≈4.	20.	—	23,913.	7843.	758.	" " "
1999J1ᴬ	≈7.	19.4	—	24,103.	7906.	759.	J. Scotti & T. Spahr, 1999

Name	Dia-meter km	Visual Magni-tude	Albedo	Distance from Planet 10³km	"	Orbital Period days	Discovery
Satellites of Saturn							
XVIII Pan[B]	≈20.	≈19.	≈0.5	134.	22.	0.577	M. Showalter, 1990
XV Atlas	30.	≈18.	0.4	137.	23.	0.601	R. Terrile, 1980
XVI Prometheus[**]	100.	≈15.	0.6	139.	23.	0.613	S. Collins, D. Carlson, 1980
XVII Pandora[**]	90.	≈16.	0.5	142.	24.	0.628	" " "
X Janus	190.	≈14.	0.6	151.	25.	0.695	A. Dollfus, 1966
XI Epimetheus[*]	120.	≈15.	0.5	151.	25.	0.695	J. Fountain, S. Larson, 1966
I Mimas[*]	390.	12.5	0.8	187.	30.	0.942	W. Herschel, 1789
II Enceladus[*]	500.	11.8	1.0	238.	38.	1.37	" " "
III Tethys[*]	1060.	10.3	0.8	295.	48.	1.888	G. Cassini, 1684
XIII Telesto	25.	≈18.	0.7	295.	48.	1.888	B. Smith et al., 1980
XIV Calypso[**]	25.	≈18.	1.0	295.	48.	1.888	D. Pascu et al., 1980
IV Dione[*]	1120.	10.4	0.6	378.	61.	2.737	G. Cassini, 1684
XII Helene	30.	≈18.	0.6	378.	61.	2.737	P. Laques, J. Lecacheux, 1980
V Rhea[*]	1530.	9.7	0.6	526.	85.	4.517	G. Cassini, 1672
VI Titan[**]	5550.[C]	8.4	0.2	1,221.	197.	15.945	C. Huygens, 1655
VII Hyperion[****]	255.	14.2	0.3	1,481.	239.	21.276	W. Bond, G. Bond, W. Lassell, 1848
VIII Iapetus[*]	1460.	11.	0.08–0.4	3,561.	575.	79.331	G. Cassini, 1671
2000S5[A]	14.	21.9	—	11,300.	1827.	449.	B. Gladman et al., 2000
2000S6[A]	10.	22.5	—	11,400.	1843.	453.	" " "
IX Phoebe[***]	220.	16.5	0.05	12,960.	2096.	550.46	W. Pickering, 1898
2000S2[A]	20.	21.2	—	15,200.	2458.	687.	B. Gladman et al., 2000
2000S8[A]	6.	23.5	—	15,500.	2506.	720.	" " "
2000S3[A]	32.	20.	—	16,800.	2717.	796.	" " "
2000S12[A]	6.	23.8	—	17,600.	2846.	877.	" " "
2000S11[A]	26.	20.4	—	17,900.	2894.	888.	M. Holman et al., 2000
2000S10[A]	8.	22.9	—	18,200.	2943.	913.	B. Gladman et al., 2000
2000S4[A]	14.	22.	—	18,250.	2951.	924.	" " "
2000S9[A]	6.	23.7	—	18,400.	2975.	935.	" " "
2000S7[A]	6.	23.8	—	20,100.	3250.	1067.	" " "
2000S1[A]	16.	21.6	—	23,100.	3735.	1311.	" " "
Satellites of Uranus							
VI Cordelia	25.[D]	24.2	<0.1	49.8	3.7	0.333	*Voyager 2*, 1986
VII Ophelia	30.[D]	23.9	<0.1	53.8	4.0	0.375	" " "
VIII Bianca	45.[D]	23.1	<0.1	59.2	4.4	0.433	" " "
IX Cressida	65.[D]	22.3	<0.1	61.8	4.6	0.463	" " "
X Desdemona	60.[D]	22.5	<0.1	62.6	4.7	0.475	" " "
XI Juliet	85.[D]	21.7	<0.1	64.4	4.9	0.492	" " "
XII Portia	110.[D]	21.1	<0.1	66.1	5.0	0.513	" " "
XIII Rosalind	60.[D]	22.5	<0.1	70.0	5.2	0.558	" " "
XIV Belinda	68.[D]	22.1	<0.1	75.3	5.6	0.621	" " "
1986U10	40.[D]	23.6	<0.1	76.4	5.8	0.638	E. Karkoschka, 1999‡
XV Puck	155.	20.4	0.07	86.0	6.5	0.763	*Voyager 2*, 1985
V Miranda	485.	16.5	0.34	129.9	9.7	1.413	G. Kuiper, 1948
I Ariel[*]	1160.	14.	0.4	190.9	14.3	2.521	W. Lassell, 1851
II Umbriel[*]	1190.	14.9	0.19	266.0	20.0	4.146	" " "

Name	Diameter km	Visual Magnitude	Albedo	Distance from Planet 10³km	"	Orbital Period days	Discovery
III Titania*	1610.	13.9	0.28	436.3	32.7	8.704	W. Herschel, 1787
IV Oberon*	1550.	14.1	0.24	583.4	43.8	13.463	" " "
XVI Caliban	60.	≈22.5	<0.1	7,169.	538.	580.	B. Gladman et al., 1997
XX Stephano	30.	≈25.	<0.1	7,920.	584.	674.	B. Gladman et al., 1999
XVII Sycorax	120.	≈21.	<0.1	12,214.	900.	1290.	P. Nicholson et al., 1997
XVIII Prospero	30.	≈24.	<0.1	16,670.	1229.	2057.	M. Holman et al., 1999
XIX Setebos	30.	≈24.	<0.1	17,810.	1313.	2271.	J. Kavelaars, 2000

Satellites of Neptune

Name	Diameter km	Visual Magnitude	Albedo	Distance from Planet 10³km	"	Orbital Period days	Discovery
III Naiad	60.	24.6	≈0.06	48.	2.3	0.3	*Voyager 2*, 1989
IV Thalassa	80.	23.9	≈0.06	50.	2.4	0.31	" " "
V Despina	150.	22.5	0.06	52.5	2.5	0.33	" " "
VI Galatea	160.	22.4	≈0.06	62.	2.9	0.43	" " "
VII Larissa**	190.	22.	0.06	73.6	3.5	0.55	" " "
VIII Proteus**	420.	20.3	0.06	117.6	5.5	1.12	" " "
I Triton*	2700.	13.6	0.8	354.	17.	5.877	W. Lassell, 1846
II Nereid***	340.	19.7	0.16	5,510.	260.	365.21	G. Kuiper, 1949

Satellite of Pluto

Name	Diameter km	Visual Magnitude	Albedo	Distance from Planet 10³km	"	Orbital Period days	Discovery
Charon*	1200.	17.	—	19.1	0.9	6.387	J. Christy, 1978

≈ Approximate value.
* Synchronous motion.
** Probably synchronous.
*** Nonsynchronous.
**** Chaotic motion. If no asterisk, rotation is unknown.
A Diameters assume an albedo (reflectivity) of 0.06.
B Orbits within the Encke gap.
C Titan's cloud-top diameter. Solid-body diameter equals 5,150 km.
D Diameter assuming the same albedo (0.07) as Puck.
† Initially detected in 1975 and then lost; recovered in 2000 by S. Sheppard and D. Jewitt.
‡ In *Voyager 2* (January 1986) images.

APPENDIX F

Further Reading

Periodicals

The principal general-interest monthly American astronomy periodicals are *Astronomy* (astronomy.com) and *Sky & Telescope* (skypub.com). One aimed specifically at amateurs is *Amateur Astronomy* (amateurastronomy.com). Some of the amateur astronomy organizations listed at the end of this appendix publish periodicals of their own.

Notable annual publications include The Royal Astronomical Society of Canada's *Observer's Handbook*, an indispensable observing guide, and Guy Ottewell's *Astronomical Calendar* (Furman University/Astronomical League), which features freehand drawings that afford fresh perspectives on sky events.

Introductions to the Constellations

H. A. Rey's *The Stars: A New Way to See Them* (Houghton Mifflin) is a charming and simple—if idiosyncratic—constellation guide. Sky Publishing's *Night Sky Guide* includes monthly star maps. Other popular introductions are *Turn Left at Orion*, by Guy Consolmagno and Dan M. Davis (Ingram); *The Constellations: An Enthusiast's Guide to the Night Sky*, by Lloyd Motz and Carol Nathanson (Doubleday); *365 Starry Nights: An Introduction to Astronomy for Every Night of the Year*, by Chet Raymo (Fireside); and Antonin Rukl's *Constellation Guidebook* (Sterling).

Star Atlases

The night sky can be thought of as layered, like an onion. The superficial layers, accessible to the naked eye, are covered by simple star charts and planispheres. Binoculars and telescopes probe to deeper layers, and the deeper one looks the more elaborate the star atlas required to keep up. The depth of each atlas is expressed in terms of the dimmest stars (and other objects) that it contains: The higher the magnitude number, the larger the atlas.

The simplest and easiest to use star charts are planispheres. A planisphere consists of a disk representing the night sky in your hemisphere with an oval window in which you rotate the disk to show the constellations at a given time and date. Light and compact, planispheres are easy to take along when backpacking or camping out, and they don't threaten to fly away on windy nights. Popular ones include *David H. Levy's Guide to the Stars*, by David H. Levy, for use at latitudes 30° to 60° North, and *The Night Sky Planisphere*, by David S. Chandler, available in three Northern Hemisphere versions, for latitude 20° to 30°, 30° to 40°, or 40° to 50°.

Among book-bound star atlases, a venerable classic, in print for nearly a century and recently updated, is Arthur Philip Norton and Ian Ridpath's *Norton's Star Atlas and Reference Handbook* (Addison-Wesley). Its dimmest stars are magnitude 6, meaning this is primarily a naked-eye atlas, but it can still be quite useful with small telescopes. Wil Tirion's *The Cambridge Star Atlas* (Cambridge) goes a bit deeper, to magnitude 6.5. *Sky Atlas 2000.0*, by Tirion and Roger W. Sinnott, covers stars to magnitude 8.5 and is available in editions with white background for indoor consultation and black background for use under the sky, making it a favorite among observers with small to medium-size telescopes.

Going deeper still, *Uranometria 2000.0*, by Wil Tirion, B. Rappaport, and G. Lovi (Cambridge), contains 332,000 stars to magnitude 9.5 along with more than 10,000 clusters, nebulae, and galaxies. It is published in two volumes, covering northern and southern skies. The largest and most expensive modern star atlas available on paper is the *Millennium Star Atlas*, by Roger W. Sinnott and Michael A. C. Perryman (Cambridge). Its three volumes contain more than a million stars to magnitude 11 and a wealth of deep-space objects.

Computer star atlases can be used to learn the sky, print custom charts, and control suitably equipped telescopes or at least show where they are pointing. Among the leading programs that I've used personally are *The Sky* (Software Bisque), which can be upgraded as you go along to offer more objects and telescope control; *Redshift* (Maris), a graphics-oriented program that can display informative stills and movies, such as the view from space of the Moon's shadow crossing Earth in any particular solar eclipse; and *Megastar* (Willmann-Bell), a sophisticated yet uncomplicated program favored by many advanced amateurs.

Internet Sky Programs

Free star-chart programs on the Web are a good place to start learning what's up on a given night and to print out your own charts. Among the more popular ones are *Planet Finder* (lightandmatter.com/area2planet.shtml), which shows planets in the sky at your location on any date; *Skyview Café* (skyview-cafe.com), with easy-to-use naked-eye sky views; *Solar System Simulator* (space.jpl.nasa.gov), from NASA's Jet Propulsion Laboratory, which presents views of any planet or satellite from any other place in the solar system; and *StarMap* (mtwilson.edu/Services/StarMap), a star-chart generator from Mt. Wilson Observatory.

Observing Guides

Some of these guides include star charts, but their main intention is to fill in the blanks, informing their readers about what's out there and how to observe it. Notable popular introductions include Patrick Moore's *Exploring the Night Sky With Binoculars* (Cambridge); Moore's *The Observer's Year: 366 Nights of the Universe* (Springer); Terence Dickinson and Alan Dyer's *The Backyard Astronomer's Guide* (Camden House); Dickinson's *Nightwatch* (Firefly); and Jay M. Pasachoff's *A Field Guide to the Stars and Planets* (Houghton Mifflin).

Stephen James O'Meara's respected *Deep Sky Companions: The Messier Objects* (Sky Publishing) features bright nebulae and star clusters, which are also the subject of essays collected in Walter Scott Houston's *Deep-Sky Wonders* (Sky Publishing). *Burnham's Celestial Handbook*, by Robert Burnham Jr. (Dover), a three-volume exposition on the facts and myths associated with each constellation, is outdated in many respects but still well worth reading.

Solar system objects are the focus of Fred William Price's *The Planet Observer's Handbook* (Cambridge); Gerald North's *Observing the Moon: The Modern Astronomer's Guide* (Cambridge); and *Introduction to Observing and Photographing the Solar System*, by Thomas A. Dobbins, Donald C. Parker, and Charles F. Capen (Willmann-Bell). Observing techniques for experienced amateurs are the subject of Gerald North's *Advanced Amateur Astronomy* (Cambridge).

Astronomy Picture Books and Introductions

Several astronomy textbooks are accessible and well written enough to be recommended to general readers, among them *Voyages Through the Universe*, by

Andrew Fraknoi, David Morrison, and Sidney Wolff (Harcourt), and *Universe*, by William J. Kaufmann, Freedman Kaufmann, and Roger A. Freedman (Freeman). Leading picture books include David Malin's lavish, large-format *The Invisible Universe* (Bulfinch); Allan Sandage's classic *Hubble Atlas of Galaxies* (Carnegie Institution), his larger and equally evocative *The Carnegie Atlas of Galaxies* (Carnegie Institution); and *One Universe: At Home in the Cosmos*, by Neil deGrasse Tyson, Charles Liu, and Robert Irion (Joseph Henry), an elegant and informative popular introduction.

On Making and Using Telescopes

For those who have the time and talent, there is no better way to learn how a telescope works—and to own one more cheaply—than to make it yourself. Introductions include *Build Your Own Telescope*, by Richard Berry (Scribner); *The Dobsonian Telescope: A Practical Manual for Building Large Aperture Telescopes*, by David Kriege and Richard Berry (Willmann-Bell); Jean Texereau's *How to Make a Telescope* (Willmann-Bell); and the venerable *Amateur Telescope Making* (Albert Ingalls, ed., 3 vols., Willmann-Bell). Advanced standard reference works include *Principles of Optics*, by Max Born and Emil Wolf (Cambridge), and *Telescopic Optics: A Comprehensive Manual for Amateur Astronomers*, by Harrie G. J. Rutten and Martin A. M. van Venrooij (Willmann-Bell). Harold Richard Suiter's *Star Testing Astronomical Telescopes* (Willmann-Bell) explains how to analyze what you see through a telescope to assess its strengths and deficits.

Amateur Astronomy Organizations

These are national organizations. To locate a nearby local amateur astronomy club, contact your local planetarium or science center or see: skypub.com/resources/directory/usa.html

American Association of Amateur Astronomers: corvus.com

American Association of Variable Star Observers: aavso.org

The American Meteor Society: amsmeteors.org

Association of Lunar and Planetary Observers: lpl.arizona.edu/alpo

The Astronomical League: astroleague.org

The Astronomical Society of Australia: atnf.csiro.au/asa_www

Astronomical Society of the Pacific: astrosociety.org

British Astronomical Association: ast.cam.ac.uk/~baa

International Supernovae Network: supernovae.net

National Deep Sky Observers Society: cismall.com/deepsky/nebulae.html

Royal Astronomical Society (U.K.): ras.org.uk

Royal Astronomical Society of Canada: rasc.ca

The Society of Amateur Radio Astronomers: bambi.net/sara.html

Star Maps

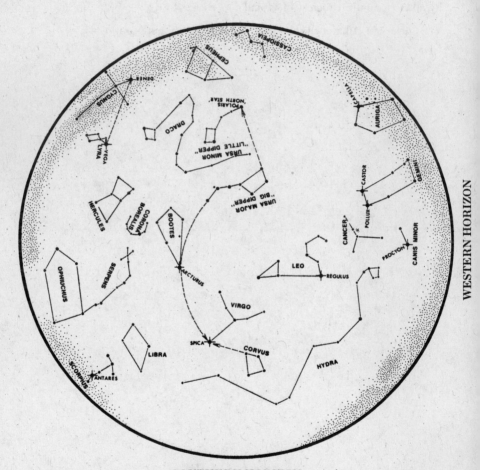

WESTERN HORIZON

SOUTHERN HORIZON

SPRING

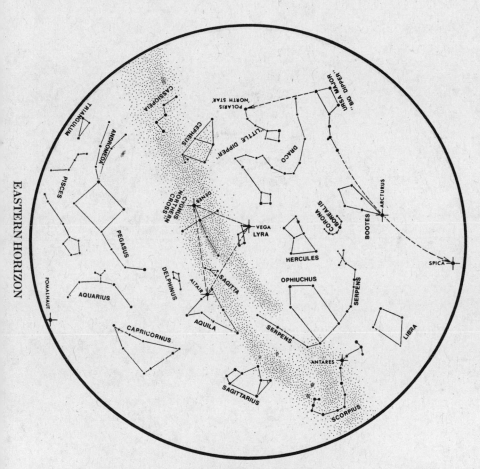

EASTERN HORIZON

SOUTHERN HORIZON

SUMMER

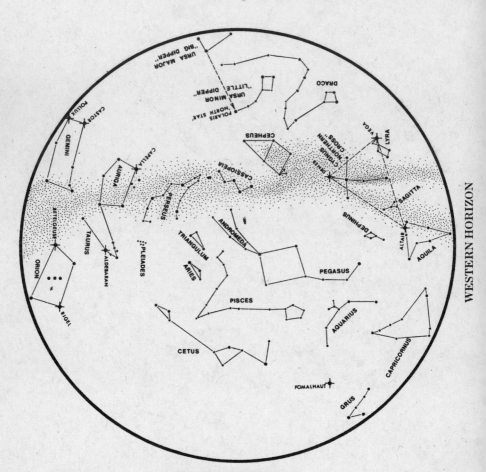

WESTERN HORIZON

SOUTHERN HORIZON

AUTUMN

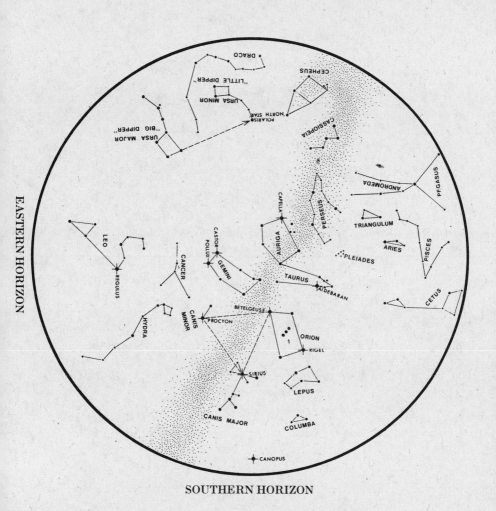

EASTERN HORIZON

SOUTHERN HORIZON

WINTER

NOTES

PREFACE

1. George E. Hale, "The Work of Sir William Huggins." *Astrophysical Journal*, April 1913, p. 145.

2. Xie Renjiang, private communication, May 9, 2001.

1. BEGINNINGS

1. V. M. Hillyer, *A Child's History of the World*. New York: Century, 1924, p. ix.

2. Ibid., p. 5.

3. Astronomers often say that observing the stars through Earth's atmosphere is like viewing a landscape from underwater. The remark is less hyperbolic than it sounds. A New Hampshire amateur astronomer, Kevin J. McCarthy, developed the curious habit of stargazing through eight feet of water, from the bottom of his swimming pool, and found the view to be better than one might expect. Refraction squeezed the 180-degree dome of the sky down to a circle only 97 degrees wide, but every star visible from up top was also visible from down below.

4. Booker T. Washington White, "Special Streamline."

5. Albert Einstein, *Autobiographical Notes*, in P. A. Schilpp, *Albert Einstein: Philosopher-Scientist*. London: Cambridge University Press, 1969, p. 5.

3. THE OZONE

1. Conflicts were being sparked by black music throughout much of the South. A flyer circulated by a white supremacist group in Tennessee read: "Stop. Help Save the Youth of America. . . . Call the advertisers of the radio stations that play this type of music and complain to them!" But the music proved irresistible.

2. In R. H. Blythe, *Zen and Zen Classics*. Tokyo: Hokuseido, 1976, vol. 2, p. 170.

3. By 2001 the project was averaging 10 teraflops, or 10,000 billion computations per second.

4. In 1997, James Cordes, T. Joseph Lazio, and Carl Sagan ("Scintillation-Induced Intermittency in SETI," astro-ph/9707039) suggested that interstellar scintillations "are very likely to allow initial detections of narrowband signals from distant sources (> 100 parsecs), while making redetections improbable." In other words, an alien signal coming from a planet several hundred light-years from Earth might be detected briefly but not be there when the SETI team rechecked it again a few minutes later, because it had in the meantime been significantly weakened by the ozonelike inconstancy of interstellar gas clouds along the signal path. They recommended that SETI projects more often reobserve places in the sky where transient signals were heard but were dismissed as false alarms when they failed to show up a second time. "We cannot rule out the possibility that [some such "false alarms"] are real, intrinsically steady ETI [extraterrestrial intelligence] signals," they wrote.

5. In some versions of this old story, Faraday's interrogator is said to have been the British prime minister. (See, e.g., John Simmons, *The Scientific 100*. Secaucus, N.J.: Carol Publishing, 1996, p. 62.) Others dismiss the tale as apocryphal.

6. "Johnny B. Goode," written and performed by Chuck Berry, from *Chuck Berry's Golden Decade*, Chess Records LP 1514D.

4. AMATEURS

1. Although the 200-inch telescope was for decades the very emblem of astronomical professionalism, its distinctive horseshoe mount, big as a building yet so smooth that it can be moved by the touch of one hand, was the invention of an amateur astronomer, H. Page Bailey. Bailey was a dentist and jack-of-all-trades who got into telescope making after a patient traded him a 15-inch mirror blank for dental care. He conceived of the horseshoe mount—an inherently stable design that permits the telescope to easily be pointed anywhere in the sky—in 1930 and built two of them, one for his personal use and the other for San Bernardino Junior College. Among the visitors who came to study his design was Russell Porter, an amateur telescope maker, illustrator, and former Arctic explorer who had been recruited to the Palomar development team. Bailey's horseshoe won out over Russell's fork design for the big mount, but Russell seems to have taken credit for it anyway, writing ambiguously in his memoirs, "Was it a coincidence that, years before, I had designed a mounting for a supertelescope such as this?" (Russell Williams Porter, *The Arctic Diary of Russell Williams Porter*. Herman Friis, editor. Charlottesville: University Press of Virginia, 1976, p. 168.) After Bailey died, unheralded, in 1962, his original mount passed through several hands, eventually winding up as a rusting hulk in a junk yard behind a Redlands, California, machine shop. Another amateur astronomer, Allan Guthmiller, renovated it in the 1990s and incorporated it into a 20-inch portable Newtonian telescope, which he would tow, uncovered, behind his truck to high-altitude observing locations. Passing motorists, gazing at this gleaming spectacle, had little way of knowing that it embodied the seed of the great Palomar telescope.

2. In John Lankford, "Astronomy's Enduring Resource." *Sky & Telescope*, November 1988, p. 482.

3. Ferguson made astronomy palatable to churchgoers by stressing that, as he saw it, "By knowledge derived from this science, not only the magnitude of the Earth is discovered . . . but our very faculties are enlarged with the grandeur of the ideas it conveys, our minds exalted above the low contracted prejudices of the vulgar, and our understandings clearly convinced, and affected with the conviction, of the existence, wisdom, power, goodness, immutability, and superintendency of the SUPREME BEING." And he was a master of the then-novel "how big it all is, how insignificant are we" school of science writing: Astronomy, he wrote, "discovers to us such an inconceivable number of suns, systems, and worlds, dispersed through boundless space, that if our Sun, with all the planets, Moon, and comets, belonging to it, were annihilated, they would be no more missed, by an eye that could take in the whole creation, than a grain of sand from the seashore—the space they possess being comparatively so small, that it would scarce be a sensible blank in the universe." (James Ferguson, *Astronomy Explained upon Sir Isaac Newton's Principles.* Philadelphia: Mathew Carey, 1806, pp. 31, 34.)

4. In George W. E. Beekman, "The Farmer Astronomer." *Sky & Telescope*, May 1990, p. 548.

5. In Allan Chapman, *The Victorian Amateur Astronomer.* New York: Wiley, 1998, p. 208.

6. Patrick Moore, "The Role of the Amateur." *Sky & Telescope*, November 1988, p. 545.

7. *Oxford English Dictionary*, second edition.

8. Chapman, *The Victorian Amateur Astronomer*, p. xi.

9. A star embedded in NGC 2261, although obscured from direct view, casts the shadows of nearby clouds onto the large, fan-shaped cloud visible from Earth. As the nearby clouds move, the shadows change, altering the brightness of the nebula.

10. This story is recounted by T. R. Williams of Rice University in his paper "The Director's Choice: Mellish, Hubble and the Discovery of the Variable Nebula," presented at AAS 197, January 7, 2001, Session 1: "Boners of the Century." Williams adds that "when Mellish discovered another comet a few weeks later, Frost delayed his notification to Harvard for several days to allow photographic confirmation of the discovery." It was so confirmed, by George Van Biesbroeck, an amateur turned professional who was observing at Yerkes.

11. John Lankford, "Astronomy's Enduring Resource." *Sky & Telescope*, November 1988, p. 483.

12. Leif J. Robinson, "Amateurs: A New Dawning." *Sky & Telescope*, November 1988, p. 453.

13. More generally, amateurism in many fields was promoted by the growing postwar prosperity of the industrialized world. My boyhood refracting telescope, which cost fifteen working days of an average American wage earner's salary in 1956, could be purchased for a fifth of that effort by 1990.

14. Don Moser, "A Salesman for the Heavens Wants to Rope You In: Astronomer John Dobson." *Smithsonian*, April 1989, p. 102.

15. Patrick Moore, *The Astronomy of Birr Castle.* London: Mitchell Beazley, 1971, p. 24.

16. Home video cameras use CCDs that can capture moments of exceptional clarity in otherwise turbulent air. Video cameras typically shoot 60 frames per second, consisting of alternating odd and even rows that are then interlaced to produce a single video frame viewed at 30 fps, a feature that a few amateurs have exploited to capture moments of especially clear "seeing" when recording bright objects like planets, the Moon, and artificial satellites. Ron Dantowitz developed this technique very effectively, taking videos through a tracking telescope at the Boston Museum of Science, where he teaches astronomy, and piecing together the sharpest parts of the clearest frames to make a high-resolution montage. His stills of the space shuttle in orbit are so detailed that one can see whether the shuttle's cargo bay doors are open or closed. When officials from the National Reconnaissance Office paid him a visit, concerned that his techniques could be used to spy *on* spy satellites, Dantowitz obligingly gave them a demonstration. Using his own database to call up its orbital elements, which were secret, he acquired and tracked a classified Lacrosse satellite for them in a matter of minutes. "You can tell it's a Lacrosse," he remarked laconically, "because of its distinct orange-red color . . . probably due to thermal insulation." He described the security officers' reaction to his feats as "stoic." (In Ron Dantowitz, "Sharper Images Through Video," *Sky & Telescope*, August 1998, p. 54.)

17. Allan Sandage, telephone interview by T.F., January 3, 2000.

18. William Blake, *The Marriage of Heaven and Hell*, 9, 7, in Geoffrey Keynes, ed., *Blake: Complete Writings.* London: Oxford University Press, 1972, p. 152.

A VISIT WITH STEPHEN JAMES O'MEARA

1. Owing, perhaps, to his amateur status, O'Meara's finding was ignored in most textbooks and scientific papers. Browsing the Web, I found such statements as, "The rotation period of Uranus could not be measured from Earth because the planet's disk is nearly featureless" (Monterey Institute of Research in Astronomy); "detailed observations of Uranus's atmospheric features have not been possible because the planet is at the resolution limit of ground-based telescopes" (Space Telescope Science Institute); and "before Voyager . . . the planet's rate of rotation could be estimated only roughly and was believed to be anywhere from 16 to 24 hours" (JPL Voyager Uranus Science Summary).

5. PROFESSIONALS

1. Joseph Patterson, "Our Cataclysmic-Variable Network." *Sky & Telescope*, October 1998, pp. 77ff.

2. Ibid.

3. Other amateurs awarded *Hubble* time included Ana Larson, to search for extrasolar planets; James Secosky, to investigate the unexplained brightening of Io when it emerges from Jupiter's shadow; and Raymond Sterner, for the study of luminous arcs connecting galaxies.

4. Paul Boltwood, "Experiences With Pro-Am Relations," in John R. Percy and Joseph B. Wilson, eds., *Amateur-Professional Partnerships in Astronomy.* San Francisco: Astronomical Society of the Pacific, 2000, pp. 193–94.

5. One study divides astronomers into five general categories and estimates that there are in North America about 5,000 professionals, 500 "master" amateurs, 40,000 "experienced amateurs," 10,000 novices, and 200,000 persons with at least a "casual interest in astronomy." Andreas Gada, Allan H. Stern, and Thomas R. Williams, "What Motivates Amateur Astronomers?" in Percy and Wilson, eds., *Amateur-Professional Partnerships*, p. 15.

6. Another survey program was TASS—The Amateur Sky Survey—which distributed simple, fixed telescopes with CCDs to volunteer observers. Although they had no moving parts and simply took images of a strip of sky that the Earth's rotation troops past each evening, a couple of dozen TASS telescopes—each consisting of little more than a camera lens and a CCD chip in a housing—were capable of making two hundred measurements on each of one hundred thousand stars annually. Although run by amateurs, TASS proved so beguiling that professional astronomers joined in as well. Future plans called for building larger instruments. "When we have them built, we are going to give them away to strangers that we meet through correspondence on the Internet," wrote TASS founder Tom Droege. "We hope they will write programs that will allow doing good science with the collected data."

A VISIT WITH JACK NEWTON

1. *Amateur Astronomy*, Winter 2000, p. 3.

6. ROCKY HILL

1. Harold Richard Suiter, *Star Testing Astronomical Telescopes: A Manual for Optical Evaluation and Adjustment*. Richmond, Va.: Willmann-Bell, 1994, p. 3.
2. Ibid., p. 2.

7. THE REALM OF THE SUN

1. In Timothy Ferris, *The Red Limit*. New York: Morrow, 1983, p. 95.
2. Certified solar filters are available in Mylar or glass. Don't use smoked glass or other homespun contraptions: They may reduce the Sun's apparent brightness enough to make it seem safe, while admitting infrared wavelengths of light that can damage the eye. Avoid filters mounted on the eyepiece itself: Concentrated solar heat can shatter the filter, in which case the observer's eye can be damaged before he or she has time to react.
3. P. Clay Sherrod, *A Complete Manual of Amateur Astronomy*. Englewood Cliffs, N.J.: Prentice-Hall, 1981, p. 101.
4. Gerald North, *Advanced Amateur Astronomy*. Cambridge, U.K.: Cambridge University Press, 1997, p. 249.
5. This stability is thought to be typical of most main-sequence, Sun-type stars, but astronomers don't know for sure. An intriguing amateur project, not yet attempted so far as I know, would be to monitor the brightness of a few hundred Sun-type stars over many years, using CCDs, to see if they generally are as stable as the Sun.
6. In Michael Maunder and Patrick Moore, *The Sun in Eclipse*. New York: Springer, 1998, p. 54.
7. In Herbert Friedman, *Sun and Earth*. New York: Scientific American Books, 1986, p. 70.
8. Terence Dickinson and Alan Dyer, *The Backyard Astronomer's Guide*. Camden East, Ont.: Camden House, 1991, p. 146.
9. William Sheehan and Thomas Dobbins, "Mesmerized by Mercury." *Sky & Telescope*, June 2000, p. 109.

8. THE MORNING AND EVENING STAR

1. Eugene O'Connor, "Chasing Venus Around the Sun." Undated report on the Web site of the Astronomical Society of New South Wales.
2. In Joseph Ashbrook, *The Astronomical Scrapbook*. Cambridge, Mass.: Sky Publishing, 1984, p. 230.
3. In David Harry Grinspoon, *Venus Revealed: A New Look Below the Clouds of Our Mysterious Twin Planet*. Reading, Mass.: Addison-Wesley, 1997, p. 48. The black drop is now thought to be produced by a combination of several effects, including the state of local seeing at the time and place of a given observation.
4. In William Sheehan and Thomas Dobbins, "Charles Boyer and the Clouds of Venus." *Sky & Telescope*, June 1999, p. 57.

5. Ibid., p. 59.

6. Ibid., p. 60.

7. In E. C. Krupp, "The Camera-Shy Planet." *Sky & Telescope*, October 1999, p. 94.

8. In Grinspoon, *Venus Revealed*, p. 49.

9. In Krupp, "The Camera-Shy Planet," p. 95.

10. In Thomas A. Dobbins, Donald C. Parker, and Charles F. Capen, *Introduction to Observing and Photographing the Solar System*. Richmond, Va.: Willmann-Bell, 1992, p. 33.

11. First sentence: Mikhail Ya. Marov and David Harry Grinspoon, *The Planet Venus*. New Haven: Yale University Press, 1998, p. 384. Second sentence: Mark A. Bullock and David H. Grinspoon, "Global Climate Change on Venus." *Scientific American*, March 1999, p. 57.

9. MOON DANCE

1. Joseph Ashbrook, *The Astronomical Scrapbook*. Cambridge, Mass.: Sky Publishing, 1984, p. 233.

2. Plutarch's account of a solar eclipse—probably that of March 20, A.D. 71—is the only one in ancient times to mention the Sun's corona, and one of only three to note that stars appear in the sky during totality. The others were Thucydides, who saw a solar eclipse in 431 B.C., and Phlegon of Tralles, whose eclipse observation probably occurred in A.D. 29.

3. Galileo Galilei, *Siderius Nuncius*, trans. A. Van Helden. Chicago: University of Chicago Press, 1989, p. 36.

4. Ibid., p. 42.

5. In Scott L. Montgomery, *The Moon and the Western Imagination*. Tucson: University of Arizona Press, 1999, p. 112.

6. In Michael J. Crowe, *The Extraterrestrial Life Debate 1750–1900: The Idea of a Plurality of Worlds from Kant to Lowell*. Cambridge, U.K.: Cambridge University Press, 1986, p. 74.

7. Ibid., p. 60.

8. Ibid., p. 63. The italics are Herschel's.

9. Ibid., p. 112.

10. Ibid., p. 207.

11. Ibid., p. 393.

12. Edgar Allan Poe, "The Unparalleled Adventure of One Hans Pfaall." In Harold Beaver, ed., *The Science Fiction of Edgar Allan Poe*, New York: Viking Penguin, 1976, p. 55.

13. Note added by Poe for publication of the story in an anthology; in Beaver, ed., *The Science Fiction of Edgar Allan Poe*, p. 58.

14. Locke's immediate target appears to have been the Reverend Thomas Dick, who argued on pluralistic religious grounds that there was life virtually everywhere in the universe. Members of the Académie des Sciences in Paris laughed knowingly at Locke's dispatches when they were read aloud there.

15. Winifred Sawtell Cameron, "Lunar Transient Phenomena." *Sky & Telescope*, March 1991, p. 265.

16. Ibid.

17. Of course, as with UFO sightings, that not all LTP reports can be explained doesn't mean there is necessarily anything to them. For a skeptical view see William Sheehan and Thomas Dobbins, "The TLP Myth: A Brief for the Prosecution." *Sky & Telescope*, September 1999, p. 118. ("TLP" is an alternative form of "LTP.") They propose that the concentration of LTP reports in a few lunar regions is due to a "feedback" effect in which observers devote "disproportionately intense scrutiny" to features where LTPs have been reported in the past. They call LTPs "as much of a flawed anachronism as the canals of Mars."

18. Alan MacRobert, "The Moon Shall Rise Again." *Sky & Telescope*, November 1988, p. 478.

19. Confusingly, this phenomenon is known observationally as the lunar *deceleration*, because the Moon, as it climbs to higher orbits, moves more slowly against the background stars—for the same reason that an airplane high in the sky moves more slowly than when seen at close range. The predicted rate of the Moon's apparent deceleration is 28 arc seconds per cen-

tury. The observed rate is 23 arc seconds. The reason for the disparity between theory and observation is not yet understood.

20. In Maurice Hershenson, ed., *The Moon Illusion*. Hillside, N.J.: Lawrence Erlbaum, 1989, p. 7.

21. Ibid., p. 11.

22. You can demonstrate this illusion in a rudimentary way with just a lightbulb and a sheet of white paper in a darkened room. Stare at the bare bulb for several seconds, then look at the paper and you will see an "afterimage" of the bulb on the paper. Then move the sheet of paper away to a greater distance, and note that the afterimage gets bigger! The brain is saying, in a sense, "Well, if it looked that big and now it turns out to be more distant than I thought, it must be *larger* than I'd thought."

23. "Maxims and Reflections," in Douglas Miller, ed. and trans., *Goethe: Scientific Studies*. New York: Suhrkamp, 1988, p. 308.

24. Bob Dylan, "License to Kill." Two weeks after the *Challenger* space shuttle exploded, on February 12, 1986, Dylan prefaced a performance of this song, in Sydney, Australia, with this statement: "Here's something I wrote a while back, about the space program. You heard about the tragedy, right? . . . The people had no right going up there. Like, there's not enough problems on Earth to solve? So I want to dedicate this song to all those poor people who were fooled into going up there." I have every respect for Dylan, the most influential artist of his time in any medium, but by the same logic the first lungfish had no right to live on land, and dinosaurs no right to fly.

25. In Christopher Dickey, "Summer of Deliverance." *The New Yorker*, July 13, 1998, p. 40.

10. MARS

1. William Herschel, "On the Remarkable Appearances at the Polar Regions of the Planet Mars, the Inclination of Its Axis, the Position of Its Poles, and Its Spheroidal Figure, with a Few Hints Relating to Its Real Diameter and Atmosphere." *Philosophical Transactions of the Royal Society* 81, pt. 1 (1781): 115; quoted in William Sheehan, *The Planet Mars: A History of Observation & Discovery*. Tucson: University of Arizona Press, 1996, p. 34.

2. This is a round number. An Earth year lasts 365.26 days; a Mars year 696.98; the average interval between oppositions is 779.74 days. This, the *synodic* period, can be as little as 764 days or as much as 810 days.

3. Vincenzo Cerulli, "Polemica Newcomb-Lowell-fotografie lunari." *Rivista di astronomia* 2 (1908), p. 13, quoted in Sheehan, *The Planet Mars*, p. 130.

4. Percival Lowell, *Mars and Its Canals*. New York: Macmillan, 1906, p. 376.

5. In Donald E. Osterbrock, "To Climb the Highest Mountain: W. W. Campbell's 1909 Mars Expedition to Mount Whitney." *Journal for the History of Astronomy* 20 (1989), p. 86.

6. Ibid., pp. 78–97.

7. E. E. Barnard, unpublished ms. at Vanderbilt University. Quoted in Sheehan, *The Planet Mars*, p. 116.

8. Bruce Murray, *Journey into Space: The First Three Decades of Space Exploration*. New York: Norton, 1989, p. 43.

9. Carl Sagan, interviewed by T.F., Ithaca, New York, 1972.

10. C. R. Chapman, J. B. Pollack, and C. Sagan, *An Analysis of the Mariner 4 Photographs of Mars*, Smithsonian Astrophysical Observatory Special Report 268. Washington, D.C.: The Smithsonian Institution, 1968.

11. A few months earlier, the astronomer Charles Capen, at Lowell Observatory, had predicted just this development. Noting that the largest dust storms evidently occur when Mars is almost at its greatest distance from the Sun, he wrote that "a vast atmospheric disturbance could interfere with . . . the first Mariner orbiter spacecraft mission," as indeed it did. An English amateur astronomer, Alan Heath, was among the first to observe it. He used a 12-inch reflector. This and a few other reports provided early warnings of what *Mariner* would—or would not—initially see at Mars.

12. Murray, *Journey into Space*, p. 65.

13. John Mellish, an amateur astronomer who was permitted to study Mars with the 40-inch Clark telescope at Yerkes Observatory in November 1915, claimed to have seen "many craters and cracks" on the red planet, adding that prior drawings made by his friend and mentor E. E. Barnard in 1892–93 at Lick Observatory similarly showed "mountain ranges and peaks and craters." Mellish's drawings went unpublished and were destroyed by a fire in his workshop, so his claim remained unverified. The science writer and amateur astronomer William Sheehan found some old Barnard drawings of Mars, which may be the ones Mellish referred to, in an attic at Yerkes. They show craterlike features, some of which are mountains but none of which are actual Martian craters. See William Sheehan, "Did Barnard & Mellish Really See Craters on Mars?" *Sky & Telescope*, July 1992, p. 23.

14. C. F. Capen, *The Mars 1964–65 Apparition*, JPL Technical Report No. 32-990. Pasadena, Calif.: Jet Propulsion Laboratory, 1966, p. 10.

15. Ibid., pp. 62, 66, 75–76, 78, 79.

16. Don Parker, private communication.

17. Stephen James O'Meara, "Observing Planets: A Lasting Legacy." *Sky & Telescope*, November 1988, p. 475.

18. Donald C. Parker and Richard Berry, "Clear Skies on Mars." *Astronomy*, July 1993, pp. 72ff.

19. Kathryn Sullivan, interviewed by T.F., Washington, D.C., March 24, 1999.

20. The inclination of a planet's axis is measured relative to an imaginary line drawn perpendicular to the plane of its orbit. A planet standing "straight up," with its equator flat along its orbital plane, would have zero axial inclination, and one lying on its side, with its poles embedded in the orbital plane, would have an inclination of 90 degrees.

21. The eccentricity of Mars's orbit is currently 0.093, and has been as high as 0.13 in the past. The eccentricity of Earth's orbit is now 0.017, and has never exceeded 0.05.

22. David Morrison, *Exploring Planetary Worlds*. New York: Scientific American Library, 1993, p. 141.

23. Ibid., p. 143.

24. Carl died on December 20, 1996, at age sixty-two. In his honor the Pathfinder lander at Ares Vallis, the first craft to land successfully on Mars since *Viking*, was named Carl Sagan Memorial Station.

A VISIT WITH JAMES TURRELL

1. Turrell was born on May 6, 1943, some fourteen months after the air raid, so the dates do not quite fit, but in telling good family stories, as Dr. Johnson said of composing lapidary inscriptions, a man is not under oath.

2. James Turrell, "Night Curtain." Internet posting under James Turrell/Roden Crater, adapted from his book *Air Mass*. London: South Bank Centre, 1993.

3. James Turrell, interviewed by T.F., Flagstaff, Arizona, December 20, 2000.

4. Ibid.

5. James Turrell, address to the University of California school of architecture, Berkeley, November 27, 2000.

6. When a Japanese television borrowed light pulses from a Turrell installation and then sped them up and put them on a Pokémon TV show, all without Turrell's knowledge or permission, hundreds of viewers became so ill from watching the program that they had to be hospitalized. Doctors disagree on the cause of this outbreak of illness, but the most likely diagnosis is photosensitive epilepsy. Such seizures can be induced by a televised strobe rate of about 32 hertz, which is quite close to the point at which the eye begins to see individual movie frames as no longer flickering but combined into a smooth continuum. The Turrell installation on which the TV effect was based isolated viewers inside a "Ganzfeld sphere," which creates a featureless visual space, like putting half Ping-Pong balls over the eyes. In that installation, however, the lights pulsed at a much slower and safer rate.

7. James Turrell, Berkeley address.

8. Jeffery Hogrefe, "In Pursuit of God's Light." *Metropolis*, August 2000.

9. Although "standstill" is used popularly, and in some archeological papers, the preferred astronomical term is lunar and solar *stillstand*.

11. STONES FROM THE SKY

1. For a summary of historical reports of meteorite deaths and injuries, see John S. Lewis, *Rain of Iron and Ice*. Reading, Mass.: Addison-Wesley, 1996, pp. 176–82.

2. "Meteorites Pound Canada." *Sky & Telescope*, September 1994, p. 11.

3. Dennis Urquhart, Research Communication of the University of Calgary, May 31, 2000.

4. Although the phrase has come to be associated with Whipple, who did often use it, when I asked him about it at Harvard in 1996 he recalled having originally employed the term "icy conglomerate" and said a newspaper reporter had come up with "dirty snowball."

5. Hughes Pack, private communication, Fall 1998.

6. Moh'd Odeh, on the Jordanian Astronomical Society Web site, www.jas.org.jo.

7. In Lewis, *Rain of Iron and Ice*, p. 50. It has been theorized that—and hotly disputed whether—this impact created Giordano Bruno, one of the geologically youngest large craters on the Moon. I think not.

8. The lunar Leonids hit the Moon two and a half hours after the same shower peaked over Europe, because the Moon on that date happened to be trailing Earth in orbit around the Sun. Since the Moon didn't enter the rich part of the Leonid stream until after Earth had left it, the terrestrial fireworks were over before the lunar fireworks began.

9. Erik Asphaug, "The Small Planets." *Scientific American*, May 2000, p. 50.

10. Determination of the structure and density of asteroids—e.g., whether they generally are solid or are more like gravel heaps—is of interest in predicting whether, when one strikes Earth, its impact more nearly resembles that of artillery shells or shotgun blasts. A few small Earth-crossing asteroids are known to be eccentrically shaped, among them 3671 Dionysus and 1996 FG3.

11. Dennis di Cicco, "Hunting Asteroids." *CCD Astronomy*, Spring 1996, p. 8.

12. Ibid., p. 11.

13. In Joseph Ashbrook, *The Astronomical Scrapbook*. Cambridge, Mass.: Sky Publishing, 1984, p. 73.

14. Comets traditionally have been named for their discoverers, plus the year and order of their discovery. Hence *IRAS-Araki-Alcock 1983d* was the fourth comet discovered in 1983, and was independently spotted by two stargazers and IRAS (the Infrared Astronomical Satellite). Under new rules set forth by the International Astronomical Union (IAU), the dates are divided into half-years, so that the first comet discovered in the first half of 1995 is designated *1995 A1*, and a letter prefix denotes whether the comet is periodic (P), long-term (C), meaning that its orbit won't bring it back for a long time, if ever, defunct (D), or possibly an asteroid (A). So *P/1996 A1 (Jedicke)* is a periodic comet, the first to be discovered in the first half of 1996. Asteroids are denoted by the year and half-month of their discovery (1982 DB was the second asteroid discovered in the second half of February 1982) plus an IAU catalog number, and a name chosen by the discoverer. Some observers who have discovered many asteroids have resorted to asking that they be named for friends, relatives, mentors, and even rock stars: Hence we have such asteroids as 17059 Elvis (1999 GX5) and 4147 Lennon (1983 AY). The IAU cautions that "names of pet animals are discouraged."

15. David Levy, "Star Trails." *Sky & Telescope*, November 1989, p. 532.

16. In Edwin L. Aguirre, "How the Great Comet Was Discovered." *Sky & Telescope*, July 1996, p. 27.

17. Alan Hale, "The Discovery of Comet Hale-Bopp." Web posting, http://galileo.ivv. nasa.gov/comet/discovery.html, September 1995.

18. In Milton Meltzer, *Mark Twain Himself*. New York: Wings Books, 1993, p. 288.

19. Bradley E. Schaefer, "Meteors That Changed the World." *Sky & Telescope*, December 1998, p. 70.

A VISIT WITH DAVID LEVY

1. David Levy, "Untitled Remarks," in William Liller, *The Cambridge Guide to Astronomical Discovery.* Cambridge, U.K.: Cambridge University Press, 1992, p. 95.

2. In Robert Reeves, "My Field of Dreams: An Interview with David H. Levy." *Astronomy,* April 1994, p. 13.

3. Levy's rebinding may itself have been suggested by Peltier, who as a boyhood stargazer checked out a favorite observing book, Martha Evans Martin's *The Friendly Stars,* from the local library so many times that the librarian eventually had it rebound in dark blue cloth and made him a gift of it. (Leslie Peltier, *Starlight Nights: The Adventures of a Star-Gazer.* Cambridge, Mass.: Sky Publishing, 1965, p. 42.)

4. Leslie C. Peltier, *Starlight Nights,* p. 137.

12. VERMIN OF THE SKIES

1. In Edwin Emerson, *Comet Lore.* New York: Schilling Press, 1910, p. 89. Digges was no gibbering spiritualist but a mathematician, experimenter—he is credited with inventing the theodolite—and, it appears, a friend of William Shakespeare's.

2. A pleasing variation has Jefferson calling them "Yankee professors." There is no evidence to support the story, which was first published in 1874, almost half a century after Jefferson's death, and has been repeated ever since. See, e.g., John F. Fulton and Elizabeth H. Thompson, *Benjamin Silliman: Pathfinder in American Science.* New York: Henry Schuman, 1947, pp. 76–78; and Silvio A. Bedini, *Thomas Jefferson: Statesman of Science.* New York: Macmillan, 1990, p. 388.

3. Aristotle, *Meteorology,* Book 1, 5ff. Samuel Taylor Coleridge seems to have had Aristotle in mind when he wrote, probably in response to his having witnessed the Leonid meteor shower of 1797, these lines of *The Rime of the Ancient Mariner:* "The upper air burst into life! / And a hundred fire-flags sheen, / To and fro they were hurried about! / And to and fro, and in and out, / The wan stars danced between. . . ."

4. Thomas Jefferson, letter to David Salmon, February 15, 1808, quoted in Bedini, *Thomas Jefferson,* p. 386.

5. Jefferson, letter to Andrew Ellicott, October 25, 1805, quoted in Bedini, *Thomas Jefferson,* p. 388.

6. In Roberta J. M. Olson, *Fire and Ice: A History of Comets in Art.* New York: Walker, 1985, p. 78.

7. In Donald K. Yeomans, *Comets: A Chronological History of Observation, Science, Myth, and Folklore.* New York: Wiley, 1991, p. 22.

8. In A. Dean Larsen, ed., *Comets and the Rise of Modern Science.* Provo, Utah: Friends of the Brigham Young University Library, 1986, p. 3.

9. Astronomers of the day—among them Halley's friend the Danzig brewer Johannes Hevelius—had established that periodic comets could be identified by determining whether their orbits were hyperbolic or parabolic. Those with hyperbolic orbits will not return; those with parabolic ("closed") orbits will. The parabolic orbit of Halley's comet takes it out past Pluto, and its orbital period varies somewhat, between seventy-four and seventy-nine years, owing mainly to the gravitational influence of Jupiter and other giant planets.

10. In Yeomans, *Comets,* p. 119.

11. Voltaire, *The Elements of Sir Isaac Newton's Philosophy,* John Hanna, trans. New York: Gryphon, 1995, p. 340.

12. Carl Sagan and Ann Druyan, *Comet.* New York: Ballantine, 1997, p. 279.

13. In Carolyn S. Shoemaker and Eugene M. Shoemaker, "A Comet Like No Other." John R. Spencer and Jacqueline Mitton, eds., *The Great Comet Crash: The Impact of Comet Shoemaker-Levy 9 on Jupiter.* Cambridge, U.K.: Cambridge University Press, 1995, p. 7.

14. Comets had been observed breaking up before, but usually because of the tidal disruptions and solar sublimation they endured when passing near the Sun. Chains of craters found on Earth and the Moon indicate impacts by fragments of broken-up comets in the past.

15. Paul Chodas, interviewed by T.F., Tucson, Arizona, October 24, 1996.

16. In Timothy Ferris, "Is This the End?" *The New Yorker*, January 27, 1997, p. 49.

17. David Morrison, interviewed by T.F., NASA Ames Research Center, October 9, 1996.

18. Posting on Len Amburgey's Web site, http://www.net1plus.com/users/lla/Index.htm.

19. In Edwin L. Aguirre, "Sentinel of the Sky." *Sky & Telescope*, March 1999, pp. 76ff.

20. James Woodford, "Outback Amateur Discovers Asteroid Threat to the World." *Sydney Morning Herald*, May 21, 1999.

21. John S. Lewis, *Rain of Iron and Ice*. Reading, Mass.: Addison-Wesley, 1996, p. 222.

22. We tend to think of planets as attracting objects rather than flinging them by the trillions into the outer reaches of the solar system, but planets are actually quite efficient pitching machines. Suppose that you were riding on a Kuiper belt comet, circa 4 billion years B.C., approaching Jupiter from behind. You probably won't actually hit Jupiter: Although it's a giant planet, it's still a small target in space. Instead, you will speed up considerably, then curve around Jupiter and fly off into space at a much faster velocity than when you arrived. Spacefarers call this "gravity assist," but the more accurate term is "angular momentum assist." What has happened is that Jupiter has slowed down a bit—just a tiny bit; its momentum is enormous—and you have sped up a lot, rather like a skater playing crack-the-whip with a Titan. You depart at the same velocity that you entered, relative to Jupiter, but your orbital velocity has increased tremendously, and when that happens to any orbiting object, it climbs to a much higher orbit. Jupiter could easily have flung five trillion comets into the extremely high orbits of the Oort cloud. Whether that is actually how the Oort cloud formed remains an open question.

13. JUPITER

1. José Olivarez, quoted in *Sky & Telescope*, December 1999, p. 120.

2. In John H. Rogers, *The Giant Planet Jupiter*. Cambridge, U.K.: Cambridge University Press, 1995, p. 8.

3. Jupiter rotates differentially, as does the Sun. Objects at the equator complete a rotation in 9 hours 50.5 minutes; at polar latitudes they take 9 hours 55.7 minutes; and radio observations indicate that the liquid depths rotate in 9 hours 55.5 minutes.

4. The pleasing practice of naming Jovian satellites after the consorts of the god Jupiter was originally proposed by Johannes Kepler to Simon Marius (1573–1624). Herschel applied the same approach to the satellites of Saturn, and it became the convention for the other planets as well.

5. J. B. Murray, "New Observations of Surface Markings on Jupiter's Satellites." *Icarus* 23, 1975, pp. 397–404.

6. Rogers, *The Giant Planet Jupiter*, p. 341.

7. S. J. Peale, P. Cassen, and R. T. Reynolds, "Melting of Io by Tidal Dissipation." *Science*, vol. 203, no. 4383, 1979, pp. 892–94.

8. In Stephen James O'Meara, "Hubble's Amateur Hour." *Sky & Telescope*, August 1992, p. 155.

9. James Secosky, "SO_2 Concentration and Brightening Following Eclipses of Io," WFPC1 program 2798, *Icarus* 111, April–June 1992.

10. "Students Receive Unique Science Lesson." *Genesee Country Express*, December 18, 1997.

11. In O'Meara, "Hubble's Amateur Hour," p. 155.

12. As misconceptions beset much of what has been written (and, alas, taught) about the Coriolis force, it may be worth taking a moment to explore it. Suppose that you fire a ballistic missile due south from New York City, so that it soars into space and then falls back to Earth twenty minutes later, 600 miles away. Were the Earth not rotating, the missile would drop harmlessly into the ocean. But as the Earth *is* rotating, toward the east, what will actually happen is that the missile will land some five degrees to the west—the Earth having rotated under it—so that it hits not the ocean but Florida. (It was to avoid mishaps of this sort that studies of ballistics led to the identification of the Coriolis force in 1835 by the French mathematician Gaspard-Gustave de Coriolis.) From the point of view of the missile, the trajectory was straight, but when observed from the ground the missile will appear to have curved westward, as if acted upon by some mysterious agency.

This Coriolis "force" causes winds to curve to their right in the Northern Hemisphere and to their left in the Southern Hemisphere, setting up large-scale atmospheric cells, or circulatory patterns, that revolve clockwise in the north and counterclockwise in the south. The oft-repeated claim that sinks drain clockwise north of the equator and counterclockwise to the south is, however, false. The Coriolis force is too weak to do that, except in carefully isolated tubs observed over long periods of time. A charlatan in Nanyuki, Kenya, a town on the equator, collected tourist money by "demonstrating" the effect in two rigged sinks, but the deception was not hard to detect, as he had happened to rig the sinks the wrong way round, so that his northern sink drained clockwise and his southern sink counterclockwise.

Hurricanes and other cyclones are low-pressure regions set spinning by impinging Coriolis winds. Counterintuitively, they rotate in the opposite direction of the Coriolis cells themselves. Imagine that the low-pressure zone is a carousel and that Coriolis winds, all curving to the right, are glancing off it from all four points of the compass: They will make the cell spin counterclockwise, just as a clockwise-rotating gear, engaged with a second gear, will turn that gear counterclockwise. High-pressure zones, however—for reasons I won't go into here—spin in the true Coriolis direction, producing anticyclones, high-pressure zones that move clockwise in the north and counterclockwise in the south.

13. Gary Seronik, "Above and Beyond." *Sky & Telescope*, May 2001, p. 130.

14. The brightest satellite, Ganymede, is yellow in color. The next brightest, Io, is usually described as white, although to my eye—perhaps influenced by post-*Voyager* knowledge of its volcanoes—it has a noticeable reddish tinge. Europa and Callisto, slightly dimmer, look white or yellow. Their approximate satellite magnitudes are Ganymede 4.6, Io 5, Europa 5.3, and Callisto 5.6. Io orbits Jupiter once every 1.8 days, Europa in 3.6 days, and Ganymede in a week. Since the innermost two orbital periods divide almost evenly into seven days, one sees these three moons in about the same positions weekly. Callisto, farther out, completes an orbit every 16.7 days.

15. John D. Bernard, "The Comet Shoemaker-Levy 9 (SL9) Collision: A Project of the Jupiter Space Station," jupiterspacestation.org/jup_sl9.html.

A VISIT WITH STUART WILBER

1. Reta Beebe, telephone interview by T.F., January 5, 2000.

14. THE OUTER GIANTS

1. In Fred W. Price, *The Planet Observer's Handbook*. Cambridge, U.K.: Cambridge University Press, 1994, p. 264.

2. The rate at which objects move in their orbits is dictated by the local strength of gravity. Since the strength of the Sun's gravitational attraction diminishes by the square of the distance, the outer planets move much more slowly than the inner ones do. The Earth orbits at a sprightly pace of nearly 30 kilometers per second, while Saturn, almost 10 times farther from the Sun, has an orbital velocity of under 10kps, and Neptune, 30 times Earth's distance fom the Sun, creeps along at only 5.4 kps. (The relationship is expressed in Kepler's third law, $R^3 = P^2$, where R is the radius of the planet's orbit and P is the length of its year.)

3. That is a probable lower limit to the rings' thickness; the upper-end estimate is about 10 kilometers.

4. Most historians give credit for the discovery of the crepe ring to William and George Bond and William Dawes in 1850, although observing notes made by the ubiquitous William Herschel describing a "quintuple belt" suggest that he detected it on the night of November 11, 1793.

5. Albert van Helden, "Saturn Through the Telescope: A Brief Historical Survey," in Tom Gehrels and Mildred Shapley Matthews, eds., *Saturn*. Tucson: University of Arizona Press, 1988, p. 32.

6. Identified by Edouard Roche in 1849, the Roche limit differs, for a given planet, depending on the structure of the satellite in question. The Roche limit for a liquid moonlet of negli-

gible mass that has the same density as the planet it orbits is 2.44 times the planet's radius, but the limit decreases for solid moons, which have greater tensile strength. A snowball moon approaching Earth will hit the Roche limit relatively soon, but science fiction movie depictions of gigantic alien spaceships hovering over terrestrial cities do not necessarily violate the Roche limit, provided that the starships are stoutly constructed.

7. Since the giant planets have captured a number of tiny, asteroidal objects, many of which doubtless remain to be found—and since some of their former satellites may have been shattered into small, orbiting fragments by impacts—it is no longer possible to declare with any great assurance how many moons each planet actually has. When I wrote the first draft of this paragraph, on a Monday, Saturn was known to have twenty-four satellites. By the end of the week, astronomers using the 3.6-meter Canada-France-Hawaii Telescope in Hawaii announced that they had discovered four more, bringing the total to twenty-eight. A few weeks later two more satellites were found, in Hawaii and at Whipple Observatory in Arizona; so Saturn by then had thirty known moons. By the time twelve additional ones were detected a few months later, most astronomers had stopped trying to estimate how many there might be. Similarly, Jupiter, which was known to have seventeen satellites when I drafted this paragraph, was up to twenty-eight by the middle of the next month, and thirty-nine a few months later. For what it's worth, other satellite totals were, for the moment, Uranus twenty-one and Neptune eight.

8. The amateur astronomer Donald Parker and colleagues estimate that Titan can be observed with almost any small telescope or even through binoculars, while Rhea and Iapetus at its western elongation can be seen with a three-inch refractor. Dione, Tethys, and Enceladus require a four-inch, Hyperion and Mimas at least eight inches of aperture. Thomas A. Dobbins, Donald C. Parker, and Charles F. Capen, *Introduction to Observing and Photographing the Solar System*. Richmond, Va.: Willmann-Bell, 1992, p. 109.

9. Minor Saturn satellites include outlying Phoebe, in retrograde orbit; Helene, which orbits between Rhea and Dione; Calypso and Telesto, between Dione and Tethys; and the inner satellites Janus, Epimetheus, Pandora, Prometheus, Atlas, and Pan. The E and G rings lie to either side of Mimas, the F ring between Pandora and Prometheus, and the A through D rings (as well as the Cassini division) inside the orbit of Pan.

10. In A. F. O'D. Alexander, *The Planet Saturn*. New York: Dover, 1962, p. 92.

11. Ibid., p. 238.

12. Further data on the opacity of Titan's atmosphere were obtained by studying the fading of radio transmissions from *Voyager 1* when it passed behind Titan, and on July 3, 1989, when amateur and professional astronomers observed an occultation of the star 28 Sagittarii by Titan, measuring the ragged curve of the star's diminishing brightness as it sank behind the clouds. Their observations of the "central flash" caused when the star was directly behind Titan, by starlight being refracted around the satellite by the lenslike effects of the atmosphere, called into question prevailing models of Titan's hazy atmosphere. One of the amateur observers, Kevin Deakes of South Yorkshire, England, noted that another benefit of the observations "is that Titan's position is now better known, which means that the forthcoming Cassini mission to Saturn and Titan can be targeted more accurately."

13. Christopher Wills and Jeffrey Bada, *The Spark of Life: Darwin and the Primeval Soup*. New York: Perseus, 2000. In Tim Flannery, "In the Primordial Soup," *New York Review of Books*, November 2, 2000, p. 56.

14. Cassini, in Alexander, *The Planet Saturn*, p. 113.

15. David Morrison, Torrence Johnson, Eugene Shoemaker, Laurence Soderblom, Peter Thomas, Joseph Veverka, and Bradford Smith, "Satellites of Saturn: Geological Perspective," in Gehrels and Matthews, eds., *Saturn*, p. 616.

16. "A knowledge of the construction of the heavens has always been the ultimate object of my observations," Herschel noted. (Michael Hoskin, "William Herschel and the Making of Modern Astronomy." *Scientific American*, February 1986, p. 106.) But he was so persistent that he found many things he hadn't been looking for. His legendarily hardworking approach to observing, attested to by many witnesses, was described by Herschel himself this way: "I employed

myself now so entirely in astronomical observations, as not to miss a single hour of star-light weather. I have many a night, in the course of eleven or twelve hours of observation, carefully and singly examined not less than 400 celestial objects . . . sometimes viewing a particular star for half an hour together, with all the various powers of my telescope." (Michael Hoskin, *William Herschel: Pioneer of Sidereal Astronomy*. London: Sheed Ward, 1959, p. 15.)

17. In E. C. Krupp, "Managing Expectations." *Sky & Telescope*, August 2000, p. 84.

18. Ibid., p. 85.

19. When Uranus approaches Neptune, pursuing its faster course along its inside orbit, the gravitational pull of Neptune accelerates it. Then, after they pass opposition—the point at which a straight line drawn from the Sun to Neptune intersects Uranus—Neptune retards Uranus's progress. The effect was pronounced in the era of Neptune's impending discovery, as opposition occurred in the year 1822.

20. Patrick Moore, *The Discovery of Neptune*. New York: Wiley, 1996, p. 17.

21. Ibid., p. 20.

22. Ibid., p. 22.

23. John Herschel, letter on "Le Verrier's Planet," in Ellis D. Miner, *Uranus: The Planet, Rings, and Satellites*. New York: Ellis Horwood, 1990, p. 25.

24. Moore, *The Discovery of Neptune*, p. 23.

25. Ibid., p. 24.

26. Quoted in *Atlas of Neptune*, p. 22, by Patrick Moore, who finds it "fascinating to think that Neptune could have been discovered almost 170 years before the much closer and brighter Uranus." Galileo's observations of Neptune were found in his logs by S. Drake and C. T. Kowal (*Scientific American*, 243, 1980, p. 74; *Nature*, 287, 1980, p. 311) and further investigated by E. Myles Standish and Anna M. Nobili (*Baltic Astronomy*, vol. 6, 1997, pp. 97–104). The latter authors note that although there is an irregularity in one Galileo drawing concerning Neptune's position, his depiction of the positions of Jupiter's satellites—the real objects of his attention—were "drawn to scale with an accuracy . . . smaller than the width of the dots with which he indicated the satellites' positions!"

27. As Herschel put it, "If we would hope to make any progress in an investigation of this delicate nature, we ought to avoid two opposite extremes, of which I can hardly say which is the most dangerous. If we indulge a fanciful imagination and build worlds of our own, we must not wonder at our going wide of the paths of truth and nature; but these will vanish like the Cartesian vortices, that soon gave way when better theories were offered. On the other hand, if we add observation to observation, without attempting to draw not only certain conclusions, but also conjectural views from them, we offend against the very end for which other observations ought to be made." *Philosophical Transactions of the Royal Society*, ci (1785), pp. 213–14.

28. J. L. E. Dreyer, ed., *The Scientific Papers of Sir William Herschel*. London: Royal Society and Royal Astronomical Society, 1912, 1, pp. 312–14.

29. Credit for the discovery of the satellites Umbriel and Ariel, which lie inside the orbit of Titania, is difficult to assign. The English amateur William Lassell is often named as their discoverer, although reasonable arguments may also be made for the Russian professional Otto Struve, while William Herschel, who also spotted some Uranian "satellites" that evidently were background stars, is claimed by some scholars to have been the first to see Umbriel. Miranda, the innermost classical satellite of Uranus, was not discovered until 1948, when Gerard Kuiper photographed it with the 82-inch reflector at McDonald Observatory in Texas. Neptune has two bright outer satellites: With pleasing symmetry, it was Lassell who discovered the larger of them, Triton, in 1846, and Kuiper the smaller one, Nereid, in 1949.

30. An exotic consequence of its radical orientation is that although Uranus rotates once every eighteen hours, each hemisphere remains in daylight for forty years and then endures a forty-year night.

31. Venus is tilted, too, by 178 degrees. Like Uranus its rotation is retrograde—which is how we know that it was knocked over more than 90 degrees—but while Uranus rotates once every eighteen hours, a day on Venus lasts 243 Earth days, suggesting that the object that hit it

came in from a direction contrary to its rotational direction, slowing it almost to a halt. Add in the theory that the Moon was formed by a major impact, and you have pretty strong evidence that would-be planets struck one another rather frequently when the solar system was young.

32. Clyde W. Tombaugh and Patrick Moore, *Out of the Darkness: The Planet Pluto*. New York: Signet, 1980, pp. 116–17.

33. In Alan Stern and Jacqueline Mitton, *Pluto and Charon: Ice Worlds on the Ragged Edge of the Solar System*. New York: Wiley, 1998, p. 19.

34. In David Levy, *Clyde Tombaugh: Discoverer of Planet Pluto*. Tucson: University of Arizona Press, 1991, pp. 69–70.

35. A French astronomer had suggested the name "Pluto" for Planet X in 1919, but nobody at the time of the discovery seems to have remembered this—certainly not Ms. Burney, who was born that year.

36. Knowing that Pluto did not fill the bill, Tombaugh searched for thirteen more years for Planet X, examining 90 million images of 30 million stars in the process. This work strongly constrained where and how large Kuiper belt objects could be, if it existed. The case was closed once the *Voyager* flyby established better values for the masses of Uranus and Neptune: When their orbits were recalculated with these new data, the perturbations thought to have been caused by Planet X disappeared.

37. "Perplexed. . . . The omission has been confusing and even distressing": Kenneth Chang, "Pluto's Not a Planet? Only in New York." *The New York Times*, January 22, 2001. "Demoted from planethood": Ira Flatow, *Talk of the Nation/Science Friday*, NPR, February 11, 2000. "The Hayden has deplanetized Pluto": Charles Osgood, CBS News *Sunday Morning*, February 18, 2001. "Pluto didn't make the cut": "Museum Explains Why Pluto Is Out of Its Orbit," *The Santa Fe New Mexican*, March 3, 2001. "In this way, you get to learn": NBC News *Saturday Today*, January 27, 2001. "Don't count planets, count families" and "We're sticking with Pluto": Chang, "Pluto's Not a Planet?" "In a different universe": "Is Pluto Really a Planet?" *The News of the World*, January 28, 2001.

38. Clyde Tombaugh, interviewed by David Levy, March 23, 1987, in Levy, *Clyde Tombaugh*, p. 182.

15. THE NIGHT SKY

1. David J. Eicher, "Warning: Globular Clusters Can Change Your Life." *Astronomy*, April 1988, p. 82.

2. Robert Henri, *The Art Spirit*. Boulder, Colo.: Westview Press, 1984, p. 79.

3. Our concentration in this book is on Northern Hemisphere constellations—which, alas, leaves out a lot—and the skies described are best seen in the spring.

4. The exceptions are Dubhe, the star in the bowl closest to Polaris, the North Star, and Benetnash, at the far end of the handle.

5. The names of most bright stars come from ancient Arabic astronomers, although some, like Sirius, come from the Greeks, and others, like Regulus, from the Romans. Following a practice introduced by Johann Bayer in his *Uranometria* atlas in 1603, prominent stars in each constellation are designated by letters of the Greek alphabet. Hence Sirius, the brightest star in Canis Major, is Alpha Canis Majoris. Several other stellar naming systems are in use, based on various catalogs. In the resulting system of multiple identities, Sirius, for instance, is also SAO 151881, GSC 5949:2767, HIP 32349, HD 48915, B-16 1591, and Flamsteed-Bayer 9-Alpha Canis Majoris. The eighty-eight constellations bear Greek and Latin names. Officially naming stars is a task handled by the International Astronomical Union, which cautions consumers that commercial organizations that charge people for "naming" stars after themselves or their friends have no professional standing. "Like true love and many other of the best things in human life," declares the IAU, in a recourse to emotive prose rare for a scientific organization, "the beauty of the night sky is not for sale, but is free for all to enjoy."

6. Robert Burnham Jr., *Burnham's Celestial Handbook*. New York: Dover, 1978, p. 1940.

7. These categories are defined by particular spectral lines: For instance, O stars display ion-

ized helium lines in their spectra, B stars have neutral hydrogen lines, and M stars show titanium oxide lines. For greater precision the categories are subdivided into numerical classes: An F5 star is halfway between F and G. In the 1920s and 1930s a second parameter was added, classifying stars by their luminosity type from I (hot supergiants) to V (cool dwarfs). The Sun's spectral classification is G2V, meaning that it's a G-type dwarf. In astrophysical parlance, however, most ordinary, main-sequence stars are dwarfs; the term doesn't mean they are terribly small.

8. In Burnham, *Burnham's Celestial Handbook*, p. 1285.

9. To extend a line of thought investigated in other contexts by the Princeton physicist J. Richard Gott, it is not surprising that we should find ourselves in a fairly typical large galaxy, since the odds favor this. Typical galaxies are by definition the most common types, and since large galaxies contain more stars than do small ones, an average observer should find himself in a rather large galaxy, just as more people are born in large cities than small ones. Relatively few observers should find themselves in extraordinarily large galaxies, as these are rare, or in dwarf galaxies, since they contain fewer stars.

10. The Milky Way and other galaxies also contain lots of so-called dark matter, the presence of which can be inferred from the gravitational dynamics of the galaxy but cannot otherwise be detected. Its composition is at present unknown, and may involve exotic subatomic particles. The dark clouds we are discussing in this chapter are not this mysterious "dark matter" but are made of gas and dust whose composition is relatively well understood.

11. In William Sheehan, *The Immortal Fire Within: The Life and Work of Edward Emerson Barnard*. New York: Cambridge University Press, 1995, p. 3.

12. In Gerrit L. Verschuur, "Barnard's 'Dark' Dilemma." *Astronomy*, February 1989, p. 32.

13. In Sheehan, *Immortal Fire Within*, pp. 9–10.

14. Ibid., p. 12.

15. Verschuur, "Barnard's 'Dark' Dilemma," p. 33, and Sheehan, *Immortal Fire Within*, p. 13.

16. Burnham, *Burnham's Celestial Handbook*, p. 1635.

16. THE MILKY WAY

1. To be fair to the ancient observers, the overabundance of bright stars near the Milky Way is not terribly striking, amounting to an excess of only about 15 to 20 percent of, for example, the 173 stars in the sky brighter than magnitude 3.0, depending on how one sorts them. But it could at least theoretically have been extracted from the data of the day. (See, e.g., D. Hoffleit and W. H. Warren Jr., *The Bright Star Catalogue*, 5th revised edition, Astronomical Data Center, NSSDC/ADC, 1991. The precise binning of stars by magnitude may contain inaccuracies due to variable stars, etc., but this has no statistically significant bearing on the results.)

2. Auguste Comte, *Cours de Philosophie Positive*, 1835.

3. A fuller account of this work may be found in my *Coming of Age in the Milky Way*. New York: Morrow, 1988, Chapter 9.

4. Stephen James O'Meara, "The Outer Limits." *Sky & Telescope*, August 1998, p. 87.

5. Walter Scott Houston, *Deep-Sky Wonders*. Selections and commentary by Stephen James O'Meara. Cambridge, Mass.: Sky Publishing, 1999, p. 2.

6. The density of the Orion nebula is typically about 1,000 atoms per cubic centimeter, much denser than interstellar space generally but still mighty rare. Dust particles in the nebula, which are the size of smoke particles and make up one percent of its mass, are separated on average by 150 meters of space.

7. In Houston, *Deep Sky Wonders*, p. 4.

8. Bok became fascinated by the night sky while camping out as a Boy Scout in Holland, "fell in love with the Milky Way," as he put it, in 1918, then turned professional under the influence of Harlow Shapley, whose measurement of globular star cluster distances revealed the place of the solar system in our galaxy. An eminently decent man of intense personal loyalties, he remained devoted to Shapley throughout his life. In 1983, months before Bok's death, I gave a talk critical of some aspects of Shapley's model of the Milky Way and was promptly taken to

task by Bok, who rose from the audience to defend his friend and mentor, then dead for a decade. I found, as many others had, that arguing with the witty and warm Bok was more pleasant than agreeing with many other scientists.

9. Since M42 and M43 were well known in Messier's time, and loom much larger in the sky than would any distant comet, scholars have wondered why Messier included them in a catalog allegedly intended to help comet seekers avoid distractions. It has been suggested that he listed them—along with M44, the Beehive cluster, and M45, the Pleiades, which seem equally unlikely to be mistaken for comets by any but the most addled observers—in order to boost the number of objects in the 1774 edition of his catalog to forty-five, thus surpassing the forty-two in his competitor Nicholas Louis de Lacaille's 1755 catalog of southern nebulae.

10. Although the mills of Orion grind exceedingly slowly, they do get results, and there are stars in the Orion nebula that have formed since Rome fell.

11. "Blue stragglers" put a fly in the ointment of this simple picture. These are blue stars that obviously are young, yet are found by the thousands in globular clusters, where virtually all the stars should be old. At least ten theories have been composed to explain where blue stragglers come from. Their scenarios range from the interactions of close binary stars to stars colliding near the cluster centers, with results that mix the stellar materials so as to produce what look like new stars.

12. Leos Ondra, "Andromeda's Brightest Globular Cluster." *Sky & Telescope*, November 1995, p. 69.

A VISIT WITH JOHN HENRY'S GHOST

1. John Henry was one of many former slaves who went to work building railroads during Reconstruction. Huddie Ledbetter told the ethnomusicologist Alan Lomax that John Henry came from Newport News, Virginia, and helped build railroads from there to Cincinnati.

2. Mississippi John Hurt, "Spike Driver Blues," 1928. Hurt's steel driver quits and runs away rather than risk dying on the job. Discreetly avoiding a confrontation with his boss ("Take this hammer and carry it to the captain / Tell him I'm gone"), he cites John Henry's example as his motivation to do something else with his life:

John Henry he left his hammer
All covered in red, all covered in red.
That's why I'm goin'.
John Henry was a steel-driving boy,
But he went down.

3. T.F. interview with Paul Davies and Steven Weinberg, San Francisco, California, May 30, 2000.

4. Ludwig Wittgenstein, *Tractatus Logico-Philosophicus*. C. K. Ogden, trans. London: Routledge, 1988, 1.1.

17. GALAXIES

1. The entire disk of the Andromeda galaxy is estimated to be 165,000 to 200,00 light-years in diameter, but I am referring here to the part of it readily seen visually through telescopes.

2. Ronald Buta, "Galaxies: Classification," in Paul Murdin, ed., *Encyclopedia of Astronomy and Astrophysics*. London: Nature Publishing Group, 2001, vol. 1, p. 861.

3. The unfortunate habit of naming these nearby galaxies after their constellations invites confusion. The Sextans dwarf elliptical, for instance, is easily conflated with Sextans A, a larger and more distant irregular, and when people say "Sculptor," they more often mean NGC 253, a major spiral ten million light-years farther away.

4. In E. J. Schreier, "NGC 5128/Centaurus A: 150 Years of Wonder." *Encyclopedia of Astronomy and Astrophysics*, p. 1831.

5. There is some evidence that M84 is actually an S0 galaxy, which is to say an armless spiral, but based on its visual appearance I'll stick with its traditional designation as an elliptical.

6. Alan Goldstein, "Explore the Virgo Cluster." *Astronomy*, March 1991, p. 73.

7. They are NGC 4438, a dusty, tidally disrupted spiral; the spirals NGC 4435, NGC 4479, and NGC 4477; the S0 galaxy NGC 4461; and the ellipticals NGC 4458 and NGC 4473.

8. In Alan M. MacRobert, "Mastering the Virgo Cluster." *Sky & Telescope*, May 1994, p. 42. Copeland wrote a number of articles on the deep sky for *Sky & Telescope*, advising in 1949 that "a larger lens or mirror is not an assured benefit. Devotion and patience are as important as light grasp."

18. THE DARK AGES

1. Harold G. Corwin Jr., "The NGC/IC Project: An Historical Perspective." NGC/IC Project Web site, updated October 12, 1999.

2. Ken Hewitt-White, "Two Galaxy Clusters in Hercules: Observing Abell 2197 and 2199 from the Cascade Mountains." *Sky & Telescope*, June 2000, p. 115.

3. Steve Gottlieb, "On the Edge: The Corona Borealis Galaxy Cluster." *Sky & Telescope*, May 2000, p. 130.

4. In Robert Roy Britt, "'Brane-Storm' Challenges Part of Big Bang Theory." Space.com Web site, April 13, 2001.

5. In Robert Burnham Jr., *Burnham's Celestial Handbook*. New York: Dover, 1978, vol. 1, p. 505.

6. Kepler, "On the New Star" (*De stella nova*). Prague: KGW, 1606, chapter 22, p. 257, line 23.

7. Reverend Robert Evans, "Visual Supernova Searching with the 40-inch Telescope at Siding Spring Observatory." *Electronic Publications of the Astronomical Society of Australia*, vol. 14, no. 2, February 1997.

8. Ibid.

9. Tim Puckett, interviewed by T.F., March 16, 2001.

10. Ibid.

11. Obscuration is doubtless the reason that no supernova has been observed in our own galaxy since the year 1680. A spiral the size of the Milky Way should produce two or three supernovae per century, but nearly all are in or near the disk, since that's where most of the stars are. Several have probably occurred during the past two centuries, but were concealed behind the clouds.

12. Carl Pennypacker, interviewed by T.F., Berkeley, California, May 5, 1994.

13. The events that immediately followed García Diaz's discovery of the supernova, on Sunday, March 28, 1993, illustrate the advantages of belonging to an amateur astronomy group and of soliciting the help of professionals. García Diaz called a fellow member of the Madrid Astronomical Association, José Carvajal Martinez, who checked the sighting with an "Asteroids appulses" computer program to verify that it was not an asteroid. While Francisco went back to observing the galaxy, Carvajal Martinez called another club member, Diego Rodriguez, who took a CCD image of the galaxy confirming that the supernova was indeed present. After a check of two digital star atlases showed no star there in the past, Rodriguez contacted Enrique Pérez, a professional astronomer who was observing that night on the 2.5-meter Isaac Newton telescope in the Canary Islands. Pérez soon was able to obtain a spectrum of the supernova, an important step toward determining what sort of star it was. At dawn the Spanish amateurs contacted Janet Mattei at AAVSO, where it was still dark, and the amateur observer Robert Evans, in Australia, where the Sun would soon set. Within hours, amateur and professional astronomers across the dark side of Earth were training their telescopes on the new supernova and obtaining useful data.

14. A. Dressler, "The Evolution of Galaxies in Clusters." *Annual Review of Astronomy and Astrophysics*, 1984, 22:212.

15. Barbara Wilson, in "Who's Afraid of Einstein's Cross," Web site posting under "Adventures in Deep Space: Challenging Observing Projects for Amateur Astronomers."

16. In John Archibald Wheeler, *At Home in the Universe*. Woodbury, N.Y.: American Institute of Physics, 1994, p. 128. See Wheeler's endnote for an extensive discussion of sources for this story.

MINERVA AT DAWN

1. Hegel, *Philosophy of Right*, Preface.

2. In Northrop Frye, *Fearful Symmetry: A Study of William Blake.* Princeton: Princeton University Press, 1969, p. 123.

3. Blake, *Jerusalem: The Emanation of the Giant Albion.* London: W. Blake, St. Molton St., 1804, p. 152.

4. In Ray Monk, *Ludwig Wittgenstein: The Duty of Genius.* New York: Macmillan, 1990, pp. 140–41.

5. Ibid., p. 572.

6. Ibid., p. 143.

GLOSSARY

Don't bear me ill will, speech, that I borrow weighty words,
then labor heavily so that they may seem light.

—Wislawa Szymborska, "Under One Small Star"

Abell cluster. A cluster of galaxies cataloged by the American astronomer George Abell.

Aberration. Optical defect that degrades an image. The main culprits are astigmatism, *coma*, spherical aberration, and *chromatic aberration*, the introduction of false colors. All involve a failure to bring light to a single focus.

Absolute magnitude. See *magnitude.*

Absorption. Blocking or dimming of light or another form of radiation by Earth's atmosphere or by gas and dust clouds in space. Absorption is constant, *scintillation* intermittent.

Absorption nebula. See *nebula.*

Accretion. Increase in the mass of an astronomical object by the addition of smaller bodies coming into contact with it.

Accretion disk. Flat, rotating zone of material, much of which is destined to be captured by the object it is orbiting, such as a *black hole.*

Achromatic (or apochromatic) lens. Lens consisting of two or more elements, each with a different refractive index, the combination of which is intended to bring light of all colors to the same focus. Goal is to eliminate *chromatic aberration.*

Adaptive optics. A system that monitors *seeing* and adjusts a telescope's focus to compensate for changes in the atmosphere, resulting in a higher *resolution* image.

Airglow. Faint light in the atmosphere produced by interactions of atmospheric atoms with particles from the Sun.

Airy disk. Disklike appearance of a star seen through an optical system, which even if perfectly made cannot render the star as a dimensionless point but must dissipate sixteen percent of the light into a series of concentric rings.

Albedo. The reflecting power of a planet or satellite. An object with an albedo of zero reflects none of the light striking it; an object with an albedo of one reflects all the light that strikes it.

ALPO. The Association of Lunar and Planetary Observers.

Altazimuth mounting. A telescope *mounting* the two axes of which are oriented to the horizon and the zenith. An ordinary altazimuth mount can thus be pointed anywhere in the sky but cannot readily track astronomical objects, as can an *equatorial mounting* equipped with a *clock drive.* Some altazimuth mounts on *Dobsonian* telescopes can, however, be fitted with motors for

341

limited tracking, and some large observatory telescopes use motorized altazimuth mounts with computer-controlled camera rotators to permit time-exposure photography.

Altitude. Angular distance of an astronomical object above the local horizon. Objects at the *zenith* have an altitude of ninety degrees; at the *horizon*, zero degrees. Compare *azimuth*.

Angstrom. Unit employed for measuring wavelengths of light. One angstrom equals one one-hundred-millionth of a centimeter.

Aphelion. Point at which a solar system object is farthest from the Sun. Compare *perihelion*.

Apogee. The point at which a satellite is farthest from the Earth. Compare *perigee*.

Apparent magnitude. See *magnitude*.

Apparition. The period during which a given astronomical object can be usefully observed. An apparition of Mars, for example, covers a period of about eight to ten months, centered on its date of *opposition*.

Appulse. The apparent close approach of one astronomical body to another on the sky, as when an *asteroid* passes near a star.

Arc minute (symbol: '). One-sixtieth of a *degree*.

Arc second (symbol: "). One-sixtieth of an *arc minute*, which, in turn, is one-sixtieth of a *degree*. "Sub arc second" *resolution*, meaning that objects smaller than one arc second can be seen through a telescope, is indicative of excellent *seeing*.

Asterism. An identifiable group of stars within a constellation (such as the Big Dipper in Ursa Major) or involving several constellations (as with the Summer Triangle, which consists of the bright stars Vega, Altair, and Deneb, in the constellations Lyra, Aquila, and Cygnus).

Asteroid. A small object orbiting the Sun; also known as a minor planet. Most asteroids are in or have come from the *asteroid belt*. Asteroids typically are dry rather than icy, but a few contain ice and resemble small *comets*.

Asteroid belt. Zone within which most asteroids are found, located between the orbits of Mars and Jupiter.

Astronomical unit (A.U.). The average distance between Earth and Sun, defined as 150 million kilometers (93 million miles, or 500 light seconds).

Aten. Class of *asteroids* whose orbits lie entirely within that of Earth.

Atom. Fundamental unit of chemical elements, composed of a *nucleus* and a cloud of one or more *electrons*.

A.U. See *astronomical unit*.

Aurora. High-altitude glow produced when particles released by solar *flares* become trapped by Earth's magnetic field and interact with atoms and molecules in the atmosphere. Called aurora borealis in the north and aurora australis in the south—or, more simply, the northern and southern lights.

Autoguider, autoguiding mechanism. A system that corrects minor errors in telescope tracking, as while taking a time-exposure image. *CCD* autoguiders attend to a single star

image in the field, automatically adjusting the tracking when the image strays to an adjacent *pixel*.

Autumn equinox. See *equinox*.

Averted vision. Practice of looking slightly away from a dim object so that its light falls on the most light-sensitive parts of the eye.

Azimuth. The angular bearing of an astronomical object, measured by degrees along the compass rose with north as 0°, east as 90°, south as 180°, and west as 270°. The position of any object in the sky can be defined at a given moment by using azimuth and *altitude*. Owing to the rotation of the Earth the azimuth and altitude of astronomical objects changes constantly, so astronomers prefer to use a sky-based charting system based on *right ascension* and *declination*.

Baily's beads. Bright spots seen near the edge of the Moon during a total solar eclipse, when sunlight shines directly through the valleys between lunar mountains at the *limb*.

Bar. Unit of atmospheric pressure. The normal air pressure at sea level on Earth is about one bar, also called "one atmosphere," equal to 14.5 pounds per square inch.

Barge. An atmospheric vortex, as seen on *Jovian* planets, that is relatively narrow in latitude.

Big bang. High-density, high-energy state of the universe near the beginning of cosmic expansion.

Big bang theory. Class of theories having to do with the structure, origin, and evolution of the expanding universe—such as those dealing with the formation of atoms from the assembling of protons and neutrons as the primordial matter cooled, and with the release of light to create the *cosmic microwave background*.

Binary star. See *multiple star*.

Binoculars. Pair of telescopes, usually small, linked together for simultaneous viewing through both eyes.

Black hole. An object—typically a collapsed star—with a gravitational field so intense that nothing, not even light, can escape it.

Blazar. A bright galaxy nucleus that varies in brightness.

Bolide. A *fireball*, or particularly bright *meteor*, especially one that explodes in the sky.

Brown dwarf. A *substellar* object, larger than a planet but lacking enough mass for thermonuclear fusion to be sustained at its core.

Cannibal galaxy. A galaxy that has incorporated the stars, dust, and gas in one or more other galaxies.

Cassegrain. See *telescope, Cassegrain*.

Cataclysmic variable star. A *variable star* that suddenly and unpredictably becomes much brighter than normal. *Novae* and *flare stars* are varieties of cataclysmic variables.

CCD. "Charge-coupled device," a light-sensitive chip used by astronomers for imaging. By extension, a camera employing such a chip.

Celestial equator. Projection of Earth's equator onto the *celestial sphere*. The celestial equator defines *declination* zero and intersects, at right angles, all lines of *right ascension*.

Celestial pole. One of two points on the sky, located at *declination* ninety degrees north and ninety degrees south, representing where Earth's axis of rotation, extended into space, encounters the *celestial sphere*.

Celestial sphere. The night sky, represented as the inside of a sphere. Coordinates on the celestial sphere are mapped by *right ascension* and *declination*.

Chromatic aberration. In optics, a fault whereby light of different colors is refracted to different focal lengths. See *aberration*.

Chromosphere. Layer of the Sun's atmosphere immediately above the *photosphere* and below the *corona*.

Circumpolar. Visible at night throughout the year, by virtue of being located close enough to the *celestial pole* to never set beneath the horizon at a given latitude.

Clock drive. A mechanism that moves a telescope or camera to compensate for the Earth's rotation.

Cloud of galaxies. See *cluster of galaxies*.

Cluster of galaxies. Gravitationally associated assembly of galaxies, typically measuring up to a few tens of millions of *light-years* in diameter. Poor clusters (like the *Local Group*) contain up to about a hundred galaxies; rich clusters (like Coma) contain thousands. "Groups" and "clouds" of galaxies are terms used, rather confusingly, to designate clusters. Also see *superclusters*.

Cluster, star. See *star cluster*.

Collimate. To align the optical components of a telescope so that the light comes to a single focus. Collimation is attained when a telescope's optical components are aligned so that the light paths are nominal in terms of its design. In the simplest case, that of a *refractor*, this means adjusting the *objective lens* so that it is perpendicular to a line running from the center of the lens to the center of the eyepiece—the telescope's optical axis.

Collimation. See *collimate*.

Coma. In optics, an *aberration* that forms a halo of light around what should be a pinpoint image. In comets, a halo of gas and dust surrounding the *nucleus*. In the sky, short for Coma Berenices, a constellation that contains the Coma cluster of galaxies.

Comet. One of a class of solar system objects that are composed of rock and ice and normally have orbits more eccentric than those of planets and *asteroids*.

Conjunction. Position of a planet or other object when it lies closest in Earth's skies to the Sun or to another planet. *Inferior planets*—Mercury and Venus, whose orbits lie inside that of Earth—are in "superior" conjunction when on the far side of the Sun and closest to it in the sky, and in "inferior" conjunction when closest to the Sun on our side. When an inferior conjunction of Mercury or Venus makes the planet visible in silhouette against the solar disk, it is said to be in *transit*.

Constellation. Astronomically, any of the eighty-eight areas into which the sky is divided. More generally, the imaginary figures in the sky made by connecting stars and contained in one of these designated areas.

Contact binary. A *multiple star* in which one star is drawing atmospheric material from another.

Coriolis force, Coriolis acceleration. Effect of a planet's rotation on circulation of its atmospheric gases. The Coriolis effect produces atmospheric "cells" that circulate clockwise in the Northern Hemisphere and counterclockwise in the Southern Hemisphere.

Corona. Outermost part of the Sun's atmosphere.

Cosmic microwave background. Radio energy suffusing the universe, thought to have originated in the *big bang* when primordial matter thinned and cooled, releasing light that the subsequent expansion of space has stretched into the lower-frequency wavelengths of microwave radio.

Cosmic ray. A high-energy particle, such as an atomic nucleus, that moves through space at nearly the velocity of light and strikes Earth's atmosphere.

Cosmology. Scientific study of the universe as a whole; or, a cosmological theory.

Culmination. The crossing, by a star or other celestial object, of the *meridian*.

Dark adaption. State in which the eye is fully sensitive to low light levels, attained by spending a prolonged period—typically twenty to as much as ninety minutes—without any significant illumination other than starlight.

Dawes's limit. The theoretical limit to a telescope's *resolving power*, equal in *arc seconds* to 4.56/d, where d is the telescope aperture in inches. Hence a ten-inch telescope is theoretically capable of resolving two objects separated by only 0.46", and the theoretical resolving power of a 20-inch telescope is 0.29". In practice, atmospheric turbulence usually limits telescope performance to less than the theoretical limits.

Declination. Distance of an object on the sky from the celestial equator, measured in degrees. North declinations are designated with a plus sign (the North *Celestial Pole* is +90°) and those to the south by a minus sign (the South *Celestial Pole* is –90°).

Degree. In astronomy, the primary unit used to measure angles on the sky. A circle of 360 degrees takes us once around the *celestial sphere*, and an arc of 90 degrees extends from the horizon to the *zenith*. Each degree is subdivided into sixty minutes of arc, and each minute into sixty arc seconds.

Dobsonian. See *telescope, Dobsonian.*

Double star. Two stars orbiting their common center of gravity. See *multiple star.*

Dwarf. In astrophysics, a *main-sequence* star with a mass up to twenty times that of the Sun. Colloquially, a small, dim star, such as a *white dwarf,* a remnant left behind when a previously main-sequence star has run low on fuel and collapsed.

Eccentricity. The amount by which an elliptical orbit differs from circularity. A perfectly circular orbit has one focus—its center—and zero eccentricity. Squash the circle to make an ellipse, and the central point splits to form two foci. The more you squash it, the farther apart the two foci are, and the greater the orbit's eccentricity.

Eclipse, lunar. Passage of the Moon through the Earth's shadow. The eclipse is total if the dark *(umbral)* part of the shadow engulfs the Moon; otherwise it is partial.

Eclipse, solar. Phenomenon seen by terrestrial observers on whom the Moon's shadow falls. Those in the central part of the shadow (the *umbra*) see a total eclipse, in which the Moon covers the entire solar disk. Those in the surrounding *penumbra* see a partial eclipse.

Eclipsing binary. A double or *multiple star* one component of which passes in front of the other as seen from Earth, altering the system's total *apparent magnitude.*

Ecliptic. The projection of the Earth's orbit onto the *celestial sphere.* Also known as the *zodiac.*

Electromagnetic spectrum. The spectrum—a plot of energy according to its wavelength—of electromagnetic energy. Visible light composes part of this spectrum. At longer wavelengths lie the regimes designated infrared light and radio; at shorter wavelengths, those of ultraviolet light, X rays, and gamma rays.

Electromagnetism. Energy produced when electrically charged particles are accelerated. Electromagnetic energy has both wave and particle properties. The particles that convey it are photons (named for light, the kind of electromagnetism perceived by the eye), and the wavelength determines how it is designated—e.g., as light or radio. See *electromagnetic spectrum.*

Electron. Negatively charged particle that, when orbiting an atomic *nucleus,* forms an atom.

Emission nebula. See *nebula.*

Epoch. Agreed-upon period of applicability for a star chart, usually altered every half-century (epoch 1950, epoch 2000, etc.). The epoch has to be defined because the *precession* of the equinoxes slowly shifts the position of celestial coordinates—*right ascension* and *declination*—against those of the stars.

Equation of time. Difference in the time measured by a sundial (apparent solar time) and by a clock (mean solar time).

Equator, celestial. See *celestial equator.*

Equatorial mounting. A telescope *mounting* that has a polar axis aligned parallel to the Earth's axis of rotation and a *declination* axis perpendicular to it, making it convenient to point the telescope using these two celestial coordinates. See *setting circle.*

Equinox. A point at which the *ecliptic* intersects the *celestial equator.* The spring equinox is designated zero hours *right ascension* (R.A.), the autumnal equinox twelve hours R.A. Compare *solstice.*

Escape velocity. The speed that an object must attain in order to depart from the gravitational influence of a given planet or other body. Escape velocity is a function of the body's mass, which determines the total strength of its gravitational force field, and its size, which determines how much of that field strength is experienced at the surface (its "surface gravity"). The escape velocity of Earth is 11.2 kilometers per second; for massive Jupiter, 60 kms; and for the Sun, 618 kms.

Extinction. Reduction in brightness of stars and planets when seen near the horizon, caused by their light having passed through much more atmosphere than when they are nearer the *zenith.* At such times they also are affected by *refraction.*

Eyepiece. Optical device placed at a telescope's focal point to magnify the image. The magnifying power is the telescope's *focal length* divided by that of the eyepiece, so the shorter an eyepiece's focal length the greater its magnifying power.

Filament. See *prominence*.

Finder telescope. Small, low-power telescope mounted on a larger one and employed to help bring objects into its field of view.

Fireball. An extremely bright (magnitude −10 or greater) *meteor*.

Flare. Brief, bright eruption in the atmosphere of the Sun or another star.

Flare star. A small, main-sequence star that suddenly (often within hours) increases in brightness and then returns to normal, a process thought to be due to the eruption of large *flares* on its surface.

Focal length. The distance between the *objective lens* or *primary mirror* of a telescope and the point at which the light rays converge at the focal point.

Focal point. See *focal length*.

Focal ratio. The *focal length* of a telescope or other optical system, such as a camera lens, divided by the diameter of its *primary* (light-gathering) *mirror* or lens. A 6-inch telescope with a 60-inch focal length has a focal ratio of 10, expressed F/10.

Force. Agency of interactions among particles. Two long-range forces are known to exist—*gravitation*, which only attracts, and *electromagnetism*, which is bipolar (positive and negative charge). There are also two short-range forces, called strong and weak, that work on the subatomic scale.

FRAS. Fellow of the Royal Astronomical Society (Great Britain).

Galactic latitude. Distance north or south of the *galactic plane*—the disk of the spiral galaxy we inhabit, projected on the celestial sphere. The galactic plane is inclined 63 degree relative to Earth's equator.

Galactic longitude. Distance eastward along the *galactic plane* from a point in Sagittarius at the galactic center.

Galactic plane. The disk of the Milky Way galaxy as projected against the sky. See *galactic latitude* and *galactic longitude*.

Galactic pole. One of two locations on the *celestial sphere* representing where the axis of rotation of the Milky Way galaxy points.

Galaxy. A large aggregation of stars, more massive than a globular *star cluster*. Almost all the stars in the universe are found in galaxies.

Galaxy cluster. See *cluster of galaxies*.

Gamma ray. Short-wavelength form of electromagnetic energy. See *electromagnetic spectrum*.

Gamma-ray burst. Brief emission of high-energy gamma rays and X rays from points in deep space.

Gibbous. Phase in which the disk of the Moon (or other satellite or planet) is seen to be more than half, but less than fully, illuminated.

Globular star cluster. See *star cluster*.

Gravitation. Mutual attraction of objects proportional to their masses and decreasing by the inverse square of the distance separating them. See *force*.

Gravity. See *gravitation*.

Greenwich mean time (GMT). See *Universal time*.

Group. A small *cluster of galaxies*, typically containing a few dozen galaxies.

Heliacal rising. Strictly, the rising of a star or planet at the same time as the Sun. Traditionally, the first time that an object can be seen rising just before the Sun.

Hertzsprung-Russell (H-R) diagram. Plot of stars in which one axis typically represents their color and the other their inherent brightness (also called *luminosity* or absolute *magnitude*). A normal star spends most of its career on the *main sequence* of the H-R diagram, then moves onto the *red giant* branch and falls to the realm of the *white dwarfs*.

Horizon. Where the sky meets the Earth, from the point of view of any given observer. In *spherical astronomy* the ideal horizon is a great circle ninety degrees from the zenith, rather as one encounters at sea. Most local terrestrial horizons differ from this ideal, owing to atmospheric refraction and the interposition of hills and other obstructions. Computerized star charts can be set to show the true horizon, taking these effects into account.

Hour, hour angle. In astronomy, the primary measurement of *right ascension*. The *celestial sphere* has a circumference of 24 hours, with each hour divided into 60 minutes and each minute into 60 seconds.

Hubble constant. See *Hubble law*.

Hubble law. That distant galaxies recede from one another at velocities directly related to their distances—the farther apart the galaxies, the faster their recession velocities. In prevailing cosmological theory, this is said to be due to large-scale expansion of cosmic space. The exact rate of expansion, known as the Hubble constant, is approximately 60 kilometers per second per megaparsec, meaning that for every *megaparsec* (3.26 million *light-years*) separating a pair of remote galaxies, their recession velocity is approximately 60 kps faster.

IAU. The International Astronomical Union, a governing body that determines astronomical nomenclature.

IC. Designates objects listed in the Index Catalogs, published in 1895 and 1908 as a supplement to the New General Catalog (or *NGC*).

Index Catalog. See *IC*.

Inferior conjunction. See *conjunction*.

Inferior planet. A planet whose orbit lies inside that of Earth.

Infrared light. Electromagnetic energy slightly longer in wavelength than visible light. See *electromagnetic spectrum*.

Interferometer. A device that typically links two separate telescopes at different locations to obtain high-resolution images as if each were part of the mirror (or dish, for a radio telescope) of one big telescope.

Interstellar. Of the regions between the stars.

Interstellar medium. Dust, molecules, and atoms adrift in interstellar space.

Ion. An atom that has fewer or more electrons than normal, resulting in its possessing a positive or negative charge. An ionized gas responds readily to magnetic fields.

Ionosphere. A layer of Earth's atmosphere, from sixty to five hundred kilometers high, that is rich in *ions* and reflects radio waves.

Jet. A long, thin filament made of *plasma*, protruding from near the poles of *protostars* and from the *accretion disks* of *black holes*.

Jovian. Class of planets comparable to Jupiter, massive and with deep, opaque atmospheres.

Jupiter. Fifth planet from the Sun; the largest planet in the solar system.

Kilometer. One thousand meters; 0.6214 miles.

Kiloparsec. One thousand *parsecs*; 3,260 *light-years*.

Kuiper belt. A zone of icy objects lying along the *ecliptic* with its inner edge beyond the orbit of Neptune. The outer limits of the Kuiper belt have not yet been charted.

Laminar flow. Smooth layer of moving air, favored by observers since its lack of turbulence tends to produce good *seeing*.

Late star. A star cooler than the Sun.

Latitude. In terrestrial coordinates, distance north or south of the equator in *degrees*; equivalent to celestial *declination*. For our galaxy, see *galactic latitude*.

LED. Light-emitting diode, a semiconductor that glows when electricity passes through it. Used by stargazers since its dim glow can show that equipment is powered up without dazzling observers.

Libration. Any of the various changes in Earth-Moon orientation by virtue of which we can see, in total, more than half the lunar surface from Earth. Physical librations are produced by genuine changes in the Moon's rate of rotation. Diurnal librations result from variations in the Moon's orbital velocity at different points in its slightly elliptical orbit.

Light bucket. A short-focus telescope (usually a large-aperture Newtonian) intended to make up in light-gathering power what it may lack in optical precision. The makers of light buckets sometimes jokingly refer to their steeply curved mirrors as "salad bowls."

Light curve. Plot of the changing brightness of an object, such as a *variable star*, over time.

Light-second. Distance light travels through space in one second—186,000 miles or 300,000 kilometers.

Light-year. The distance light travels through space in one year, about six trillion miles.

Limb. The apparent edge of the disk of the Sun, Moon, or a planet or satellite.

Limb darkening. Decrease in brightness of the solar disk (or a planetary disk) toward the *limb*, caused by the fact that light coming from near the edges of the disk passes through more upper atmosphere on its way to our eyes.

Local Group. The group of galaxies to which the Milky Way and the Andromeda galaxy belong.

Longitude. In terrestrial coordinates, distance west from Greenwich, England; the celestial equivalent is *right ascension*. For the Milky Way galaxy, see *galactic longitude*.

Lookback time. The age of the light reaching us from a distant object. Owing to lookback time, when we observe, say, light from a *quasar* that is three billion light-years distant, we are seeing the quasar as it looked three billion years ago.

LTP. *Lunar Transient Phenomenon.*

Luminosity. The intrinsic brightness of a star; its absolute *magnitude*. The luminosity of stars differs greatly, from dwarfs ten thousand times dimmer than the Sun to massive young stars a million times brighter than the Sun.

Lunar. Of or pertaining to Earth's Moon.

Lunar month. The interval (29 days 12 hours 44 minutes, or 29.53 days) between two successive new Moons, also known as the synodical month. Because the relative position of the Sun, which illuminates the Moon, changes as the Earth moves in its orbit, the lunar month is not the same as the Moon's orbital period of 27.32 days. Since the Moon always keeps the same face toward us, its rotation period (a lunar "day") is the same as its orbital period.

Lunar Transient Phenomenon (LTP). Short-lived sign of activity, such as a flare, haze, or other change in brightness or color, observed on the Moon. Also known as Transient Lunar Phenomenon (TLP).

M. Designates objects in the *Messier Catalog.*

Magellanic Cloud. One of the two largest satellite galaxies of the Milky Way.

Magnetopause. The surface at which a planet's magnetic field is balanced by the pressure of the *solar wind*.

Magnetosphere. The spheroidal zone inside a planet's *magnetopause.*

Magnitude. An astronomical brightness scale in which the brightest stars are about 1st magnitude and the faintest stars visible to the unaided eye are about 6th magnitude, with each integer denoting a difference of 2.5 times. Hence a magnitude 3 star is 2.5 times brighter than a magnitude 4 star. Extremely bright objects are assigned negative numbers: The full Moon is magnitude −13, and Jupiter at its brightest is magnitude −2.9. A large telescope in a long CCD exposure can detect objects as dim as magnitude 25 or 26. All these references are to apparent magnitude—that is, what we see in the skies of Earth. Absolute magnitude refers to the apparent magnitude that each object would have if viewed from a distance of 10 parsecs (32.6 light-years). The apparent magnitude of a given star thus results from its absolute magnitude and its distance (plus any dimming introduced by interstellar matter). The apparent magnitude of the Sun, for instance, is −27, since it is so close to Earth, but its absolute magnitude is a modest 4.8.

Main sequence. The path occupied by most normal stars on the *Hertzsprung-Russell diagram*, a plot of stars by their colors and luminosities (or by related properties).

Maksutov telescope. See *telescope, Maksutov.*

Mare. Large, dark lava plain on the Moon; plural "maria."

Maria. See *mare.*

Mars. Fourth planet from the Sun. Mars is Earthlike in that it has polar caps, seasons, and an atmosphere, but it is smaller and colder than Earth and its atmosphere is much thinner.

Mass. Quantity of matter in an object. Since mass is directly related to the force of its gravitational field at a given distance, the mass of a planet, for instance, can be determined from the rate at which a satellite orbits it.

Megaparsec. One million *parsecs;* 3.26 million *light-years.*

Mercury. Planet nearest the Sun. Mercury is small, hot, and heavily cratered.

Meridian. For any given observer, a line drawn across the sky running north to south and passing through the *zenith.* Stars and other celestial objects are highest in the local sky when crossing the meridian, a situation known as *culmination.*

Messier Catalog. List of relatively bright, nebulous objects (including nebulae, star clusters, and galaxies) that might be mistaken for comets, compiled by the seventeenth-century French comet hunter Charles Messier. (For a list of Messier objects, see Appendix D.)

Meteor. See *meteoroid.*

Meteorite. A *meteoroid* that hits Earth.

Meteoroid. Umbrella term for small chunks of dust, rock, or ice that enter Earth's atmosphere from space. Called meteors when they streak across the sky, meteorites if they hit land or sea.

Micrometeorite. A tiny *meteorite,* typically the size of a grain of dust or sand.

Milky Way. The spiral galaxy to which the Sun and its planets belong, along with more than a hundred billion other stars. Also, the glowing river of light in the sky produced by *stars* and *nebulae* in the galaxy's disk.

Minor planet. See *asteroid.*

Minute of arc. See *arc minute.*

Molecule. Fundamental unit of chemistry, comprising two or more *atoms.*

Mounting, mount. The mechanical system that supports a telescope and permits it to be pointed at various parts of the sky.

Multiple star. Two or more stars bound together gravitationally, so that they orbit their mutual center of gravity. When limited to two components, may be called a binary or double star.

Nadir. For any given observer, a point directly below, toward the center of the Earth. The nadir lies 180 *degrees* from the *zenith.*

Near Earth Asteroid (NEA). See *Near Earth Object.*

Near Earth Object (NEO). An asteroid or comet with an orbit that passes close to Earth's.

Nebula. From the Latin for "cloud," a diffuse patch of interstellar gas and dust. Emission, or "bright," nebulae glow like neon lights, re-emitting energy injected into them by nearby stars. Absorption, or "dark," nebulae do not glow, but may be seen in silhouette. Parts of many bright nebulae also display absorption features, as when cooler gas in the foreground absorbs light at certain wavelengths emitted by the brighter nebula behind. Reflection nebulae are dark clouds illuminated by reflected, rather than absorbed and re-emitted, starlight. Also see *planetary nebula.*

NEO. See *Near Earth Object.*

Neptune. Eighth planet from the Sun, *Jovian* in structure. The outermost of the solar system planets if one classifies *Pluto* as a *Kuiper belt* object.

Neutron. Neutrally charged particle found in the nucleus of an atom.

Neutron star. See *star*.

New General Catalog. See *NGC*.

Newtonian telescope. See *telescope, Newtonian*.

NGC. Designation of objects listed in the New General Catalog of Nebulae and Clusters of Stars, originally published in 1888 and based on the work of William and John Herschel.

Node. One of the two points at which the orbit of an object (such as the Moon or a planet) intersects the plane of the *ecliptic*. When the object is moving south to north it is said to be passing the ascending node; when north to south, the descending node.

Northern lights. See *aurora*.

Nova. A star that brightens suddenly, often becoming visible to the naked eye when it was too dim to previously be seen without a telescope. Hence the name, meaning a "new" star. Compare *supernova*.

Nuclear fusion. Interaction in which atomic nuclei are forged together, creating new nuclei and releasing energy. See *star*.

Nucleus. Of an atom, the proton (for hydrogen) or protons and neutrons (for the heavier elements) around which the atom's *electrons* swarm. Of a *comet*, the solid object that ejects gas and dust to form the comet's *coma* and *tail*. Of a *galaxy*, the bright, central region.

Objective lens. The large light-gathering lens of a refracting telescope or binoculars.

Oblateness. The ratio between the equatorial and polar radius of a planet. Fluid, fast-rotating planets are more oblate than are solid, more slowly rotating planets like Earth.

Obliquity. The angle between a planet's axis of rotation and a line drawn perpendicular to the plane of its orbit—that is, the angle between the orbital plane and the planet's equator. It is this tilt of its axis, and not distance from the Sun, that causes the seasons on Earth.

Observable universe. See *universe*.

Occultation. Passage of an apparently large astronomical object, such as the Moon, in front of another, apparently smaller, more distant one, such as a star.

Oort cloud. A large, spherical aggregation of *comets*, centered on the Sun.

Open star cluster. See *star cluster*.

Opposition. Position of a planet or other solar system body when it lies on the opposite side of our skies from the Sun.

Orbit. The recurrent path pursued by an object around another, more massive object. The Earth orbits the Sun, which orbits the center of the Milky Way galaxy, which orbits the center of gravity of the *Local Group* of galaxies.

Parallax. Displacement in the apparent position of a star in the sky when viewed from two or more different locations, used as a method of measuring stellar distances through triangulation. The distances of nearby stars can be so measured by observing them from opposite sides of the Earth, while that of somewhat more distant stars can be obtained by making two observations, six months apart, from the opposite sides of Earth's orbit.

Parsec. Unit of distance measurement employed in deep-space studies, equal to 3.26 *light-*

years. A kiloparsec is 1,000 parsecs (3,260 light-years); a megaparsec is 1 million parsecs (3.26 million light-years).

Penumbra. In a solar eclipse, the lighter, outer Moon shadow, within which observers see a partial eclipse. On the Sun itself, the outer, lighter parts of *sunspots*.

Penumbral. See *penumbra*.

Perigee. The point in its orbit at which a satellite is nearest to the Earth.

Perihelion. Point at which a planet or other solar system object comes closest to the Sun.

Perturbation. Of a solar system object, deviation from the orbit it would pursue if unaffected by anything other than the Sun. The orbits of comets suffer extreme perturbations when a comet passes near a massive planet. Smaller perturbations can be discerned in the orbits of planets, caused by the gravitational attraction of other planets.

Phase. The portion of a planet or satellite that is seen to be illuminated from Earth at a particular time. In astronomical parlance, what is commonly called a "half" Moon is actually a quarter Moon. The lunar phases are: new, waxing crescent, first quarter, waxing gibbous, full, waning gibbous, last quarter, waning crescent.

Photometer. Instrument employed for doing *photometry*.

Photometry. Measurement of the apparent brightness of a star or other object.

Photon. Subatomic particle that conveys light and other forms of electromagnetic energy.

Photosphere. The visible surface of the Sun.

Pixel. One among the thousands or millions of light-sensitive dots on a *CCD* camera.

Planet. An astronomical object that orbits a star and is less massive than a *star* but more massive than are *asteroids, comets*, and *Kuiper belt* objects.

Planetarium. A device that re-creates the appearance of the night sky. More generally, the building or domed theater housing such a device.

Planetarium program. A computer program that displays the astronomical sky.

Planetary nebula. Glowing shell (or shells) of gas ejected by an unstable star, so named because many have disklike shapes superficially resembling the disks of planets when viewed through a telescope.

Planetesimal. An object in the primordial *solar nebula* with a diameter of less than ten kilometers. Planets are thought to have formed from the accretion of planetesimals.

Plasma. State of matter in which, rather than atoms, there are just free atomic nuclei and electrons—*ions*. Characteristic of stars and ionized nebulae, plasma interacts strongly with electromagnetic fields.

Plate tectonics. Phenomenon in which a planet is sufficiently geologically active to move parts of the crust (the "plates") over time. Plate tectonics on Earth produces continental drift.

Pluto. An icy object orbiting the Sun with an orbit intersecting that of Neptune. Originally assumed to be a planet, Pluto is perhaps better classified as a *Kuiper belt* or *trans-Neptunian object*.

Position angle. The location on the sky of one astronomical object relative to another, measured by drawing a line from the North *Celestial Pole* through the first object to the second. The angle is computed as with a compass rose; hence, if the second object is due east of the first its P.A. is 90 degrees, and if due west, 270 degrees. Position angles aid observers in locating stars in binary systems and recently discovered *supernovae*.

Power. Extent to which an *eyepiece* magnifies an image. As in human affairs more generally, too much power is not necessarily a good thing: For a given telescope observing a given object under a particular set of atmospheric conditions there is an ideal range of powers, best ascertained by trying several eyepieces, and this is seldom the highest available power.

Precession. The slow, conical wobble of a planet's axis of rotation. Since Earth's precession causes a gradual change in the position of the *equinoxes*, the phenomenon is sometimes called "precession of the equinoxes."

Primary mirror. The large, light-gathering mirror in a reflecting telescope.

Prominence. An eruption of hot gas and plasma from the Sun's surface. Prominences are described as such when seen at the edge of the Sun, where they are silhouetted against the space beyond. Prominences seen against the disk itself are called filaments.

Proper motion. Change of a star's position on the sky caused by a combination of its actual velocity through space and that of the solar system.

Proton. Positively charged particle found in the *nucleus* of an *atom*.

Protostar. A *star* in the process of forming.

Pulsar. A neutron *star* that emits pulses of radio energy.

Quantum. Fundamental unit of energy.

Quantum flux. Random variations in energy levels at the subatomic level, equated with the temporary appearance of "virtual particles" out of the vacuum.

Quasar. A bright object at the center of a galaxy, probably powered by energy released from material spiraling into the vicinity of a *black hole*. Rare in the universe today, bright quasars were commonplace when the universe was younger. See *lookback time*.

Radio astronomy. The study of astronomical phenomena in the wavelengths of radio, such as the 21-centimeter-wavelength radio noise emitted by hydrogen atoms in space. See *electromagnetic spectrum*.

Rayleigh scattering. Diffusion of light by small particles in the atmosphere. Rayleigh scattering is inversely proportional to the fourth power of light's wavelength. Hence blue light, which is short in wavelength, is much more scattered by Earth's atmosphere than is red light—which is why the sky is blue.

Red giant. A star, late in its evolutionary career, the outer atmosphere of which has expanded to form a large and relatively cool (hence ruddy) outer envelope.

Reddening. Effect of interstellar dust on starlight, which dims the star's *apparent magnitude* and, since blue light is more scattered than red, makes it look redder than it really is. Not to be confused with *redshift*.

Redshift. Displacement of the *spectral lines* of a distant galaxy toward the red, low-frequency, end of the *spectrum*, thought to be caused by the expansion of cosmic space.

Reflecting telescope. See *telescope, reflecting.*

Reflection nebula. See *nebula*.

Reflector. See *telescope, reflecting.*

Refracting telescope. See *telescope, refracting.*

Refraction. Bending of light, as by its passing through a lens.

Refractor. See *telescope, refracting.*

Resolution. Amount of detail attainable with a given telescope. See *resolving power.*

Resolving power. The smallest angle between two points (such as components of a double star) that a given telescope can distinguish. The theoretical limit for a given aperture, known as *Dawes's limit*, is attainable only with excellent optics and seeing conditions.

Retrograde. Property of moving opposite to the normal direction in an orbit, or spinning in the opposite of normal direction on an object's axis of rotation. Most solar system satellites orbit their planets in the same direction; those that go the other way round have retrograde orbits. Most planets rotate in the same direction; the exceptions have retrograde rotation. A third meaning applies to the apparent but not real backward motion of planets against the sky: Owing to Earth's orbital motion, planets sometimes appear to stall in their progress along the *ecliptic* and move backward—in retrograde motion—for a while before resuming their normal courses.

Right ascension. Distance of an object on the sky, measured eastward from the spring *equinox* in hours, with 24 hours constituting a circle around the *celestial sphere.*

Ritchey-Chretien telescope. See *telescope, Ritchey-Chretien.*

Roche limit. The minimum distance at which a satellite can orbit a planet without being torn apart by differential gravitational forces.

Roll-off roof. Observatory design in which the roof rolls away on wheels to expose the telescope to the sky.

Saturn. Sixth planet from the Sun. Cooler and somewhat smaller than Jupiter, Saturn has spectacular rings.

Schmidt-Cassegrain telescope. See *telescope, Schmidt-Cassegrain.*

Scintillation. Variation in the apparent brightness of an image (in light or another wavelength of electromagnetic radiation, such as radio) caused by changes in Earth's atmosphere or in dust and gas clouds in space. Compare *absorption.*

Season. Recurring change in prevailing climate in one hemisphere of a planet, such as Earth or Mars, caused primarily by a tilt in the planet's axis. See *solstice.*

Second of arc. See *arc second.*

Secondary mirror. In a reflecting telescope, any mirror in the light path other than the *primary mirror*. In a Newtonian telescope, the secondary reflects light to the side of the tube; in a Cassegrain telescope, light reflected from the secondary goes straight back, and passes through a central hole in the primary mirror. See *telescope.*

Seeing. The steadiness of the air through which observations are being made, usually described in terms of resolution—e.g., the closest binary stars that can be resolved. For a given optical system, the overall quality of an image on a given night is dependent on atmospheric seeing and *transparency.*

Setting circle. One of a pair of dials (or computer readouts, in which case one speaks of "digital setting circles") that show where a telescope is pointing by specifying the celestial coordinates *right ascension* and *declination*.

Shooting star. A *meteor*.

Solar flare. See *flare*.

Solar nebula. The cloud of gas and dust from which the Sun and its planets formed, 4.5 billion years ago.

Solar prominence. See *prominence*.

Solar system. The Sun and everything that is gravitationally bound to it, including *planets*, *asteroids*, *Kuiper belt* objects, and *comets*.

Solar telescope. A telescope designed specifically for observing the Sun.

Solar wind. Ionized hydrogen and helium gas blown outward from the Sun.

Solstice. One of the two points on the *ecliptic* midway between the spring and autumn *equinoxes*. At summer solstice the Sun stands farthest north of the *celestial equator* and hence shines most directly down on Earth's Northern Hemisphere, which is what causes northern summer. At winter solstice the Sun is farthest south of the celestial equator, and its relatively indirect rays produce northern winter. Compare *equinox*.

Southern lights. See *aurora*.

Spectra. See *spectrum*.

Spectral line. See *spectrum*.

Spectrogram. See *spectroscope*.

Spectrograph. See *spectroscope*.

Spectroheliogram. See *spectrohelioscope*.

Spectroheliograph. See *spectrohelioscope*.

Spectrohelioscope. An optical system that rejects all but a narrow band of spectral light, making possible solar observations that otherwise could be obtained only during a total solar eclipse. A device that records such observations is called a spectroheliograph, and the resulting recording is called a spectroheliogram.

Spectroscope. A device that enables observation of an object's *spectrum*. A spectrograph records the spectrum, and this recording is called a spectrogram.

Spectroscopic binary. A double star that cannot be resolved ("split") in a telescope, but the duplicity of which can be discerned by studying its spectroscopic lines, which may reveal both the differing compositions of the two stars and their motions as they orbit their common center of gravity.

Spectroscopy. The study of spectra. See *spectrum*.

Spectrum. An image or plot that breaks down the light from an object into frequencies, in something like the way piano keys break music into distinct notes. The spectral lines denote particular transitions in atoms, and hence can be studied to determine the composition of the object.

Spherical aberration. See *aberration*.

Spherical astronomy. Mapping and analysis of the sky as if it were the inside of a sphere, as when aiming a telescope or navigating a ship by means of celestial coordinates. See *celestial sphere.*

Spring equinox. See *equinox.*

Standard candle. An astronomical object of known brightness, the distance of which can therefore be accurately determined by comparing its absolute with its apparent *magnitude.* Standard candles employed in measuring the distances of galaxies include Cepheid *variable stars* and *supernovae.*

Star. A gravitationally bound ball of *plasma* with sufficient mass to sustain *nuclear fusion* at the core. Exceptions include neutron stars, which are extremely condensed objects resembling giant atomic nuclei, the somewhat less compressed white dwarfs, and brown dwarfs, which may include objects with so little mass—not much more than a dozen Jupiters— that they can manage only a few kinds of nuclear fusion, and those intermittently.

Star cluster. A gravitationally associated group of stars. Globular star clusters are large, massive, and old. Open star clusters are smaller, much less densely populated, and generally younger.

Star, variable. Any star whose brightness changes periodically. All normal stars pulsate and change in brightness somewhat, but the term variable is applied to those where the changes are noticeable.

Substellar. Lacking sufficient mass to sustain thermonuclear reactions at the core.

Summer solstice. See *solstice.*

Sun. Star orbited by the Earth.

Sunspot. A relatively dark, cool patch on the Sun's glowing surface (or *photosphere*), caused by local magnetic storms. Viewed with a proper filter, sunspots are readily visible with small telescopes or the unaided eye.

Supercluster. An array of thousands of galaxies, typically measuring 150 to 300 million *light-years* in diameter and containing a hundred or more clusters of galaxies.

Supergiant. An extremely high-luminosity star, absolute *magnitude* typically −5 to −8.

Superior conjunction. See *conjunction.*

Superior planet. See *conjunction.*

Supernova. An exploding star. Supernovae are a hundred times more luminous than *novae,* and unlike novae do not recur.

Synchronous rotation, synchronous motion. Situation in which a satellite (such as Earth's Moon) rotates on its axis in the same period of time that it orbits its planet, so that it always presents the same face to the planet. Also known as gravitational lock or captured rotation.

Synodic period. From the Greek for "meeting," the interval between *conjunctions* of planets.

Tail. Long streamers produced by dust and gas ejected from the *nucleus* of a *comet* and swept outward by sunlight and the *solar wind.*

Telescope. A device that gathers light (or another form of electromagnetic energy) and brings it to a focus.

Telescope, Cassegrain. A design in which light reflected from the *primary mirror* is bounced directly back by a *secondary mirror* and passes through a hole in the primary before coming to focus. Said to have been invented by the French sculptor Guillaume Cassegrain.

Telescope, Dobsonian. Inexpensive form of Newtonian telescope, invented by John Dobson.

Telescope, Maksutov. A design in which the light first passes through a spherically figured corrector lens and then bounces off a spherical mirror. Most Maksutovs use a reflective spot on the back of the corrector plate to reflect the light back through a hole in the mirror, Cassegrain-style. This folded design is suitable for portable telescopes, since it packs a long focal length in a compact package.

Telescope, Newtonian. A design in which the light from the *primary mirror* is bounced off to the side of the tube by a single *secondary mirror.* Invented by Isaac Newton.

Telescope, reflecting. A telescope that uses a *primary mirror* to gather light.

Telescope, refracting. A telescope that uses a single or compound *objective lens* to gather light to a focus.

Telescope, Ritchey-Chretien. A photographic reflector currently preferred for *CCD* imaging, using a Cassegrain design but with exotically figured mirrors to deliver sharp images across a wide field. Designed by the American optician George Willis Ritchey and the French optical designer Henri Chrétien.

Telescope, Schmidt-Cassegrain. A design, combining innovations by Guillaume Cassegrain and the astronomer Bernhard Schmidt, in which the folded Cassegrain optical path is combined with a corrector lens at the starlight end of the tube. The main distinction between Schmidt-Cassegrain and Maksutov telescopes lies in the figure of the corrector plate, which is simpler for Maksutovs.

Terminator. The line dividing the sunlit from the dark sides of planets and satellites.

Terrestrial. Of or pertaining to Earth.

Theodolite. A surveyor's tool, usually incorporating a small telescope, employed to measure horizontal and vertical angles.

Tide. On Earth, a change in the elevation of the seas (and, to a lesser extent, the land) resulting from the gravitational pull of the Moon and (to a lesser extent) the Sun.

TLP. See *Lunar Transient Phenomenon (LTP).*

TNO. See *Trans-Neptunian Object.*

Transient Lunar Phenomena (TLP). See *Lunar Transient Phenomenon (LTP).*

Transit. Passage of a small astronomical object across the disk of a larger one, such as that of Mercury or Venus across the Sun, or a satellite across Jupiter or Saturn. See *conjunction.*

Trans-Neptunian Object (TNO). One of a class of generally icy solar system bodies found mostly beyond the orbit of Neptune.

Transparency. The clarity of air through which observations are being made—e.g., its absence of clouds, dust, and water vapor. The overall quality of the air at a given time and place is described in terms of its transparency and *seeing.*

Ultraviolet. Light shorter in wavelength than is visible light, but longer than X rays. See *electromagnetic spectrum*.

Umbra. In a solar eclipse, the dark inner shadow of the Moon, within which terrestrial observers see a total eclipse. On the Sun, the dark core of a *sunspot*, surrounded by a lighter (and hotter) *penumbra*.

Umbral. See *umbra*.

Universal Time (UT). The time at the prime meridian, which passes through Greenwich, England, and marks zero degrees *longitude*. Employed by astronomers to avoid their having repeatedly to convert from local times at various locations.

Universe. The set of all past, present, and potential future phenomena observable from Earth. The observable universe is the subset of all past and present such phenomena.

Uranus. Seventh planet from the Sun. *Jovian* in composition, Uranus is relatively devoid of upper-atmosphere features.

Variable star. See *star, variable*.

Venus. Second planet from the Sun. Venus is almost a twin of Earth in terms of its size and mass, but is much hotter and has a denser atmosphere.

White dwarf. A hot but dim *star* with a mass about that of the Sun but much higher in density.

Winter solstice. See *solstice*.

X ray. Electromagnetic energy with wavelengths between those of ultraviolet light and gamma rays. See *electromagnetic spectrum*.

Year. The time (365.256 days) it takes Earth to complete one orbit around the Sun.

Zenith. For any given observer, the point directly overhead in the sky. Opposite of the *nadir*.

Zodiac. The band to which planets are confined in Earth's skies—i.e., the *ecliptic*. Traditionally (e.g., astrologically) there are twelve zodiacal constellations, although on modern star charts the ecliptic passes through thirteen (the other one is Ophiuchus). "Zodiac" is Greek for "circuit of animals," since most of the zodiacal constellations depict animals. They are Capricornus the goat, Pisces the fishes, Aries the ram, Taurus the bull, Cancer the crab, Leo the lion, and Scorpius the scorpion—plus, if you like, Sagittarius, who was half man and half horse.

Zodiacal light. Sunlight scattered off dust particles, visible on dark nights as a cone or band of light along the *ecliptic*.

INDEX

ABOUT THE AUTHOR

TIMOTHY FERRIS has been stargazing since 1956 and writing about astronomy since 1960. His bestselling books include *The Whole Shebang* and *Coming of Age in the Milky Way*, which have been translated into fifteen languages and were named by *The New York Times* as two of the leading books published in the twentieth century. A former newspaper reporter and editor of *Rolling Stone* magazine, Ferris has published more than two hundred articles and essays in periodicals including *The New Yorker*, *Harper's*, *Life*, *National Geographic*, *Nature*, *Newsweek*, *Time*, *Reader's Digest*, *The Nation*, *The New Republic*, and *The New York Review of Books*. He wrote and narrated the PBS television specials "The Creation of the Universe" and "Life Beyond Earth," and produced the Voyager phonograph record, an artifact of human civilization launched aboard the twin *Voyager* interstellar spacecraft in 1977 and now exiting the solar system. A consultant to NASA on the long-term goals of space exploration, Ferris most recently served on the space agency's Near-Earth Object Steering Group.

Professor Ferris has taught in five disciplines at four universities, and is currently on the faculty at the University of California, Berkeley. He has received the American Institute of Physics science writing medal, the American Association for the Advancement of Science prize, and a Guggenheim Fellowship. His books have been nominated for the National Book Award and the Pulitzer Prize.

To learn more about Professor Ferris, visit his Web site at timothyferris.com.